Automatic Computational Techniques in Civil and Structural Engineering

Automatic Computational Techniques in Civil and Structural Engineering

by E. Litton
B.Sc., M.Sc., C.Eng., M.I.C.E., M.I.Struct.E.

Head of Department of Civil Engineering, Dundee
College of Technology

Crosby Lockwood   London

Granada Publishing Limited
First published in Great Britain 1973 by
Crosby Lockwood
Park Street St Albans and
3 Upper James Street London W1R 4BP

ISBN 0 258 96820 6

Typeset in Great Britain by
Santype Limited (Coldtype Division)
Salisbury Wiltshire England
Printed in Great Britain by
Fletcher & Son Ltd, Norwich

A/624.028

Contents

12: Conclusion 360

Appendix 1: Program DCTCES 3:
Elastic Analysis of Various Single-span Beams with either a Concen-
trated or Uniformly Distributed Load—Details of Formulae
Used 361

Appendix 2: Program DCTCES 4:
Elastic Analysis of a Single Bay, Pitched Roof, Portal Frame with
Pinned Bases—Details of Formulae Used 365

Outline

The purpose of this book is to introduce the reader to the concept of the development and use of automatic computational techniques in civil and structural engineering, endeavouring to present a general philosophy and policy rather than merely solving a collection of miscellaneous problems. It does not claim to be a detailed treatise of specific methods of structural analysis, which, in any case, are fully described elsewhere.

The book is intended for two types of readers, namely qualified civil and structural engineers and research workers, who have little or no experience in using computers for analysis, design, or to a lesser extent, data-processing techniques in the laboratory or on site, and for undergraduates in those universities and colleges that are orientating their engineering syllabuses towards a modern approach involving considerable use of computers and ancillary equipment. (The demand from this latter source will certainly grow.)

The author has long felt that the first step taken into the field of automatic computational techniques by an engineer who is experienced in traditional methods of structural analysis, design, and laboratory procedure is an important one. In their haste to deal with their own narrow, specialized and rather advanced topics of study it is, however, one that is usually neglected or glossed over by many other writers. The result is that the beginner may well obtain an unbalanced view of the subject as a whole and be discouraged from further study because of the speed and disarray with which the complicated new ideas have been delivered. The book therefore appraises the situation and 'sets the stage' for further serious study by the reader of his own particular interest.

Care has been taken throughout to show both the practising engineer and the student how to set about applying the automatic computational techniques to everyday engineering problems they encounter in the office, on site or in the laboratory, and suitable illustrative examples are provided.

The author's own special interest is in structural engineering and most of the techniques described have been derived while he has been working on such problems. Nevertheless, he considers that, from this nucleus, the step into other branches of civil engineering is not a large one, and the same basic approach may therefore be followed. Some examples of these applications are briefly outlined.

A list of references and a bibliography appear at the end of chapters.

Foreword

There is still time enough to prepare for the arrival of the fateful year '1984', and it is certainly not my wish that engineers should drift into the way of life predicted by Orwell merely by default, either by just not seeing the danger or by not taking appropriate steps in time to control or harness the blessing (or curse!) of automation as applied to the activities of the professional civil engineer. I firmly believe that the trend towards automation is indeed a blessing, and should be regarded as an opportunity for the engineer to extricate himself from the drudgery of irksome, routine tasks, thereby releasing himself for effort and thought on a higher and more productive plane. He should then be able to open new doors and extend new boundaries for knowledge and achievement, so providing fulfilment for the more creative and intellectual side to his character.

This book provides a means of approaching the whole problem of automation 'my way', and embodies years of personal experience grappling with the various aspects of the subject, so that the treatment accorded is at times rather subjective. For this I make no apology.

I have drawn from many sources for my material and wish to thank all who helped in any way by supplying me directly or indirectly with specific information. In particular I would thank:

the Governors of Dundee College of Technology for providing me with the opportunities, facilities and material within the College from which I have been able to draw as required during the preparation of this work;

my colleague Mr J. P. Cole for his close co-operation in providing much of the material for Chapter 8;

my former colleague Mr J. R. Thorpe for supplying me with descriptive notes on the Finite Element Method and for providing me with Program No. DCTCEL 1, to which reference is made in Chapter 9;

Dr R. Webster, Head of the Computer Unit, Dundee College of Technology, and his staff, for providing me with such a sympathetic computing service over the past few years during the preparation of this book;

Mr N. Craven, the College Librarian, for his assistance in the compilation of the lists of references and bibliographies;

Mr J. S. Roper, Senior Lecturer in Computing at the University of Durham, for collaborating with me over many years on computing matters (some of which are referred to in Chapter 9) and more recently, for writing Algol versions of the former Autocode programs (Program Nos. DCTCES 5 and 6) which are referred to in Chapter 3;

Mr K. Lowe, Liverpool Polytechnic, for much valuable assistance in computing matters generally and in particular for assisting me with the preparation of Program Nos. DCTCES 3 and 4 in Chapter 3;

Mr T. V. Thompson (now retired, but formerly Chief Design Engineer) and Mr K. G. Heward, Technical Manager, Redpath Dorman Long, Ltd., Middlesbrough, for their close co-operation and professional advice during the period of my consultancy with the Company, and for their subsequent help in providing me with some of the material used in Chapter 9;

International Computers, Ltd., and Elliott Automation Ltd., for providing me with library programs and other material over several years, and for allowing me to reproduce various extracts from this information in these pages;

Messrs G. Maunsell and Partners, Consulting Engineers, for providing me with full details of Program No. A27 and for permitting me to reproduce the relevant material on this program as given in Chapter 4;

Dr P. D. Forbes and his colleagues at the National Engineering Laboratory, East Kilbride, for providing me with computer calculations for the problem analysed by the Finite Element Method, of which the main details are recorded in Chapter 4;

the Clyde Port Authority for permission to reproduce the photographs of Fig. 9.21 and the accompanying notes on the Clyde Tidal Model in Chapter 9;

W. Szeto, Esq., Architect, Hong Kong, for permission to reproduce the photographs of the multi-storey building in Fig. 9.2;

British Olivetti Ltd., for permission to reproduce the photograph of the Olivetti Programma 101 in Fig. 6.11;

the Solartron Electronic Group Ltd., for permission to reproduce the photograph of the Digital Voltmeter in Fig. 8.1.

Finally, but by no means least, I wish to thank my wife for her encouragement, assistance and patient forbearance at all times during the preparation of this book.

E. LITTON
Monifieth
Angus

1
The Problem

INTRODUCTION

This is the Age of Automation and its influence on civil engineering is readily apparent. However, full automation in civil engineering is a concept well nigh impossible to envisage, and may inculcate sinister forebodings in the minds of many engineers such as:

a false impression that the ultimate goal is an age in which engineers have made themselves redundant by their very acts of devising methods for the automatic design, analysis, and possibly even the construction of works; a vision similar to George Orwell's *1984*, in which all engineers become morons carrying out monotonous routine duties; or concern regarding the danger of a departure from reality leading to a loss of 'feel' or of a practical understanding of the problem so that incorrect solutions emerge with disastrous consequences.

Such misgivings might tempt experienced engineers to hold fast to the time-proven ways. Nevertheless, the author refuses to see the lot of the professional engineer as a penance in which dreary calculations have to be carried out with only a brain, a pair of hands, pencil, paper, slide rule, patience and determination! Despite the misgivings listed, engineers should ensure that, by assigning more of the tedious work routine to computers, robots and similar devices, they can thereby be freed to tackle more projects, to consider more possibilities, and so to increase production. Moreover, the practical significance of the results can be confirmed, for example, by making loading tests on full-scale structures and models as a complementing standard procedure.

Automation should be regarded as a goal to be sought rather than to be fully achieved, with successive improvements replacing one another, so tending towards the limit which is full automation. Provided that sufficient engineers work to this concept when dealing with their own particular problems, it will be possible in the case of a design project, say, to piece together the individual contributions to form a complete process which can be programmed for use on a computer. This is the opportunity that awaits the modern engineer, and it is hoped that this book will help him to gain it.

When first using automatic computational techniques there is an immediate danger of not seeing the wood for the trees i.e., of not appreciating the basic issues as a whole before grappling with the intricacies of the independent technologies which abound in this work. The civil or structural engineer need not become expert in all the basic technologies in the first instance—rather should he start with a view of the complete picture as a manager in control, capable of evaluating the accuracy and significance of the results obtained and acting on them. Naturally, he must know something of the individual subjects used, though he should guard against being side-tracked into a detailed study of the basic skills at too early a stage: he can soon satisfy himself as to the validity and economic efficiency of each process by using spot checks or trial and adjustment tests. Once the overall system has been established and is functioning reasonably the engineer can study the component skills in greater detail personally or through a team of suitably qualified staff.

When the engineer decides to use automatic computational techniques, he is confronted simultaneously with three fields of study:

> the actual engineering problem involving theory, design, analysis, experimental work and the like;
> mathematics;
> computer technology.

These are shown in Fig. 1.1. The basic problem in civil or structural engineering can be tackled either by using traditional methods with the computer replacing the slide rule, or by modern methods specially devised to make the most effective use of the computer. Both traditional and modern methods generally require mathematical theory outside (or stretching!) the knowledge of the engineer, and the topics are listed in Fig. 1.1. The use of the computer itself is the final stage in the process and involves an understanding of those items

1.	2.	3.
PROCESS OF SOLVING THE ACTUAL ENGINEERING PROBLEM.	MATHEMATICS.	COMPUTER TECHNOLOGY.
Theory.	Matrix Algebra.	Programming.
Design.	Numerical Analysis.	Use of Library &
Analysis.	Theory of Errors.	Special Programs.
Experimental Work.	Statistics.	Computer Operation.
	Solution of Equations.	General Computer
Either Based On:—	Binary Algebra.	Appreciation.
Traditional Methods,	Boolean Algebra.	Use of Ancillary
or Modern Methods	etc.	Equipment.
		Use of Programs written at other Centres or for other Computers, etc.

FIG. 1.1

tabulated under Computer Technology, of which some may be entirely new to the reader.

Once the brief fundamentals in these three fields of study have been grasped, the engineer can become a computer user, devoting his time and effort to suit his own particular circumstances. His progress in the use of automatic computational techniques can then be quite rapid.

The Evolution of Traditional Methods

The traditional method of solving a typical engineering problem prior to the introduction of the electronic digital computer was initiated by a mathematician or an engineer who endeavoured to obtain a solution based on strict scientific reasoning and with no regard to the resulting calculations. The subsequent methods of solution dealt mainly with devices to simplify these calculations while taking account of additional refinements in theory to a lesser extent. However, this process was terminated by the introduction of the electronic computer, which provides the opportunity for a fresh approach to the problem.

This pattern of development is well illustrated in the elastic analysis of frames and beams. The better-known methods of analysis (in chronological order) are as follows.

Method	Attributed to	Approx. Date
Theorem of Three Moments	Clapeyron	1857
Moment of Area	Mohr	1874
Strain Energy	Castigliano and others	1879
Slope Deflection	Mohr	1892-93
	Maney and Wilson	1915
Direct Integration	Macaulay	1919
Moment Distribution	Hardy Cross	1930
Column Analogy	Hardy Cross	1932
Semi-graphical Integration	Robertson	1949
Kani	Kani	1957

This list is not complete, but does show the development of the various methods over the years. In broad terms, each successive method was produced because of shortcomings in the preceding one, either restrictions in the use of the method or difficulties in the calculations caused by the volume of the arithmetical work which would have had to be carried out without the help of computers or calculating machines.

Notes on the Methods

The bibliography for this chapter gives full details of the various methods, only the computational processes being considered here.

The Theorem of Three Moments

This method involves the solution of linear simultaneous equations. Rounding-off errors can accrue, and the calculations can become quite extensive when dealing with a large number of unknowns. Ill-conditioned equations can also be formed in particular numerical examples; in such cases care and accuracy are specially required. (See chapters 3 and 5).

Moment-of-Area

Non-uniform members can also be dealt with by this method, which again generally reduces to the solution of linear simultaneous equations (see above).

Strain Energy

These methods, such as those of Castigliano, can be used to analyse most structural frames and beams. However, the calculations are arduous, and rounding-off and other errors can readily occur. There is no 'glimpse' of the solution during the calculations, so that the final answer has to be accepted at its face value.

Slope Deflection

This method gives deflections and rotations as well as moments and forces. Once more, the calculation reduces to the solution of linear simultaneous equations with the associated difficulties. This technique is the basis of the Stiffness Method—a modern process devised specially for the electronic digital computer. (See chapters 2 and 4).

Direct Integration

This method is particularly suited for analysing beams. Since it is based on the calculus, the magnitude and location of maximum and minimum slopes and deflections can be conveniently obtained. As with strain energy methods, the arithmetic involved can be arduous and errors easily made.

Moment Distribution

This is a method of successive approximation, and represents the first attempt to overcome the computational difficulties experienced in all the previous techniques. At any stage of the calculation an approximate answer is available, thus giving the engineer the opportunity of using his judgement, or 'feel', in appraising the validity of the work. The arithmetic is considerably simplified from that of previous methods, and the process is a major contribution to modern structural analysis, rendering complicated frames more capable of solution by design office staff than hitherto.

However, each frame has generally to be given special treatment, and difficulties can arise when dealing with unusual frames. The arithmetic can again become arduous in the case of large frames subjected to several loading cases, though various research workers have been and still are extending and improving the method to overcome such snags.

Column Analogy

This method uses the parallel which exists between the actual problem and an equivalent short column problem, which can be readily solved. In certain cases, this technique is quite ingenious and can greatly simplify the calculation, though, as in the case of strain energy methods, there is no 'feel' in the process and the results must be accepted at their face value.

Semi-Graphical Integration

This method gives a neat solution to the analysis of portal frames having sloping members and fixed or pinned bases. The calculations are usually tabulated and carried out in an orderly manner. The technique is essentially a unit load method and is the basis of the Flexibility Method—a modern procedure devised with the computer in mind. (See chapters 2 and 7).

Kani's Method

Here the starting point is from the slope deflection equations, which are solved by iteration. The calculation is laid out in a distinctive way, akin to that used in moment distribution, though it avoids the inconvenience of the additions and subtractions needed in that method. The arithmetic is simple and once again there is 'feel' in the method. Unusual frames can, however, cause difficulties and the solution in such cases, albeit very skilful, can tend to become rather ponderous.

This completes the review of the process of developing solutions to engineering problems before the invention of the computer; these are termed *traditional methods*. Most of them reveal the considerable resourcefulness of engineers in dealing with the burden imposed on them from calculations involving the solution of linear simultaneous equations. Thus Southwell overcame this mathematical problem by devising his Method of Relaxation, which of course is now widely used in science and engineering.

The Development and Use of the Computer

The ever-increasing magnitude of arithmetical calculations in more recent years has acted as a stimulus for the invention of all sorts of calculating devices. The introduction of the modern electronic digital computer (just after World War 2) solved most arithmetical difficulties so that the chronological development of traditional engineering methods could in fact be disregarded and the position re-appraised. Thus the engineer could use any method in a sequence of development to give him, above all, the information he required according to his terms of reference, rather than being primarily concerned with the amount of arithmetical work. (Examples of terms of reference are: Are deflections required? Is rib shortening important? Has the frame irregular features? Are there several load cases?) Thus many traditional methods of solving engineering problems, if still valid, can be gainfully used by the engineer who has access to a digital computer.

Since computers have become widely available, research workers' priorities have changed, as they can now make a completely fresh approach to time-honoured problems and obtain results in a more efficient way. For example, again in the elastic analysis of structures, the search in the computer age is essentially to find a single method which can be used on all types of frames, no matter how complicated or large, giving deflections and rotations as well as forces and moments to a high degree of accuracy in the minimum possible time. This type of approach is alluded to here as the use of *modern methods* (see chapters 2, 4 and 7). A limitation is the size of existing computers though clearly any new exceptional method devised requiring a larger machine would provide an incentive for manufacturers to supply the necessary product.

Thus the computer can be used in two ways.

With traditional methods: Processes developed before the invention of the computer can be employed directly as the basis for a computer program. Superseded or rejected techniques may come back into favour

With modern methods: Such methods are derived specially for the computer and are essentially a fresh approach.

Summary of the Problem

The problem is shown in Fig. 1.2. Such a chart is termed a *flow diagram*. Starting from the basic problem in civil or structural engineering, which could be in theory, design, experimental work or a combination of these three, the engineer can choose to use either traditional or modern methods. He should assess the practical feasibility of each alternative according to his own circumstances and then make his selection. Either alternative normally involves a fresh look at mathematics with the computer specially in mind.

The next stage in the flow diagram is programming (Fig. 1.2). Computer programs must either be obtained or specially written to carry out the calculations. The engineer may well have access to several computers, and their accessibility and availability will, in no small measure, dictate to him which alternative he should adopt in his own particular circumstances.

The preparation of the numerical data to suit the program and computer is the next step, leading to teleprinting, tape or card punching, etc., to present the data in the form of input required by the selected computer. This data is then fed into the computer, the calculations performed and the output produced in a form as prearranged in the program.

This completes the process. However, the engineer might well be displeased with the results obtained, and the entire process would then have to be repeated with different data until he was satisfied.

Experimental work (Fig. 1.2) brings the added problem of data logging and processing. This leads to a study of electronic equipment and instrumentation with the computer specially in mind. The engineering data could be recorded by such equipment directly on punched tape or card to be used with suitable programs as input data to the computer, so avoiding the need for any additional preparation of data, teleprinting, or the like. The engineer seeking to solve his problems with speed, accuracy and economy may well try to commit his organisation to the development of such a system.

Finally it should be possible to link together many of the processes shown in Fig. 1.2, and the limit of this enterprise, albeit never fully achieved, is full

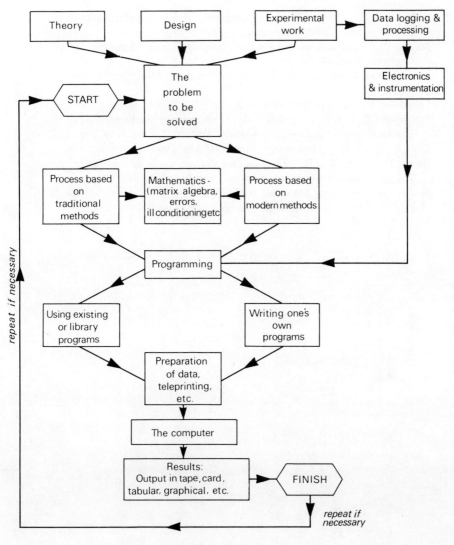

FIG. 1.2

automation. The various aspects of this system are discussed in greater detail later.

Bibliography

Pippard, A. J. S. and Baker, J. F. *Analysis of Engineering Structures*. 4th ed. Arnold. 1968
Steel Designers' Manual. 4th ed. Crosby Lockwood. 1972
Hoff, N. J. *Analysis of Structures*. Wiley. 1956

Robertson, R. G. 'Semi Graphical Integration Applied to the Analysis of Rigid Frames.' *Structural Engineer*, Nov. 1949

Naylor, N. 'Side Sway in Symmetrical Building Frames.' *Structural Engineer*, April 1950

Cross, H. and Morgan, N. D. *Continuous Frames of Reinforced Concrete.* Wiley

Lightfoot, E. *Moment Distribution*, Spon. 1961

Southwell, R. V. 'Stress Calculations in Frameworks by the Method of Systematic Relaxation of Constraints.' Parts I, II, Proc. Roy. Soc. 1935

Kani, G. *Analysis of Multistorey Frames.* Ungar. 1957

Binah, D. Advances in Deformation Distribution Methods for Elastic Frame Analysis. M.Sc. Thesis. University of London. 1961

Morley, A. *Strength of Materials.* 9th ed. Longmans. 1940

Morley, A. *Theory of Structures.* 3rd ed. Longmans. 1948

2
The Role of the Engineer

An Approach and a Philosophy

The previous chapter described how the electronic digital computer has freed the engineer to delve back into the past and rediscover many of the ingenious techniques derived over the years, yet left unused because of these computational difficulties until now. This approach, termed throughout the book as the use of traditional methods, can be very useful when dealing with a wide variety of practical problems.

The other approach, already outlined briefly, is to disregard all previous developments of methods and to start afresh, devising techniques with the computer specially in mind. This alternative, here called the use of modern methods, has received and is receiving the greater attention by research workers in the computer age, and offers a vast field for exploration and discovery.

But the engineer should always bear in mind that there is scope for development in both approaches and he should ask himself the following questions:

Is there a need for me or my organisation to build up a private computer system, with due regard to time and cost?
What computers in actual operation are available for me to use now or as required?
Where are these computers located geographically?
What is their storage capacity?
What is the normal time from the placing of a request to the receiving of the output?
What are the maximum and minimum hours of machine time permitted by any single user on any given occasion?
Can I, or my own operators, have personal access to the computer system?
What facilities are there for using library programs, for the writing of special programs?
What is the cost of computer time and of the various services provided?
What facilities are there for preparing and checking input data?
What about irregular, intermittent demands i.e. what computers are particularly readily accessible and available to me or my organisation when engaged on, say, trial and error calculations which could involve an unknown number of further runs on the computer as a result of the first output, until acceptable results are achieved?

Each engineer will obviously have different circumstances giving a variety of answers to such questions. The key question is: in my special circumstances, what is the best approach to my own computational problems?

9

For example, the use of a large-store fast computer such as Atlas to solve a comparatively small problem involving, say, half an hour of machine time, is wasted if all information has to be sent to, and returned from, the computer centre by post, with the additional delay of a reservation time of, say, four days, when there is the alternative of using a smaller, slower, but readily accessible computer taking, say, three hours of machine time with two hours travelling and half-a-day reservation time, together with the advantage of personal attendance.

Again it is unrealistic for a practical engineer, seeking speedy solutions to his immediate problems, to try to study and use an elaborate modern method of analysis which has not yet been programmed, or which can only be run on a few comparatively inaccessible computers.

Who Does What?

Adam Smith's principle of the division of labour is an important factor in organising the work. (See also chapter 11). Who shall undertake the various duties indicated in Fig. 1.2?

A civil or structural engineer should be in overall control, as a manager or section leader. His clear grasp of the problem should enable him to evaluate the significance of the results obtained and to act on them.

His team should consist of:
a civil or structural engineering assistant
a mathematician/programmer
an electronics engineer (for data logging, etc.)
a secretary/teleprinter and computer operator
a messenger/clerk
a draughtsman.

The exact composition of the team depends on considerations of economics, time, whether the staff are full-time or part-time (possibly being engaged in more conventional duties as well), the size of the problem and supply of work, training the staff, and the particular circumstances of the organisation, e.g. equipment owned, access to computer, etc.

The main duties of the various categories of staff should be as follows.

Civil or Structural Engineering Assistant
He and the manager should be responsible for appraising the basic engineering problem, converting it into a form capable of being processed by the team, and for supplying all details and information regarding the engineering theory and methods to be used in structural analysis, design, interpretation and the processing of experimental results. He will prepare flow diagrams from which the computer programs will be written, and will be available for consultation by the programmer during the writing and testing of these programs. He will instruct the electronics engineer as to what information is required from tests and experiments.

Mathematician/Programmer
He will advise the engineering assistant when required on mathematical techniques, and perhaps even prepare subroutines on mathematical matters for inclusion in the flow diagrams. His responsibilities include writing and testing of

the computer programs, and operating the computer as required. He should prepare full details of all programs written.

Electronics Engineer
If data logging devices are used to record experimental data such an engineer will design and assemble the circuits, obtain the results, and maintain the equipment.

Secretary/Teleprinter and Computer Operator
Her duties are to punch data and program tapes and cards, to operate the computer, and to provide a secretarial service to the team.

Messenger/Clerk
A messenger may be required to deliver the input data to and obtain the output from the computer centre, and can also assist in the clerical and miscellaneous work of the team.

Draughtsman
A draughtsman may be needed in certain projects to help the engineering assistant with his work.

The employment of a mathematician/programmer in the team is recommended (rather than suggesting that this work be assigned to an engineer) because the author considers the writing of substantial programs to be a task for a specialist.

A typical pattern in UK universities and colleges has been for the Department of Mathematics to assume control, to buy a computer, and then to staff the unit with mathematicians—often on a part-time basis, since they are normally engaged in teaching conventional mathematics as well: such staff then organise computer programming courses. Usually these courses are attended enthusiastically by students and lecturers alike from the Engineering Departments. When the courses end, the organiser naively expects the participants to go forth, write programs, and trouble him no more! (At this point the author would nevertheless like to record his appreciation to mathematicians for helping him in this way, even although he does not agree with this pattern of development!)

Clearly an engineer who spends most of his working day writing programs is no longer an engineer. Certainly he can write simple useful programs, but it is seldom long before he is faced with the need to write a vast, intricate program, and it is at this stage that the process breaks down. This is a task for an expert—a full-time computer programmer/mathematician—and the services of such a person should be available in each installation. Nevertheless, it is desirable that all engineers attend a programming course, not so much to study programming as such, but rather to appreciate what is involved in the process of writing programs.

Making the Most of Existing Resources and Methods

In the early 1950s, manufacturers began to sell computers and ancillary equipment (i.e. hardware—see Chapter 6) in earnest to customers who, at that time, were unreceptive, ill-prepared and uneducated for the computer age. These difficulties obliged manufacturers to write and collect a wide variety of programs on topics of potential interest over the whole field of science and engineering, as a means of advertising their computers.

Thus when an organisation bought a computer system (the hardware), it acquired a comprehensive array of special or library programs (software) at no extra cost. Manufacturers also built up their own teams of trained personnel and undertook to write special programs, and train their clients' staff on request— though on a satisfactory commercial basis.

It is in this way that many programs on various procedures in civil and structural engineering have been written. These library programs have been listed by the manufacturers and, more recently, by research establishments and engineering institutions, and prove most useful to the aspiring computer user if readily obtainable. The engineer can often use a library program merely by following the instructions in the accompanying booklet without necessarily understanding the method. However, in such circumstances, spot checks are recommended. Nowadays manufacturers are not so inclined to provide as many programs for their new range of computers as formerly. Indeed this has prompted a critic to suggest that manufacturers should supply their hardware free and charge only for the software!

Finally, textbooks, books of tables and formulae, and the like provide an extensive source of existing information and methods which the engineer can use when preparing his own programs. This procedure is illustrated in the examples in Chapters 3 and 9.

The Use of Methods Based on Modern Computational Processes

The library programs now available from an ever-increasing variety of sources, e.g. consulting engineers, industry, universities, colleges and government research organisations, as well as from computer manufacturers and engineering institutions, are often based on modern computational methods. Again, the engineer can use these to solve his immediate problems without necessarily understanding the methods employed though, in such cases, certain safeguards should be adopted. Indeed, from such a start the engineer may begin to appreciate the usefulness of computers and the associated computational processes, so encouraging him to promote such methods within his organisation. He can therefore play a significant part in establishing the system and can demonstrate, educate, train and use. The subjects to be covered using this approach have already been outlined in Chapter 1 and are more fully dealt with later.

3
Structural Analysis—Applications of Automatic Techniques Using the Traditional Methods

Outline

The usefulness of the computer in structural analysis can be quickly made apparent to the engineer by demonstrations in which numerical examples are evaluated using standard library programs. For example, it is possible to obtain beam deflections or bending moments in a portal frame, etc., using programs which are already available. The engineer can prepare input data for his own numerical problems, using the relevant library program to produce the required output information. Details of this process are generally given in a booklet or brochure provided with the program tapes or cards.

Provided the numerical data associated with any particular problem is within the specification of the library program and the capability of the computer, the user needs no special knowledge of the engineering method used (see Chapter 2), or of programming or mathematics (i.e. numerical methods and associated techniques); he merely follows simple instructions to obtain his solution. In this way the engineer uses the computer as a tool, just as the public use telephones, television sets, and other electrical devices. However, if the library program used is based on, say, formulae or a method of analysis with which the user is already familiar, the process can be even more readily appreciated and a comparison made with the corresponding manual process (i.e. using only hand multiplication, logarithmic tables, a slide rule, or a desk calculating machine).

This is the approach used in this chapter, and the examples given illustrate how some typical well-known problems in structural analysis can be solved using a computer. In all these cases full details of the programs are given so that the reader might use these with his own data on a suitable computer. They are written in the programming language known as ALGOL, although a knowledge of programming is not necessary at this stage. (An appropriate background on computer appreciation and programming is given in Chapter 6). However, those who have no previous experience of programming should examine the program 'print-outs' to familiarise themselves with the appearance, expressions, and type of notation used, so that they can begin to grasp a mental picture of what is involved in compiling a program. The examples are presented in increasing complexity, and it is important to realise that much general knowledge on the use of computers can be acquired by their systematic study.

Examples

Example 1:   Simply-supported Beam Carrying a Uniformly-distributed Load

Find the reactions, the maximum moment, deflection and slope of a simply-supported beam of constant cross-section carrying a uniformly-distributed load (see Fig. 3.1).

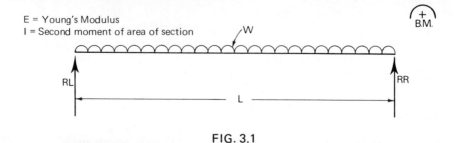

E = Young's Modulus
I = Second moment of area of section

FIG. 3.1

Span = L and Total applied load = W
Reaction at left-hand support = RL = $\frac{1}{2}W$
Reaction at right-hand support = RR = $\frac{1}{2}W$
Maximum bending moment (at midspan) = M_{max} = $WL/8$
Maximum deflection (at midspan)   = Δ_{max} = $5WL^3/(384EI)$
Maximum slope (at the supports)   = i_{max} = $WL^2/(24EI)$

This is a simple problem and would certainly not require the use of a computer to obtain the solution for one particular span and load, though such a means could well be justified for a large number of different spans and loads. However, the point of this first example is to introduce the concept and use of a computer program, and this is best done by means of a simple illustration.

It seems reasonable to deduce that if a computer were used, then the numerical values of L, W, E and I associated with a particular problem should be supplied in some way as input to the computer, and the required numerical values of RL, RR, M_{max}, i_{max}, and Δ_{max} obtained as output. In fact, such is the process, and the series of detailed instructions to be inserted in the computer in order to achieve this aim is the computer program. (Programs are written in different codes or 'languages', and details of these are given later.)

A program capable of solving this problem is Program DCTCES 1 given below. This program is written in ALGOL, a language that is described in Chapter 6. Full details of the program are as follows.

Program DCTCES 1:   Simply-supported beam carrying uniformly-distributed load.

```
CES 1;
"BEGIN"  "REAL" W, L, E, I, A, B, C, D;
         "READ" W, L, E, I;
         A: = W/2;
```

$$B: = W * L/8;$$
$$C: = W * L \uparrow 2/(24 * E * I);$$
$$D: = 5 * W * L \uparrow 3(384 * E * I);$$
"PRINT" A, B, C, D;

"END";

The entire program consists of the letters, numbers and symbols, starting with CES 1 (which is an abbreviated title) and finishing with "END";. This is known as the *program 'print-out'*.

The program is now prepared in the form necessary for input to the particular type of computer to be used. The more common types of input at the present time are punched cards or tape in an appropriate binary-type code, and these can be prepared using a device such as a teleprinter or a card punch. Such machines generally possess a typewriter keyboard and produce punched tapes or cards together, sometimes, with a print-out of the program or data being processed. This optional facility provides a means of checking the input tapes or cards for errors. As an example, the paper tape that is produced when punching Program DCTCES 1, using a teleprinter coded for use with an Elliott 4100 computer system, is shown in Fig. 3.2.

Particular numerical examples can now be chosen for use with the program. A complete set of input data consists of four numerical quantities for W, L, E, and I respectively. Then, when $W = 23$ tons, $L = 120$ inches, $E = 13{,}400$ tons/in^2, and $I = 200$ in^4, the input data consists of the four numbers

    23
    120
    13400
    200

An input data tape or a set of data cards should now be prepared for these numbers using the same teleprinter or device as employed for processing the program.

The computer to be used must now be prepared to receive the program and the data. This consists essentially of clearing all the stores and inserting an appropriate set of basic instructions that enable the computer to understand the language used and to carry out the necessary computational processes required. In this particular case the 'appropriate basic set of instructions' is the Algol Compiler which, in the form of a punched tape, a set of punched cards, or a magnetic tape, should be inserted into the computer. (Compilers are basic requirements in the use of any computer and are generally provided by the computer manufacturers. Further details are given in Chapter 6.)

The computer is now ready to receive the program and the data, and these should next be inserted. The calculation is carried out and the answers provided as output in the form of punched tape or cards that should be decoded using the same teleprinter or device, so obtaining a print-out. If a computer has a line printer attached, the print-out is obtained direct.

The output consists of four numerical quantities, RL, M_{max}, i_{max}, and

FIG. 3.2
PUNCHED PAPER TAPE FOR PROGRAM NUMBER DCTCES 1 USING
8 CHANNEL ELLIOTT 4100 PAPER TAPE CODE

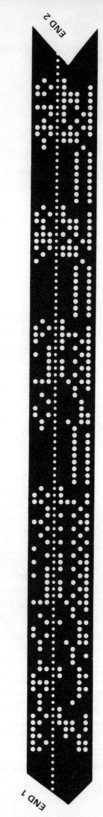

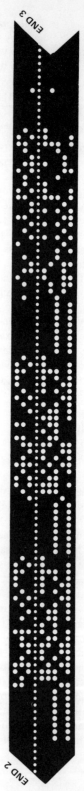

(*NOTE*: THIS IS, IN FACT, A SINGLE TAPE SHOWN HERE IN TWO PARTS
TO FACILITATE THIS PHOTOGRAPHIC REPRODUCTION.)

16

Δ_{max}, respectively. With the numerical data specified the output is

> 11.500000
> 345.00000
>   .00514925
>   .19309701

The answer therefore is

> $RL = RR = 11.500000$ tons
> M_{max} = 345.00000 ton. inches
> i_{max} = 0.00514925 radians
> Δ_{max} = 0.19309701 inches

and these figures can now be rounded off to realistic practical values.

The units for both input and output are consistent; i.e., all forces are in tons and all units of length are in inches. Program DCTCES 1 can therefore be used with data expressed in any units or any dimensional system, provided that this is similar for both input and output.

Consider the following numerical example in SI units.

$W = 100$ kN; $L = 10,000$ mm, $E = 200$ kN/mm^2, and $I = 3 \times 10^9$ mm^4.

In this case the input to the computer is

> 100
> 10000
> 200
> 3×10^9

and the output from the computer is

> 50
> 125000
>   .00069444
> 2.17014

The answer is therefore

> $RR = RL = 50$ kN
> M_{max} = 125,000 kNmm
> i_{max} = 0.00069444
> Δ_{max} = 2.17014 mm

and these figures can again be rounded off to realistic practical values. All forces are expressed in kN and all lengths in millimetres. It is more usual to express moments in kNm so that $M_{max} = 125,000$ kNmm $= 125$ kNm.

The reader should now carry out the above two numerical problems for himself. An experienced computer operator will be able to give the necessary assistance in punching tapes and cards and in operating the computer (though it is recommended that the engineer should then learn how to do such things himself).

Notes on Example 1

1.   Program DCTCES 1 can be readily modified to accept input in any units and provide output in any other units, though the modified program can then only be used for such units and no others. Another method of dealing with the various versions of SI units is to convert the input data into pure SI units. The output is then also in pure SI units and can subsequently be modified to the more usual multiples employed in structural engineering.

2.   This introductory basic program accepts four numbers as input and gives four others as output. When the data are as severely restricted as this, it is possible to confuse the various numbers, and errors can easily be made. This can be overcome by preparing standard format sheets for input data (especially in the cases where this is extensive) for use with the various programs, and by providing a more elaborate form of output which describes more clearly the meaning of the answers given. This is done in later examples.

Example 2:   Elastic Analysis of a Built-in Beam Carrying a Concentrated Load

Find the reactions, the fixed-end moments, the bending moments and the deflection at the load point P and the mid-point C, as well as the magnitude and location of the maximum deflection of any point on the span of a built-in beam (See Fig. 3.3).

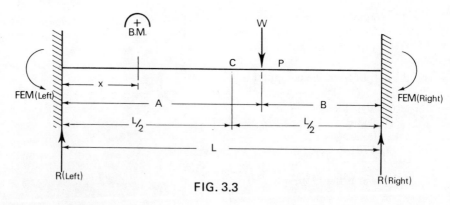

FIG. 3.3

Referring to Fig. 3.3:

Fixed end moment at left-hand
    support $= \text{FEM(LEFT)} = WB^2A/L^2$

Fixed end moment at right-hand
    support $= \text{FEM(RIGHT)} = WA^2B/L^2$

Vertical reaction at right-hand
    support $= \text{R(RIGHT)} = [WA + \text{FEM(RIGHT)}]/L$

Vertical reaction at left-hand
    support $= \text{R(LEFT)} = W - \text{R(RIGHT)}$

Bending moment at load point
$P = M[P]$ $= \text{FEM(LEFT)} - A \cdot \text{R(LEFT)}$

Deflection at P $= WA^3 B^3 / (3EIL^3)$

When $A \geqslant L/2$, deflection at C $= \dfrac{WB^2 [3LA - (L + 2A) \cdot L/2]}{24EIL}$

and when $A < L/2$, deflection at C $= \dfrac{WA^2 [3LB - (L + 2B) \cdot L/2]}{24EIL}$

When $A \geqslant L/2$, maximum deflection of any point on span $= 2WA^3 B^2 / [3EI(3L - 2B)^2]$

where x $= 2LA/(L + 2A)$

and when $A < L/2$, maximum deflection $= 2WA^2 B^3 / [3EI(3L - 2A)^2]$

where x $= L^2 / (3L - 2A)$

Program DCTCES 2 has been written (in Algol) to solve this problem allowing for all possible values of A. Once more the entire program consists of the letters, numbers and symbols given on pp. 20-21. Again the reader is not necessarily expected to be capable of understanding the preparation of a program at this stage. The print-out is given in full so that the engineer can note the work involved in writing such a program and can use it with his own data in due course.

This program is used in the same way as that given previously. Here, however, the input data consists of six numbers which are respectively $E, L, A, B, I,$ and W, where E is Young's Modulus, I is the second moment of area of a section of the beam, and $L, A,$ and B are the dimensions specified in Fig. 3.3. Also this time the six values must be preceded by a descriptive heading contained between the symbols / and \. This enables any input data to be identified later.

Then, when $E = 13{,}400$ tons/in^2, $L = 120$ in, $A = 80$ in, $B = 40$ in, $I = 200$ in^4, and $W = 15$ tons, with a descriptive heading: TRIAL DATA 1 12/3/70, the input is as follows:

/ TRIAL DATA 1 12/3/70 \
13,400
120
80
40
200
15

Note that it is also permissible to separate each number with a comma, so that the input data could equally well be:

/ TRIAL DATA 1 12/3/70 \
13400, 120, 80, 40, 200, 15,

```
CES2;

      "COMMENT"  DCTCES2 ELASTIC ANALYSIS OF BUILT-IN BEAM WITH
                 CONCENTRATED LOAD;

"BEGIN" "INTEGER" P;
        "INTEGER" "ARRAY" Q[1:100];
        P:=1;
        INSTRING(Q,P);
        P:=1;
        PUNCH(4);   SAMELINE;
"BEGIN" "REAL" E,L,A,B,I,W,X;
        "REAL" "ARRAY" M[1:4],D[1:3],R[1:2];
        "READ" E,L,A,B,I,W;
        M[1]:=W*(B↑2)*A/(L↑2);
        M[2]:=W*(A↑2)*B/(L↑2);
        R[2]:=(W*A+M[2]-M[1])/L;
        R[1]:=W-R[2];
        M[4]:=M[1]-A*R[1];
        D[2]:=W*A↑3*B↑3/(3*E*I*L↑3);
        "IF" A "GE" L/2 "THEN"
```

```
"BEGIN"  M[3]:=M[1]-R[1]*L/2;
         D[1]:=W*B↑2*(3*L*A-L*(L+2*A)/2)/(24*E*I*L);
         D[3]:=2*W*A↑3*B↑2/(3*E*I*(3*L-2*B)↑2);
         X:=2*L*A/(L+2*A);
"END"
"ELSE"
"BEGIN"  M[3]:=M[1]-R[1]*L/2+W*(L/2-A);
         D[1]:=W*A↑2*(3*L*B-L*(L+2*B)/2)/(24*E*I*L);
         D[3]:=2*W*A↑2*B↑3/(3*E*I*(3*L-2*A)↑2);
         X:=L↑2/(3*L-2*A);
"END";
"PRINT"  ''L'',  DCTCES 2:  ELASTIC ANALYSIS OF BUILT-IN BEAM',
         ' WITH CONCENTRATED LOAD','L4'';
OUTSTRING(Q,P);
"PRINT"  ''L4'';
"PRINT"  'E= ',E,''S3'','L= ',L,''S3'','A= ',A,''S3'',
         'B= ',B,''S3'','I= ',I,''S3'','W= ',W,''S3'',
         ''L6'','FEM(LEFT)= ',M[1],''L'',
         'FEM(RIGHT)= ',M[2],''L'','M[C]= ',M[3],''L'',
         'M[P]= ',M[4],''L'','R(LEFT)= ',R[1],''L'',
         'R(RIGHT)= ',R[2],''L'',
         'DEFLEXION AT C= ',D[1],''L'',
         'DEFLEXION AT P= ',D[2],''L'',
         'MAXIMUM DEFLEXION= ',D[3],''L'',
         'WHERE   X = ',X;
"END";

"END";
```

The output from the computer this time is:

DCTCES 2: ELASTIC ANALYSIS OF BUILT-IN BEAM WITH
CONCENTRATED LOAD.

TRIAL DATA 1 12/3/70

E = 13400.000	L = 120.00000	A = 80.000000
B = 40.000000	I = 200.00000	W = 15

FEM(LEFT)	= 133.33333
FEM(RIGHT)	= 266.66667
M[C]	= −100.00000
M[P]	= −177.77778
R(LEFT)	= 3.8888889
R(RIGHT)	= 11.111111
DEFLEXION AT C	= .03731343
DEFLEXION AT P	= 0.3537866
MAXIMUM DEFLEXION	= .03898873
WHERE X	= 68.571429

This output is virtually self-explanatory and is a distinct improvement on the
restricted output given in Example 1. It consists of the following:

the program number and title
the input data descriptive heading
the input data in readily identifiable form as a check
the output data in readily identifiable form.

The units used in the output are consistent with those in the input where loads
and forces are in tons and dimensions in inches. Thus in the output, moments
are in ton. inches, forces are in tons, and deflections and dimensions are in inches.

As before any system of units may be used for the input data and the output
will be in the corresponding consistent units.

Consider the following example in SI units.

Title: DATA 2   SI UNITS   JULY 70

$E = 200 \text{ kN/mm}^2$, $L = 15$ m, $A = 10$ m, $B = 5$ m, $I = 4 \times 10^9 \text{ mm}^4$,
and $W = 80$ kN.

This data could either be converted into pure SI units (i.e. N and m only), or
loads and forces could be expressed in kN and lengths in millimetres. Using
the latter alternative the input data is

/ DATA 2   SI UNITS   JULY 70 \

200
15000

10000
5000
4×10^9
80

and the output from the computer is

DCTCES 2: ELASTIC ANALYSIS OF BUILT-IN BEAM WITH
CONCENTRATED LOAD

DATA 2   SI UNITS   JULY 70

E = 200	L = 15000	A = 10000
B = 5000	I = 4×10^9	W = 80
FEM(LEFT)		= 88888.889
FEM(RIGHT)		= 177777.78
M [C]		= −66666.667
M [P]		= −118518.52
R(LEFT)		= 20.740741
R(RIGHT)		= 59.259259
DEFLEXION AT C		= 1.3020833
DEFLEXION AT P		= 1.2345679
MAXIMUM DEFLEXION		= 1.3605442
WHERE X		= 8571.4286

Here, the moments are in kNmm, forces are in kN, and deflections and dimensions are in mm. It is simple to convert moments into the more usual kNm units.

Notes on Example 2

1.   The numerical output in the above problems can now be rounded off to realistic practical values.
2.   The reader should develop the practice of keeping the input, output, and the corresponding diagram for one set of data together in his calculations for future reference. Computer tapes or cards should always be properly labelled and filed from the outset, and all material associated with unsuccessful operations should be destroyed. This process is made most effective by using a diary, log book, or register, and it is most important that the staff concerned (see Chapter 2) should use a satisfactory filing system.
3.   As far as possible computer programs should be written in a flexible manner, so that different data can be readily used. For example, even if it is only required to multiply 2 by 4, it is better to write a program which will multiply x by y.
4.   All reasonable possibilities in the choice of input data should be allowed for when preparing a program (rather than only making provision for the numerical data to hand at the time). Thus in Example 2 provision is made for the value of *A* being greater than, equal to, or less than, that of *B*.

Example 3:   Elastic Analysis of Various Single-Span Beams with Either a
Concentrated or a Uniformly-Distributed Load

Find the reactions, fixed-end moments, and bending moments and deflections at
various points on a beam carrying either a concentrated or a uniformly-
distributed load in which the ends are either free, pinned, or fixed (see Fig. 3.4).

In this example there are 18 possible conditions of support, for all of which
provision must be made in the program. It is convenient to specify the various
conditions by means of the integers P, Q, and R. P indicates the nature of the
applied loading, and can have the values of 1 or 2.

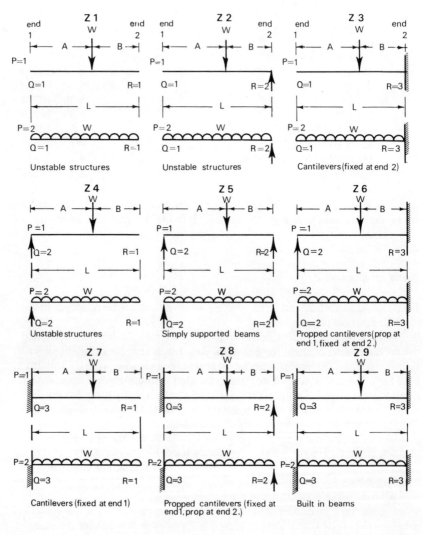

FIG. 3.4

When *P* is 1 a concentrated load *W* is applied at a point on the span; when *P* is 2 a uniformly-distributed load *W* is applied over the entire span. *Q* and *R* indicate the nature of the support at ends 1 and 2 respectively.

When *Q* or *R* is 1, the respective end is unsupported; when *Q* or *R* is 2, the respective end is pinned; and when *Q* or *R* is 3, the respective end is fully fixed.

There are nine different types of beams, *Z*1 to *Z*9 inclusive, and these are described in conventional structural engineering terminology.

Thus when *P* = 1 or 2:

For $Z = 1, Q = 1, R = 1$ title: Unstable structure
For $Z = 2, Q = 1, R = 2$ title: Unstable structure
For $Z = 3, Q = 1, R = 3$ title: Cantilever (fixed at end 2)
For $Z = 4, Q = 2, R = 1$ title: Unstable structure
For $Z = 5, Q = 2, R = 2$ title: Simply-supported beam
For $Z = 6, Q = 2, R = 3$ title: Propped cantilever (fixed at end 2, prop at end 1)
For $Z = 7, Q = 3, R = 1$ title: Cantilever (fixed at end 1)
For $Z = 8, Q = 3, R = 2$ title: Propped cantilever (fixed at end 1, prop at end 2)
For $Z = 9, Q = 3, R = 3$ title: Built-in beam

Program DCTCES 3 has been written in Algol to solve this problem (pp. 26-37).

Input: The input consists of nine numbers which are respectively *P, Q, R, L, A, B, W, E, I*, where *E* is Young's Modulus, *I* is the second moment of area of the uniform section of the beam, and the other quantities are given in Fig. 3.4. (Nine numbers must always be specified as input: see notes below).

Output: The output consists of:

(a)   The number and title of the program.
(b)   A print-out of the input data, *P, L, A, B, W, E* and *I* as a check.
(c)   The label and description of the structure.
(d)   The computed values of:

REACTION AT END 1
REACTION AT END 2
F.E.M. AT END 1
F.E.M. AT END 2
B.M. AT LOAD POINT
MAX. POSITIVE B.M.
MAX. NEGATIVE B.M.
DISTANCE OF MAX. POSITIVE B.M. FROM END 1
DISTANCE OF MAX. NEGATIVE B.M. FROM END 1
DISTANCE OF MAX. DEFLEXION FROM END 1
DEFLEXION OF END 1
DEFLEXION OF END 2
DEFLEXION OF LOAD POINT (if applicable)
MAX. DEFLEXION ON SPAN

and (e) The statement – END OF PROGRAM.

Print-out of Program DCTCES 3: Elastic analysis of various single span beams
carrying either a concentrated or uniformly-distributed load

```
CES3;

"COMMENT" DCTCES 3 ELASTIC ANALYSIS OF VARIOUS SINGLE SPAN BEAMS
         WITH EITHER A CONCENTRATED OR UNIFORMLY DISTRIBUTED
         LOAD;

"BEGIN" "REAL" L,A,B,W,E,I,F1,F2,M1,M2,M3,M4,M5,X1,X2,X3,D1,D2,D3,D4;
        "INTEGER" P,Q,R;
        PUNCH(4);  SAMELINE;
        "PRINT" ''L'',''DCTCES 3 ELASTIC ANALYSIS OF VARIOUS SINGLE SPAN''
        , BEAMS'',''L''WITH EITHER A CONCENTRATED OR UNIFORMLY'',
        , DISTRIBUTED LOAD'',''L3''; 

        "READ" P,Q,R,L,A,B,W,E,I;
        "PRINT" ''L'',''P= '',SAMELINE,DIGITS(1),P,''L'',''L= '',
        ALIGNED(6,3), L,''L'',''A= '',A,''L'',''B= '',B,''L'',
        ''W= '',W,''L'',''E= '',SCALED(5),E,''L'',''I= '',I;

        "IF" Q=1 "AND" R=1 "THEN" "GOTO" Z1;
        "IF" Q=1 "AND" R=2 "THEN" "GOTO" Z2;
        "IF" Q=1 "AND" R=3 "THEN" "GOTO" Z3;
        "IF" Q=2 "AND" R=1 "THEN" "GOTO" Z4;
        "IF" Q=2 "AND" R=2 "THEN" "GOTO" Z5;
        "IF" Q=2 "AND" R=3 "THEN" "GOTO" Z6;
        "IF" Q=3 "AND" R=1 "THEN" "GOTO" Z7;
        "IF" Q=3 "AND" R=2 "THEN" "GOTO" Z8;
        "IF" Q=3 "AND" R=3 "THEN" "GOTO" Z9;
        "PRINT" ''L'',''DATA WRONG'';
        "GOTO" FIN;
```

```
Z1:"PRINT" ''L'',''Z1 UNSTABLE STRUCTURE'';
    "GOTO" FIN;

Z2:"PRINT" ''L'',''Z2 UNSTABLE STRUCTURE'';
    "GOTO" FIN;
Z4:"PRINT" ''L'',''Z4 UNSTABLE STRUCTURE'';
    "GOTO" FIN;
Z3:"PRINT" ''L'',''Z3 CANTILEVER (FIXED AT END 2)'';
    "IF" P=1 "THEN"
"BEGIN" F1:=0.0;
    F2:=W;
    M1:=0.0;
    M2:=W*B;
    M3:=0.0;
    M4:=M2;
    M5:=.50;
    X1:=L;
    X2:=.50;
    X3:=0.0;
    D1:=W*B↑2*(L+A/2)/(3*E*I);
    D2:=0.0;
    D3:=W*B↑3/(3*E*I);
    D4:=D1;
    "GOTO" Y1;

"END";
```

27

```
F1:=0.0;
F2:=W;
M1:=0.0;
M2:=W*L/2;
M3:=.50;
M4:=M2;
M5:=.50;
X1:=L;
X2:=.50;
X3:=0.0;
D1:=W*L↑3/(8*E*I);

D2:=0;
D3:=.50;
D4:=D1;
"GOTO" Y1;
Z5:"PRINT" ''L'',`Z5 SIMPLY SUPPORTED BEAM';
"IF" P=1 "THEN"
"BEGIN" F1:=W*B/L;
F2:=W*A/L;
M1:=0.0;
M2:=0.0;
M3:=-W*A*B/L;
M4:=.50;
M5:=M3;
X1:=.50;
X2:=A;
```

```
"IF" ABS(A-L/2) "GE" ∎-6 "THEN"
X3::=SQRT(A*(L+B)/3) "ELSE"
X3::=SQRT(B*(L+A)/3);
D1::=0.0;
D2::=0.0;
D3::=W*A↑2*B↑2/(3*E*I*L);
"IF" ABS(A-L/2) "GE" ∎-6 "THEN"
D4::=W*A*B*(L+B)*SQRT(3*A*(L+B))/(27*E*I*L)   "ELSE"
D4::=W*A*B*(L+A)*SQRT(3*B*(L+A))/(27*E*I*L);
"GOTO" Y1;

"END";
    F1::=W/2;
    F2::=F1;
    M1::=0.0;
    M2::=0.0;
    M3::=∎50;
    M4::=∎50;
    M5::=-W*L/8;

    X1::=∎50;
    X2::=L/2;
    X3::=X2;
    D1::=0.0;
    D2::=0.0;
    D3::=∎50;
    D4::=5*W*L↑3/(384*E*I);
    "GOTO" Y1;
```

```
Z6:"PRINT" ''L'','Z6 PROPPED CANTILEVER (FIXED AT END 2, ',
           'PROP AT END 1)';
"BEGIN" "IF" P=1 "THEN"
        F1:=W*Bↆ2*(2*L+A)/(2*Lↆ3);
        F2:=W*A*(3*Lↆ2-Aↆ2)/(2*Lↆ3);
        M1:=0.0;
        M2:=W*A*B*(L+A)/(2*Lↆ2);
        M3:=-W*A*Bↆ2*(2*L+A)/(2*Lↆ3);
        M4:=M2;
        M5:=M3;
        D1:=0.0;
        D2:=0.0;
        D3:=W*Aↆ2*Bↆ3*(3*L+A)/(12*E*I*Lↆ3);
        "IF" ABS(B-A*SQRT(2))<ₒ-6 "THEN"
        D4:=W*Lↆ3/(102*E*I)    "ELSE"
        "IF" B>A*SQRT(2) "THEN"
        D4:=W*A*Bↆ3*(L+A)ↆ3/(3*E*I*(3*Lↆ2-Aↆ2)ↆ2)    "ELSE"
        D4:=W*A*Bↆ2*SQRT(A/(2*L+A))/(6*E*I);
        X1:=L;
        X2:=A;
        "IF" ABS(B-A*SQRT(2))<ₒ-6 "THEN"
        X3:=A    "ELSE"
        "IF" B>A*SQRT(2) "THEN"
        X3:=L-2*L*M2/(W*A+M2)    "ELSE"
        X3:=SQRT(L*(W*A*B-L*M2)/(W*B-M2));
        "GOTO" Y1;

"END";
```

```
F1:=3*W/8;
F2:=5*W/8;
M1:=0.0;
M2:=W*L/8;
M3:=-50;
M4:=M2;
M5:=-9*W*L/128;
X1:=L;
X2:=3*L/8;
X3:=0.4215*L;
D1:=0.0;
D2:=0.0;
D3:=-50;
D4:=W*L↑3/(185*E*I);
"GOTO" Y1;
Z7:"PRINT" ''L'','Z7 CANTILEVER (FIXED AT END 1)'';
"IF" P=1 "THEN"
"BEGIN" F1:=W;
F2:=0.0;
M1:=W*A;
M2:=0.0;
M3:=0.0;
M4:=M1;
M5:=-50;
X1:=0.0;
X2:=-50;
X3:=L;
```

```
D1:=0.0;
D2:=W*A↑2*(L+B/2)/(3*E*I);
D3:=W*A↑3/(3*E*I);
D4:=D2;
"GOTO" Y1;

"END";

F1:=W;
F2:=0.0;

M1:=W*L/2;
M2:=0.0;
M3:=.5Ø;
M4:=M1;
M5:=.5Ø;
X1:=0.0;
X2:=.5Ø;
X3:=L;
D1:=0.0;
D2:=W*L↑3/(8*E*I);
D3:=.5Ø;
D4:=D2;
"GOTO" Y1;
Z8:"PRINT" '','L'','Z8 PROPPED CANTILEVER (FIXED AT END 1, ',',
          'PROP AT END 2)';

    "IF" P=1 "THEN"
"BEGIN" F1:=W*B*(3*L↑2-B↑2)/(2*L↑3);
        F2:=W*A↑2*(2*L+B)/(2*L↑3);
```

```
M1:=W*A*B*(L+B)/(2*L↑2);
M2:=0.0;
M3:=-W*A↑2*B*(2*L+B)/(2*L↑3);
M4:=M1;
M5:=M3;
X1:=0.0;
X2:=A;
"IF" ABS(A-B*SQRT(2))<=-6 "THEN"
X3:=A   "ELSE"
"IF" A>B*SQRT(2) "THEN"
X3:=2*L*M1/(W*B+M1)   "ELSE"
X3:=L-SQRT(L*(W*A*B-L*M1)/(W*A-M1));

D1:=0.0;
D2:=0.0;
D3:=W*A↑3*B↑2*(4*L-A)/(12*E*I*L↑3);
"IF" ABS(A-B*SQRT(2))<=-6 "THEN"
D4:=W*L↑3/(102*E*I)   "ELSE"
"IF" A>B*SQRT(2) "THEN"
D4:=W*A↑3*B*(L+B)↑3/(3*E*I*(3*L↑2-B↑2)↑2)   "ELSE"
D4:=W*A↑2*B*SQRT(B/(2*L+B))/(6*E*I);
"GOTO" Y1;

F1:=5*W/8;
F2:=3*W/8;
M1:=W*L/8;
M2:=0.0;
```

"END";

```
M3:=-50;
M4:=M1;
M5:=-9*W*L/128;
X1:=0.0;
X2:=5*L/8;
X3:=0.5785*L;
D1:=0.0;
D2:=0.0;
D3:=-50;
D4:=W*L↑3/(185*E*I);
"GCTO" Y1;
Z9:"PRINT" ''L'',''Z9 BUILT-IN BEAM'';
"IF" P=1 "THEN"
"BEGIN" F1:=W*B↑2*(L+2*A)/L↑3;
F2:=W*A↑2*(L+2*B)/L↑3;
M1:=W*A↑2*B/L↑2;
M2:=W*A↑2*B/L↑2;
M3:=-2*W*A↑2*B↑2/L↑3;
"IF" A<B "THEN"
M4:=W*A↑2*B/L↑2    "ELSE"
M4:=W*A↑2*B/L↑2;

M5:=-2*W*A↑2*B↑2/L↑3;
"IF" ABS(A-B)<■-6 "THEN"
X1:=-2  "ELSE"
"IF" A>B "THEN"    X1:=L;
X1:=0  "ELSE"    X1:=L;
X2:=A;
```

```
"IF" ABS(A-B)<=-6 "THEN"
X3:=L/2 "ELSE"
"IF" A>B "THEN"
X3:=2*L*A/(L+2*A)    "ELSE"
X3:=L↑2/(L+2*B);
D1:=0.0;
D2:=0.0;
D3:=W*A↑3*B↑3/(3*E*I*L↑3);
"IF" ABS(A-B)<=-6 "THEN"    "ELSE"
D4:=W*L↑3/(192*E*I)    "ELSE"
"IF" A>B "THEN"
D4:=2*W*A↑3*B↑2/(3*E*I*(L+2*A)↑2)    "ELSE"
D4:=2*W*A↑2*B↑3/(3*E*I*(L+2*B)↑2);
"GOTO" Y1;
"END";

F1:=W/2;
F2:=F1;
M1:=W*L/12;
M2:=M1;
M3:=.50;
M4:=M1;
M5:=-W*L/24;
X1:=-2;
X2:=L/2;
X3:=X2;
D1:=0.0;
D2:=0.0;
```

35

```
D3:=.50;
D4:=W*L↑3/(384*E*I);
"GOTO" Y1;

Y1:"PRINT" ''L3'',''REACTION AT END 1 = '',ALIGNED(6,3),F1,''L'',
          ''REACTION AT END 2 = '',F2,''L'',''F.E.M. AT END 1 = '',
          M1,''L'',''F.E.M. AT END 2 = '',M2,''L'',
          ''B.M. AT LOAD POINT = '';
"IF" ABS(M3-.50)<.-6 "THEN"
"PRINT" ''N.A.''  "ELSE"
"PRINT" ALIGNED(6,3),M3;
"PRINT" ''L'',''MAX. POSITIVE B.M. = '';
"IF" ABS(M4-.50)<.-6 "THEN"
"PRINT" ''N.A.''  "ELSE"
"PRINT" ALIGNED(6,3),M4;
"PRINT" ''L'',''MAX. NEGATIVE B.M. = '';
"IF" ABS(M5-.50)<.-6 "THEN"
"PRINT" ''N.A.''  "ELSE"
"PRINT" ALIGNED(6,3),M5;
"PRINT" ''L'',''DISTANCE OF MAX. POSITIVE B.M. FROM END 1 = '';
```

```
"IF" ABS(X1-50)<=-6 "THEN"
"PRINT" 'N.A.'  "ELSE"
"IF" ABS(X1+2)<=-6 "THEN"
"PRINT" '0,L'  "ELSE"
"PRINT" ALIGNED(6,3),X1;
"PRINT" ''L'','DISTANCE OF MAX. NEGATIVE B.M. FROM END 1 = ';
"IF" ABS(X2-50)<=-6 "THEN"
"PRINT" 'N.A.'  "ELSE"
"PRINT" ALIGNED(6,3),X2;
"PRINT" ''L'','DISTANCE OF MAX. DEFLEXION FROM END 1 = ',
        ALIGNED(6,3),X3,''L'','DEFLEXION OF END 1 = ',
        ALIGNED(3,6),D1,''L'','DEFLEXION OF END 2 = ',D2,''L'',
        'DEFLEXION OF LOAD POINT = ';
"IF" ABS(D3-50)<=-6 "THEN"
"PRINT" 'N.A.'  "ELSE"
"PRINT" ALIGNED(3,6),D3;
"PRINT" ''L'','MAXIMUM DEFLEXION ON SPAN = ',ALIGNED(3,6),D4;
"PRINT" ''L3'','END OF PROGRAM';
FIN:"PRINT"
"END";
```

37

The output always includes the print-out of the above descriptions alongside their numerical values.

Numerical Illustration

Consider the beam shown in Fig. 3.5. Then, for $I = 12 \times 10^9$ mm^4 and $E = 200$ kN/mm^2, the input (for P, Q, R, L, A, B, W, E, I respectively) is as follows.

> 1
> 3
> 2
> 90000
> 60000
> 30000
> 100
> 200
> 12×10^9

and the output from the computer is as follows.

DCTCES 3: ELASTIC ANALYSIS OF VARIOUS SINGLE-SPAN BEAMS WITH EITHER A CONCENTRATED OR UNIFORMLY-DISTRIBUTED LOAD

> P = 1
> L = 90000.00
> A = 60000.00
> B = 30000.00
> W =   100.00
> E = 2×10^2
> I = 1.2×10^{10}
> Z8 PROPPED CANTILEVER (FIXED AT END 1, PROP AT END 2)

REACTION AT END 1	= 48.148
REACTION AT END 2	= 51.852
F.E.M. AT END 1	= 1.3333×10^6
F.E.M. AT END 2	= 0.000
B.M. AT LOAD POINT	= 1.5556×10^6
MAX POSITIVE B.M.	= 1.3333×10^6
MAX. NEGATIVE B.M.	= 1.5556×10^6
DISTANCE OF MAX. POSITIVE B.M. FROM END 1	= 0.000
DISTANCE OF MAX. NEGATIVE B.M. FROM END 1	= 60000.000
DISTANCE OF MAX. DEFLEXION FROM END 1	= 55384.615
DEFLEXION OF END 1	= 0.000000
DEFLEXION OF END 2	= 0.000000
DEFLEXION OF LOAD POINT	= 277.777778
MAXIMUM DEFLEXION ON SPAN	= 284.023669

END OF PROGRAM

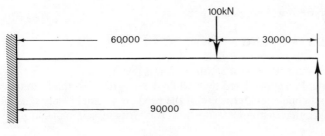

FIG. 3.5

In the above results, reactions are in kN, moments in kNmm, dimensions and deflections in mm, and the numerical values can be rounded off to practical values.

Notes on Example 3

1.   The formulae used to obtain the above expressions are given in Appendix 1. These have been obtained from such works as References 1-3 and in some instances from first principles.

2.   If, by mistake, the input data had been given as

    1
    1
    1
    90000
    60000
    30000
    100
    200
    12×10^9

the output would contain the heading:

     Z1 UNSTABLE STRUCTURE

and no moments or reactions would be given.

3.   If a particular item in the output is not applicable, the letters N.A. will be printed, e.g.

     B.M. AT LOAD POINT = N.A.

in the case of a uniformly-distributed load.

4.   Nine numbers must always be given as input. Therefore, with a uniformly distributed load, where the dimensions A and B are not relevant, insert $A = 0$ and $B = 0$.

5.   In the case of certain obvious errors in input data (e.g. putting Q or $R = 4$,

say), the output will contain the phrase:

DATA WRONG

6.   As before, any form of consistent units may be used.
7.   It should be appreciated that, while the program and the formulae are rather lengthy and complicated, the use of the program is simple; only nine quantities need be specified as input and the output is virtually self-explanatory.

Example 4:   Elastic Analysis of a Single Bay, Pitched Roof, Portal Frame With Pinned Bases.

Find the reactions, and bending moments at the joints, on a pitched-roof portal frame with pinned bases when subjected to any combination of the loads shown in Fig. 3.6. The fourteen cases of applied loading cover the more common types of loads which may be applied to a building constructed from such frames; namely, dead, imposed, wind (pressure and suction), vertical crane and crane surge loads. By superposition a typical practical loading system can be subdivided into a sum of any combination of the basic cases shown in Fig. 3.6, and the total reactions and bending moments obtained by addition.
     Program DCTCES 4 has been written (in Algol) to obtain the solutions in this way (pp. 42-47).

Input: The input consists of the following items which must be drawn up in the order specified
(a) A descriptive heading (contained between the symbols / and \) to identify the job, the frame, and/or the loading case.
(b) N, the number of basic loading cases used to make up the total loading, this integer being in the range 1 to 14.
(c) N numbers. These integers are the identification numbers of the N basic loading cases specified in Fig. 3.6.
(d) Eight numbers specifying the dimensions of the frame, namely L, h, f, I_1, I_2, a, b and c, which must be given in this order. Where a, b, and c do not apply, insert $a = 0$, $b = 0$, and $c = 0$, so that eight numbers are always specified under this item.
(e) N numbers, being the individual applied loads in all N basic loading cases. These are the values of either W or P shown in Fig. 3.6 and are positive when acting as shown. One number must be given for each basic loading case specified for a particular problem, and these must be presented in the order given in (c) above.

Output: The output from the computer is as follows
(a) The number and title of the program.
(b) The descriptive heading.
(c) The computed values of H_A, H_E, V_A, V_E, M_E, M_C, and M_D
(The above descriptions are also printed out together with their numerical values).

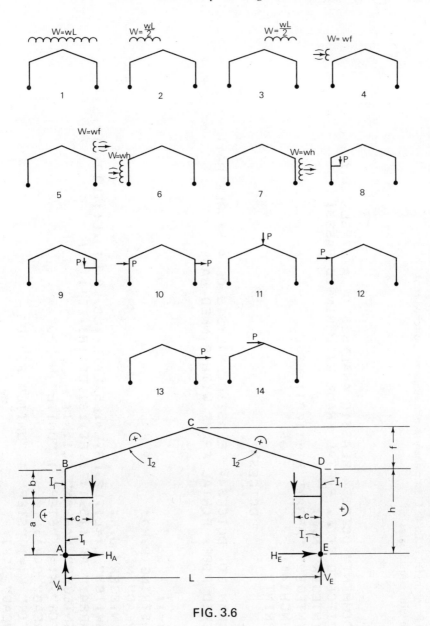

FIG. 3.6

Numerical Illustration

Consider the frame in Fig. 3.7. The loading is due to wind + crane + imposed loads, and the reactions at A and E (i.e. H_A, V_A, H_E, and V_E), and the bending moments at B, C, and D (i.e. M_B, M_C, and M_D) induced from this combined loading are required.

Print-out of Program DCTCES 4: Elastic analysis of a single bay, pitched roof, portal frame with pinned bases

```
CES4;
    "COMMENT" DCTCES4 ELASTIC ANALYSIS OF A SINGLE BAY,
    PITCHED ROOF, PORTAL FRAME WITH PINNED BASES;
"BEGIN" "INTEGER" N,M;
    "INTEGER" "ARRAY" P[1:100];
    PUNCH(4);
    "PRINT" '

                DCTCES4

PROGRAM NO. DCTCES4.  ELASTIC ANALYSIS OF A SINGLE BAY,
PITCHED ROOF, PORTAL FRAME WITH PINNED BASES

';
    M:=1;
    INSTRING(P,M);
    "READ" N;
"BEGIN" "INTEGER" J,K;
    "SWITCH" L:=L1,L2,L3,L4,L5,L6,L7,L8,L9,L10,L11,L12,L13,L14;
    "ARRAY" A[1:15,1:7],B[1:6],D[1:8],E[1:14];
    "INTEGER" "ARRAY" C[1:N];
    "FOR" K:=1 "STEP" 1 "UNTIL" N "DO"
    "READ" C[K];
    "FOR" K:=1 "STEP" 1 "UNTIL" 8 "DO"
    "READ" D[K];
```

```
"FOR" K:=1 "STEP" 1 "UNTIL" N "DO"
"READ" E[C[K]];
B[1]:=D[5]*D[2]/(D[4]*(D[1]+2/4+D[3]+2)+0.5);
B[2]:=D[3]/D[2];
B[3]:=1+B[2];
B[4]:=2*(B[1]+1)+B[3];
B[5]:=1+2*B[3];
B[6]:=B[4]+B[3]*B[5];
"FOR" K:=1 "STEP" 1 "UNTIL" 15 "DO"
"FOR" J:=1 "STEP" 1 "UNTIL" 7 "DO"
A[K,J]:=0.0;
"FOR" K:=1 "STEP" 1 "UNTIL" N "DO"
"BEGIN" "GOTO" L[C[K]];
L1:A[1,1]:=E[1]*D[1]*(3+5*B[3])/(16*D[2]*B[6]);
A[1,2]:=-A[1,1];
A[1,3]:=E[1]/2;
A[1,4]:=E[1]/2;
A[1,5]:=D[2]*A[1,1];
A[1,6]:=A[1,1]*(D[2]+D[3])-1/2*A[1,3]*D[1]+1/8*E[1]*D[1];
A[1,7]:=-A[1,1]*D[2];
"GOTO" M1;
L2:A[2,1]:=E[2]*D[1]*(3+5*B[3])/(16*D[2]*B[6]);
A[2,2]:=-A[2,1];
A[2,3]:=3*E[2]/4;
A[2,4]:=E[2]/4;
A[2,5]:=D[2]*A[2,1];
A[2,6]:=A[2,1]*(D[2]+D[3])-1/2*A[2,3]*D[1]+1/4*E[2]*D[1];
A[2,7]:=-A[2,1]*D[2];
"GOTO" M1;
```

```
L3:A[3,1]:=E[3]*D[1]*(3+5*B[6])/(16*D[2]*B[6]);
   A[3,2]:=-A[3,1];
   A[3,3]:=E[3]/4;
   A[3,4]:=3*E[3]/4;
   A[3,5]:=D[2]*A[3,1];
   A[3,6]:=A[3,1]*(D[2]+D[3])-1/2*A[3,3]*D[1];
   A[3,7]:=-A[3,2]*D[2];
   "GOTO" M1;
L4:A[4,1]:=-E[4]*D[3]*(B[5]+B[3])/(8*B[6]*D[2])-E[4]/2;
   A[4,2]:=-A[4,1]-E[4];
   A[4,3]:=-E[4]*(D[2]+D[3]/2)/D[1];
   A[4,4]:=-A[4,3];
   A[4,5]:=D[2]*A[4,1];
   A[4,6]:=A[4,1]*(D[2]+D[3])-1/2*A[4,3]*D[1]+1/2*E[4]*D[3];
   A[4,7]:=-A[4,2]*D[2];
   "GOTO" M1;
L5:A[5,1]:=+E[5]*D[3]*(B[5]+B[3])/(8*B[6]*D[2])-E[5]/2;
   A[5,2]:=-A[5,1]-E[5];
   A[5,3]:=-E[5]*(D[2]+D[3]/2)/D[1];
   A[5,4]:=-A[5,3];
   A[5,5]:=D[2]*A[5,1];
   A[5,6]:=A[5,1]*(D[2]+D[3])-1/2*A[5,3]*D[1];
   A[5,7]:=-A[5,2]*D[2];
   "GOTO" M1;
L6:A[6,2]:=-E[6]*(2*(B[4]+B[5])+B[1])/(8*B[6]);
   A[6,1]:=-A[6,2]-E[6];
   A[6,3]:=-E[6]*D[2]/(2*D[1]);
   A[6,4]:=-A[6,3];
```

```
  A[6,5]:=D[2]*A[6,1]+E[6]*D[2]/2;
  A[6,6]:=A[6,1]*(D[2]+D[3])-1/2*A[6,3]*D[1]+E[6]*(D[3]+1/2*D[2]);
  A[6,7]:=-A[6,2]*D[2];
  "GOTO" M1;
L7:A[7,1]:=-E[7]*(2*(B[4]+B[5])+B[1])/(8*B[6]);
  A[7,2]:=-A[7,1]-E[7];
  A[7,3]:=-E[7]*D[2]/(2*D[1]);
  A[7,4]:=-A[7,3];
  A[7,5]:=D[2]*A[7,1];
  A[7,6]:=A[7,1]*(D[2]+D[3])-1/2*A[7,3]*D[1];
  A[7,7]:=-A[7,2]*D[2]-1/2*E[7]*D[2];
  "GOTO" M1;
L8:A[8,1]:=E[8]*D[8]*(B[4]+B[5]-B[1]*(3*(D[6]*D[6])/(D[2]*
  D[2])-1))/(2*B[6]*D[2]);
  A[8,2]:=-A[8,1];
  A[8,4]:=E[8]*D[8]/D[1];
  A[8,3]:=E[8]-A[8,4];
  A[8,5]:=D[2]*A[8,1]-E[8]*D[8];
  A[8,6]:=A[8,1]*(D[2]+D[3])-1/2*A[8,3]*D[1]+E[8]*(1/2*
  D[1]-D[8]);
  A[8,7]:=-A[8,2]*D[2];
"GCTO" M1;
L9:A[9,1]:=E[9]*D[8]*(B[4]+B[5]-B[1]*(3*(D[6]*D[6])/
  (D[2]*D[2])-1))/(2*B[6]*D[2]);
  A[9,2]:=-A[9,1];
  A[9,3]:=E[9]*D[8]/D[1];
  A[9,4]:=E[9]-A[9,3];
```

```
        A[9,5]:=D[2]*A[9,1];
        A[9,6]:=A[9,1]*(D[2]+D[3])-1/2*A[9,3]*D[1];
        A[9,7]:=-A[9,2]*D[2]-E[9]*D[8];
        "GOTO" M1;
L10:    A[10,1]:=-E[10];
        A[10,2]:=-E[10];
        A[10,3]:=-2*E[10]*D[6]/D[1];
        A[10,4]:=-A[10,3];
        A[10,5]:=D[2]*A[10,1]+E[10]*D[7];
        A[10,6]:=A[10,1]*(D[2]+D[3])-1/2*A[10,3]*D[1]+E[10]*(D[3]+D[7]);
        A[10,7]:=-A[10,2]*D[2]-E[10]*D[7];
        "GOTO" M1;
L11:    A[11,1]:=E[11]*D[1]*B[5]/(4*D[2]*B[6]);
        A[11,2]:=-A[11,1];
        A[11,3]:=E[11]/2;
        A[11,4]:=E[11]/2;
        A[11,5]:=D[2]*A[11,1];
        A[11,6]:=A[11,1]*(D[2]+D[3])-1/2*A[11,3]*D[1];
        A[11,7]:=-A[11,2]*D[2];
        "GOTO" M1;
L12:    A[12,2]:=-E[12]*(B[4]+B[5])/(2*B[6]);
        A[12,1]:=-A[12,2]-E[12];
        A[12,3]:=-E[12]*D[2]/D[1];
        A[12,4]:=-A[12,3];
        A[12,5]:=D[2]*A[12,1];
        A[12,6]:=A[12,1]*(D[2]+D[3])-1/2*A[12,3]*D[1]+E[12]*D[3];
        A[12,7]:=-A[12,2]*D[2];
        "GOTO" M1;
L13:    A[13,1]:=-E[13]*(B[4]+B[5])/(2*B[6]);
        A[13,2]:=-A[13,1]-E[13];
```

```
A[13,3]:=-E[13]*D[2]/D[1];
A[13,4]:=-A[13,3];
A[13,5]:=D[2]*A[13,1];
A[13,6]:=A[13,1]*(D[2]+D[3])-1/2*A[13,3]*D[1];
A[13,7]:=-A[13,2]*D[2];
"GOTO" M1;
L14:A[14,1]:=-E[14]/2;
A[14,2]:=-E[14]/2;
A[14,3]:=-E[14]*(D[2]+D[3])/D[1];
A[14,4]:=-A[14,3];
A[14,5]:=D[2]*A[14,1];
A[14,6]:=A[14,1]*(D[2]+D[3])-1/2*A[14,3]*D[1];
A[14,7]:=-A[14,2]*D[2];
M1:"END";
    M:=1;
    OUTSTRING(P,M);
    "FOR" K:=1 "STEP" 1 "UNTIL" 7 "DO"
    "FOR" J:=1 "STEP" 1 "UNTIL" 14 "DO"
    A[15,K]:=A[15,K]+A[J,K];
    SAMELINE;
    "PRINT" ,
HA= , ,A[15,1],
HE= , ,A[15,2],
VA= , ,A[15,3],
VE= , ,A[15,4],
MB= , ,A[15,5],
MC= , ,A[15,6],
MD= , ,A[15,7];
"END";
"END";
```

47

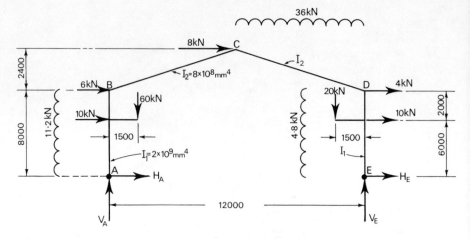

FIG. 3.7

The input is as follows.

/ PORTAL FRAME FIG. 3.7. WIND + CRANE + SUPER \

9
3   6   7   8   9   10   12   13   14
12000   8000   2400
2×10^9 8×10^8
6000   2000   1500
36   11.2   4.8   60   20   10   6   4   8

and the output is:

DCTCES 4 ELASTIC ANALYSIS OF A SINGLE BAY, PITCHED ROOF, PORTAL FRAME WITH PINNED BASES

PORTAL FRAME FIG. 3.7 WIND + CRANE + SUPER

HA	=	−18.982214
HE	=	−35.017786
VA	=	35.066667
VE	=	80.933333
MB	=	−177057.71
MC	=	−7735.0206
MD	=	210942.29

In these results, reactions are in kN and moments in kNmm, the figures obtained then being rounded off to practical values.

Notes on Example 4

1.   The program can be readily modified to give a more elaborate output including, for example, a print-out of the dimensions of the frame.

2.   The sign conventions for all applied loads, reactions and bending moments are shown in Fig. 3.6. These are positive when as shown and negative when acting in the other direction or sense.

3.   The reactions and bending moments due only to any individual component of the applied loading can be obtained by putting $N = 1$ and modifying the rest of the input data accordingly.

4.   The formulae used in Program DCTCES 4 are given in Appendix 2. These have been obtained from References 1-3 and, in some instances, from first principles.

Example 5:   Elastic Analysis of Multistorey Frames

An elastic analysis of a stiff-jointed frame in which all members are either horizontal or vertical and form a regular continuous structure between the outer limits is required (e.g. the frame in Fig. 3.8).

On this occasion an existing library program[4] was available. It had been specially written in an obsolescent programming language and use must be made of this software if at all possible. This is a typical problem confronting the engineer, and it is important to realise that a solution along the above lines can frequently prove difficult. Reference has already been made to the reluctance of computer manufacturers to maintain an extensive supply of software, and in

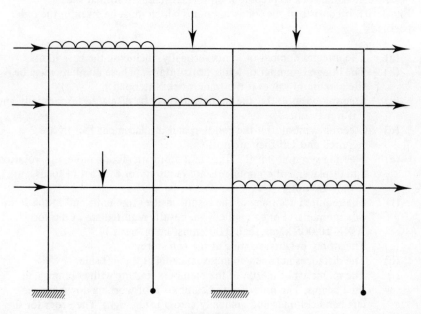

FIG. 3.8

such circumstances the onus is on the user to modify existing material to his own requirements. The various ways this can be done are briefly discussed in Chapter 6.

In this particular case, it was finally decided to rewrite the existing program in Algol. Program DCTCES 5[5]: Analysis of a framework by Kani's Method, was therefore prepared by J. S. Roper. (Because of shortage of space it is not possible to include the print-out of this extensive program here, though this can be obtained from the program brochure[5].)

Use and features of Program DCTCES 5[5]

The method of structural analysis used is Kani's Iterative Method[6], brief notes on which are given in Chapter 1. This program has the same basic features and does the same task as Program LC 2[4]. (Reference should be made to the brochure on LC 2 for any further details.)

The program calculates the end moments on members of a rectangular or step-shaped, plane, rigid-jointed frame with up to j joints. (j depends on the type and size of computer used. For example, with an Elliott 803 computer with 4096 words of store, $j = 50$.)

The user must specify whether or not each of the base joints is fixed or pin-jointed (all other joints being fixed), and whether or not the frame may be displaced horizontally. Calculations are made for a series of loadings acting either on the beams or on one outer edge of the frame. The first loading is the dead load and subsequent loadings are alternative live loads. Part of the output consists of the maximum effects of any combinations of these loads.

Input: The input for this program is both lengthy and detailed, and it is therefore advantageous to prepare it on special standard format sheets (Fig. 3.10). It consists of the following items which must be given in the order specified.

(a)   The integral number of joints vertically, including the base joints.
(b)   The integral number of joints horizontally. (These numbers must be the maximum values if the framework is stepped.)
(c)   Steering symbol: 0 if the framework may be displaced horizontally and 1 if it may not.
(d)   Steering symbol: 0 if the rotation and displacement factors are required and 1 if they are not.
(e)   Steering symbol: 0 if the fixed end moments, fixing moments, rotation and displacement components are required for each set of loads, and 1 if they are not.
(f)   The required accuracy of the results in the same units and in the form of a moment. (For example, if the results required are in the form X.XX ±0.005 kNm, then 0.005 must be inserted.)
(g)   The storey heights, starting at the top storey.
(h)   The distances between columns, starting at the left side.
(i)   The moments of inertia of the members starting with columns and then beams. The order for the columns is of working down for the left-hand columns and gradually across to the right. The order for the beams is of across the top row from left to right and gradually working down. Zero must be inserted for all the members which are missing from a

stepped framework. (The program, in fact, assumes that all frameworks are perfectly rectangular.)

(j)   Steering symbols, one for each base joint in order from left to right, each being 0 if the joint is fixed and 1 if it is pin-jointed.

(k)   Steering symbol as follows: 0 if the loading is only vertical; 1 if the loading is only horizontal; or 2 if both types of loading are used. This symbol applies to the set of loads immediately following it. Other sets of loads may begin with another value.

(l)   For every beam, the number of loading cases applied to it. The loading on each beam must be specified in terms of the imposed loading types in Fig. 3.9. These may have a negative value for the force or force per unit length, so that many configurations of loads are possible. These numbers are written in an order depending on the position of the beam. The order is from left to right for each level starting at the top and working down. Zero must be inserted for each beam with no load acting on it (and for imaginary beams in stepped frameworks).

(m)   The loading-case specifications. This is a list of dimensions and forces W (or w or M), a, b, c, y, for each of the loading cases acting on the beams. The numbers described in (l) above and the order in which these cases are written should be consistent. For example, if the numbers given under section (l) were 3, 2, 5, then the first three loading cases would be taken to be acting on the first beam, the next two on the second, the next five on the third and so on. W, a, b, c, and y are defined for each case in Fig. 3.9. Zero should be inserted for any value of a, b, or c not applicable to the loading case.

(n)   The horizontal forces at the joints. These are written in order starting at the top level and working down. Zero must be inserted for joints where no force is acting. Positive values are taken to be acting from left to right.

(o)   The integer 777, a steering symbol to indicate the end of a set of loads. Sections (k) to (o) may be repeated for as many sets of loads as are required.

(p)   Finally an additional integer 777 should be inserted.

Depending on the value of the steering symbol described in (k), the horizontal forces or the vertical forces may be missed out completely. If its value is 1 and it is not the first set of loads, then the forces are taken to be wind forces.

Output: This begins with a restatement of the input data with headings and is followed by the numerical values of the calculated moments called for by the various steering symbols. These moments are suitably labelled in the output and are given in the sequence of the joint numbering which is from left to right. The storeys are numbered 0, 1, 2 . . . from the base upwards. At any joint four values of moments are given in the order as shown in Fig. 3.11. (When less than four members meet at a joint, four values are still quoted, zero being inserted for a non-existent member.) Finally the maximum and minimum values are given for all moments due to the addition of the dead load with the most adverse combination of the live loads. When obtaining these maximum and minimum effects, the horizontal live load is taken to be due to wind capable of acting in both directions.

LOADING CASES

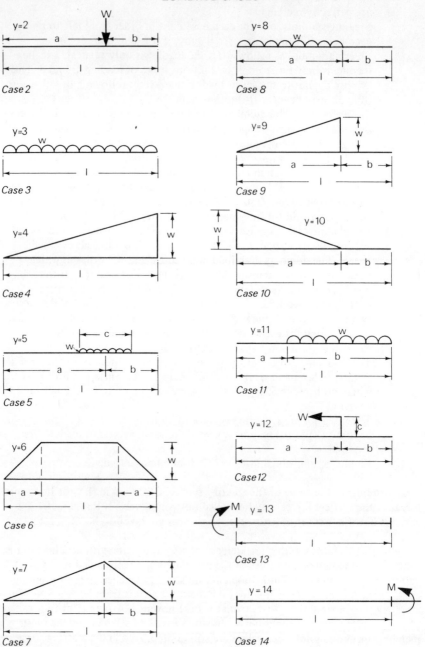

FIG. 3.9

Program no. DCTCES 5 - Input data   Sheet 1

Framework analysis by Kani's method

Steering symbols

Number of joints vertically	Number of joints horizontally	Displacement horizontally	Output of factors	Output of moments	Accuracy

Heights of columns

Distance between columns

Moment of inertia

Steering symbols for base joints

Diagram indicates the order
in which data is to be written.
No. of joints vertically ·· 4
No of joints horizontally ·· 4

FIG. 3.10

53

Program no. DCTCES 5 - Input data   Sheet 2

Loading data

This sheet may be repeated for as many sets of loads as are required

Steering symbol for wind loading

No. of loading cases on the beams

Loading on the beams

w or W or M	a	b	c	y

Wind forces at the joints

777		777	*delete if inappropriate*

FIG. 3.10 (*continued*)

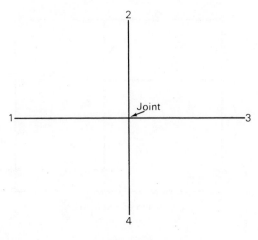

FIG. 3.11

Limitations and accuracy
(a)   All members of the framework must be either horizontal or vertical and must form a continuous structure between the outer limits. However, the outer limits may be stepped in any way. (Certain structures with beams and columns missing from the inside of the structure have been successfully analysed by this program, though this may not work in all cases.)
(b)   All joints must be rigid except the base joints, each of which may be either rigid or hinged.
(c)   The cross-section of each member must be constant, though these need not be the same for each member.
(d)   Loads may only be applied to beams.
(e)   Only frameworks with up to j joints may be analysed by this program. (In this case of the Elliott 803 computer with 4096 words of store, $j = 50$. In other cases j depends on the type and size of computer used and can be determined either by trial or by establishing what storage remains once the program has been inserted.)
(f)   When analysing a stepped framework imaginary members must be specified to complete a rectangular structure. The limit of j joints referred to in (e) must include these imaginary members.
(g)   The accuracy of the calculation is chosen by the user.
(h)   Any system of units may be used, and a separate system can be used for moments of inertia. The results are given in the same system used for input data.

Illustration
As an illustration consider the frame in Figure 3.12.
It is required to find the moments at the joints due to the different load conditions and then obtain the maximum and minimum moments which can be sustained on application of either (or neither) of the live loads together with the dead load.

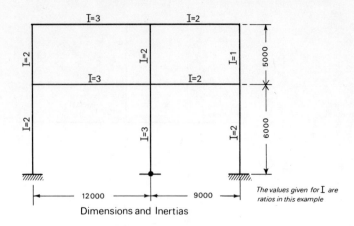

Dimensions and Inertias

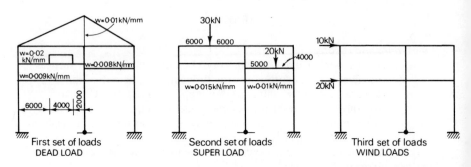

FIG. 3.12

The input data is shown in Fig. 3.13 on standard format sheets. All forces and loads are expressed in kN and all lengths in mm, so that all output moments will be in kNmm. The inertias of the members are expressed as ratios. In punching the data only the letters, numbers and symbols contained in the rectangular boxes should be included, the descriptive words around these boxes being ignored.

The output given in Fig. 3.14 is in accordance with the details given above.

Notes on Example 5

1.   By employing the other steering symbols in (d) and (e) of the input details above, the rotation and displacement factors, fixed end moments, etc., could have been obtained.

2.   The maximum deviation is given for all iterations accomplished in the actual output. However only the value associated with the last iteration for each load case is given here because of shortage of space (e.g. there are 23 iterations for the dead load case).

Program no. DCTCES 5 - Input data   Sheet 1

Framework analysis by Kani's method

Steering symbols

Number of joints vertically	Number of joints horizontally	Displacement horizontally	Output of factors	Output of moments	Accuracy
3	3	O	I	I	0·05

Heights of columns

5000	6000				

Distance between columns

12000	9000				

Moment of inertia

2	2	2	3	I	2
3	2	3	2		

Steering symbols for base joints

O	I	O			

Diagram indicates the order in which data is to be written.
No. of joints vertically ·· 4
No of joints horizontally ·· 4

FIG. 3.13

57

Program no. DCTCES 5 - Input data   Sheet 2a

Loading data

This sheet may be repeated for as many sets of loads as are required

Steering symbol for wind loading

O

No. of loading cases on the beams

I	I	2	I																			

Loading on the beams
w or W or M

w or W or M	a	b	c	y
0·01	0	0	0	4
0·01	9000	0	0	10
0·009	0	0	0	3
0·02	8000	4000	4000	5
0·008	0	0	0	3

Wind forces at the joints

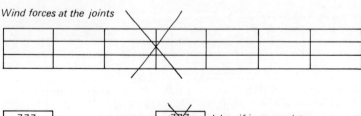

777

777	delete if inappropriate

FIG. 3.13 (*continued*)

Program no. DCTCES 5 - Input data   Sheet 2b

Loading data

This sheet may be repeated for as many sets of loads as are required

Steering symbol for wind loading

O

No. of loading cases on the beams

I	O	I	2																			

Loading on the beams

w or W or M	a	b	c	y
30	6000	6000	0	2
0·015	0	0	0	3
0·01	0	0	0	3
20	5000	4000	0	2

Wind forces at the joints

777

| ~~777~~ | *delete if inappropriate*
|---|

FIG. 3.13 (*continued*)

Program no. DCTCES 5 - Input data Sheet 2c

Loading data

This sheet may be repeated for as many sets of loads as are required

Steering symbol for wind loading

I

No. of loading cases on the beams

Loading on the beams
w or W or M

Wind forces at the joints

10	20					

777		777	delete if inappropriate

FIG. 3.13 (*continued*)

FIG. 3.14 — SHEET 1
PROGRAM NO. DCTCES 5:— NUMERICAL EXAMPLE
PRINT-OUT OF INPUT DATA
(As punched from Input Data Sheet — see Fig. 3.13)

```
3,3,0,1,1,0.05,
5000,6000,
12000,9000,
2,2,2,3,1,2,
3,2,3,2,
0,1,0,
0,
1,1,2,1,
0.01,0,0,0,4,
0.01,9000,0,0,10,
0.009,0,0,0,3,
0.02,8000,4000,4000,5,
0.008,0,0,0,3,
777,
0,
1,0,1,2,
30,6000,6000,0,2,
0.015,0,0,0,3,
0.01,0,0,0,3,
20,5000,4000,0,2,
777,
1,
10,20,
777,
777,
```

FIG. 3.14 — SHEET 2
PROGRAM NO. DCTCES 5:— NUMERICAL EXAMPLE
OUTPUT FROM COMPUTER

```
NO. OF JOINTS IN VERTICAL DIRECTION   3

NO. OF JOINTS IN HORIZONTAL DIRECTION   3


MEASURES OF THE SYSTEM

STOREY HEIGHTS

  5000.0000
  6000.0000

DISTANCES BETWEEN COLUMNS

  12000.000
  9000.0000

MOMENTS OF INERTIA OF COLUMNS

  2.0000 ⌑+00
  2.0000 ⌑+00
  2.0000 ⌑+00
  3.0000 ⌑+00
  1.0000 ⌑+00
  2.0000 ⌑+00
```

FIG. 3.14 — SHEET 2 (*continued*)

MOMENT OF INERTIA OF BEAMS

```
3.0000 ₑ+00
2.0000 ₑ+00
3.0000 ₑ+00
2.0000 ₑ+00
```

BASE JOINTS(LEFT TO RIGHT)

0FIXED

1PIN JOINTED

2FIXED

L OADCASEVERTICAL L OAD

	W	A	B	C	Y
SPAN NO 1					
	.0100	.0000	.0000	.0000	4
SPAN NO 2					
	.0100	9000	.0000	.0000	10
SPAN NO 3					
	.0090	.0000	.0000	.0000	3
	.0200	8000	4000	4000	5
SPAN NO 4					
	.0080	.0000	.0000	.0000	3

```
 1 ITERATION ACCOMPLISHED MAX. DEVIATION    36234.0182

 2 ITERATION ACCOMPLISHED MAX. DEVIATION     9220.1568
```
— — — — — — — — —

— — — — — — — —
```
22 ITERATION ACCOMPLISHED MAX. DEVIATION        0.0771

23 ITERATION ACCOMPLISHED MAX. DEVIATION        0.0448
```

MEMBER END MOMENTS

STOREY 2

N	1	0.000	0.000	-43734.197	43734.210
N	2	76076.225	0.207	-44424.716	-31651.716
N	3	15695.771	-1.682	-0.841	-15692.403

STOREY 1

N	1	2.096	83284.475	-147236.028	63947.361
N	2	226471.370	-66577.399	-93847.367	-66046.576
N	3	29059.942	-13097.105	0.000	-15962.837

FIG. 3.14 — SHEET 2 (*continued*)

```
                        STOREY   0
  N  1          0.000      29008.636       0.000     -29008.636
  N  2          0.000          0.000       0.000          0.000
  N  3          0.000     -10946.462       0.000      10946.462
```

L OADCASEVERTICAL L OAD

```
                W           A           B           C           Y
```

SPAN N O 1

```
                30.00       6000        6000        .0000       2
```
SPAN N O 3

```
                .0150       .0000       .0000       .0000       3
```
SPAN N O 4

```
                .0100       .0000       .0000       .0000       3
                20.00       5000        4000        .0000       2
```

```
    1 ITERATION ACCOMPLISHED MAX. DEVIATION    33793.3153

    2 ITERATION ACCOMPLISHED MAX. DEVIATION     6494.9374
    _  _  _  _  _  _  _  _  _  _  _  _  _  _  _  _
    _  _  _  _  _  _  _  _  _  _  _  _  _  _  _  _

   23 ITERATION ACCOMPLISHED MAX. DEVIATION        0.0770

   24 ITERATION ACCOMPLISHED MAX. DEVIATION        0.0448
```

MEMBER END MOMENTS

```
                        STOREY   2
  N  1          0.000          0.000    -40680.054      40680.068
  N  2      41172.351         -0.639     -1700.389     -39471.322
  N  3       7243.885          1.457         0.728      -7246.793


                        STOREY   1
  N  1          2.038      74800.559   -137355.967      62551.332
  N  2     188844.604     -46393.765   -113447.179     -29003.633
  N  3      61985.239     -22368.684         0.000     -39616.556


                        STOREY   0
  N  1          0.000      28576.461       0.000     -28576.461
  N  2          0.000          0.000       0.000          0.000
  N  3          0.000     -22507.483       0.000      22507.483
```

L OADCASEH ORIZONTAL L OAD

H ORIZONTAL F ORCES AT THE J OINTS

```
       10.000      20.000
```

FIG. 3.14 — SHEET 2 (*continued*)

```
 1 ITERATION ACCOMPLISHED MAX. DEVIATION    38571.4286

 2 ITERATION ACCOMPLISHED MAX. DEVIATION    22180.2778
 —  —  —  —  —  —  —  —  —  —  —  —  — -
 —  —  —  —  —  —  —  —  —  —  —  —  — —
26 ITERATION ACCOMPLISHED MAX. DEVIATION       0.0534

27 ITERATION ACCOMPLISHED MAX. DEVIATION       0.0310
```

MEMBER END MOMENTS

```
                     STOREY   2
N   1          0.000        0.000      10814.826    -10814.817
N   2      10972.595        0.594       8458.709    -19431.897
N   3       7023.759        0.503          0.251     -7024.762


                     STOREY   1
N   1          0.925     2000.785      28985.096    -30987.731
N   2      23283.882   -15990.669      21489.144    -28782.337
N   3      27349.249     1261.402         -0.000    -28610.651


                     STOREY   0
N   1          0.000   -46403.868         0.000     46403.868
N   2          0.000       -0.000         0.000         0.000
N   3          0.000   -45215.329         0.000     45215.329
```

MAXIMUM MEMBER END MOMENTS

```
                     STOREY   2
N   1          0.000        0.000     -32919.371     95229.094
N   2     128221.171        0.801     -35966.007    -12219.819
N   3      29963.415        0.278          0.139     -8667.641


                     STOREY   1
N   1          5.060   160085.819    -118250.932    157486.424
N   2     438599.856   -50586.730     -72358.223    -37264.239
N   3     118394.431   -11835.703          0.000     12647.814


                     STOREY   0
N   1          0.000   103988.965         0.000     17395.232
N   2          0.000        0.000         0.000         0.000
N   3          0.000    34268.866         0.000     78669.274
```

MIN.   MEMBER END MOMENTS

```
                     STOREY   2
N   1          0.000        0.000     -95229.077     32919.393
N   2      65103.630       -1.025     -54583.813    -90554.935
N   3       8672.012       -2.185         -1.092    -29963.958
```

FIG. 3.14 – SHEET 2 (*continued*)

			STOREY 1		
N	1	1·171	81283·690	-313577·091	32959·630
N	2	203187·488	-128961·833	-228783·689	-123832·546
N	3	1710·693	-36727·191	-0·000	-84190·044

			STOREY 0		
N	1	0·000	-17395·232	0·000	-103988·965
N	2	0·000	-0·000	0·000	0·000
N	3	0·000	-78669·274	0·000	-34268·866

Example 6:   Calculation of Bending Moments and Shearing Forces in a
Continuous Beam

The bending moments and shearing forces at various points along a continuous
beam on rigid supports by an elastic analysis, due to the application of a dead
load and various live loads, are required. Typical beams and loadings are
illustrated in Fig. 3.15.

Again, use is made of an existing library program[7] written in Autocode (see
chapter 6) for an older computer, and as in the previous example, this program
was rewritten in Algol and modified by J. S. Roper, resulting in Program
DCTCES 6[8]: Calculation of Bending Moments and Shearing Forces in a
Continuous Beam. (Again because of the shortage of space it is not possible to
include the print-out of this extensive program here, though this can be obtained
from the program brochure[8].)

Use and Features of Program DCTCES 6
This program has essentially the same basic features and does the same task as
Program Number LC3[7]. (Reference should be made to the brochure on LC3

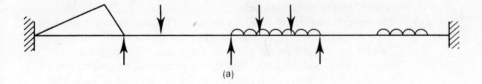

(a)

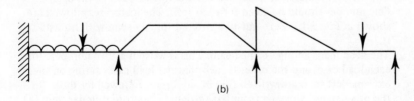

(b)

FIG. 3.15

for any additional details.) The spans may have different moments of inertia, but each span must itself be uniform. The extreme ends of the beam may either be fixed, simply supported, or any combination of the two. The effects of live load may be calculated if required, and the resulting maximum and minimum values are then given for the most adverse conditions of such loading. The process used involves solving the various equations obtained from the Theorem of Three Moments for pairs of adjacent spans, together with two equations for the end conditions, to obtain the support moments. These, together with the bending moments and shearing forces on simply-supported beams due to the loading, give the complete solution.

Input: Again, the input data is fairly complicated, and it is advantageous to write this on the special sheets (Fig. 3.16). The input consists of the following items which must be in the order specified.

(a)   The number of supports. Fixed ends are regarded as supports for this purpose.

(b)   A steering symbol which is an integer between 0 and 3 to indicate the end conditions, being
0 for both ends fixed
1 for left end pinned, right end fixed
2 for left end fixed, right end pinned
3 for both ends pinned.

(c)   A steering symbol: 0 when intermediate results are required, or 1 when intermediate results are not required. This facility is only available on Program LC2. Here, either 0 or 1 gives the same output.

(d)   A steering symbol: 0 when live loads are to be used, or 1 when there are no live loads.

(e)   The lengths of the spans starting at the left and working to the right.

(f)   The moments of inertia of the spans in the same order as the lengths. These can be expressed in any consistent units, quite apart from the units of length used elsewhere.

(g)   The integral number of divisions for which the moments are required to be output. One integer must be given for each span. If, for example, 2 and 4 are entered for the first two spans and their lengths are respectively 0 and 50 units, then the moments will be printed out for points at the following distances from the left support of each span.
Span 1: distances 0   5   10
Span 2: distances 12.5   25   37.5   50

(h)   The integral number of dead load cases acting on each of the spans. One number should be given for each span. The load cases allowed are shown in Fig. 3.17. Zero must be entered for any span with no load on it.

(i)   The dead loads acting on the beam. This is written as five numbers for each load case, and the order is such that the load cases acting on the extreme left span are written first in any order, followed by those for the other spans, working from left to right. Therefore if under item (h) the first two numbers were 2, 3, then the details of the load cases would be in such an order that the first two act on the first (left-hand

Program no. DCTCES 6   Input data sheet

B.M.S. & S.F.'s in a continuous beam

No. of supports	Steering symbol for end condition	Steering symbol for output	Steering symbol for live loads
4	☒ 1 ☒ ☒	☒ 1	0 ☒

Span	1	2	3	4	5	6	7	8	9
L_i	6	9	7·5						
J_i	8	30	15						
N_i	2	2	2						

Dead load
Numbers of load cases

I	I	I						

M or W or w	a	b	c	y
0·7	0	0	0	2
0·7	0	0	0	2
0·7	0	0	0	2

Live load
Numbers of load cases

I	2	I						

M or W or w	a	b	c	y
I	0	0	0	2
0·5	0	0	0	2
25	5	4	0	I
2	1·5	6	0	4

777

FIG. 3.16

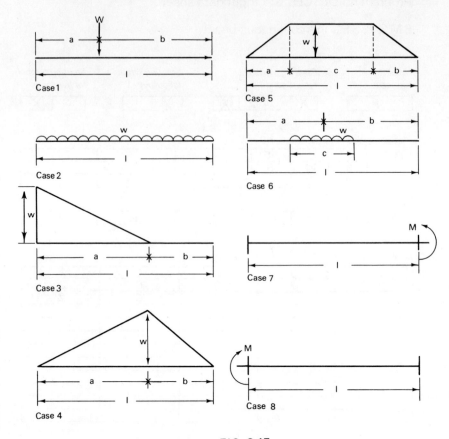

FIG. 3.17

span) and the next three act on the second span, and so on. The five numbers for each span are defined:

w, (or W or M) is the load (either a load per unit length, a concentrated load or a moment)
a, b and c are dimensions (see Fig. 3.17)
y is the loading case number.
Zero must be entered for any dimension which does not apply.

(j)   If 0 is entered for item (d) then the details of live loads are entered exactly as described in (h) and (i) above.
(k)   777 is used as an end of data signal.

Output: This begins with a restatement of the input data with headings, followed by the calculated values of shearing forces and bending moments called for by the various steering symbols. These values are fully described in the print-out. Finally the maximum and minimum values of bending moments and shearing forces, due solely to the live loading, are given. The specified live loadings may, or may not, act on each span and all such combinations are considered.

Limitations
The following limitations must be observed.
  (a)   The only loading cases which may be used are those shown in Fig. 3.17.
  (b)   The minimum number of supports allowed depends on the end conditions of the beam:
      (i) both ends fixed,   minimum number of supports = 2
      (ii) left end pinned, right end fixed,   „   „   „   = 2
      (iii) left end fixed, right end pinned,   „   „   „   = 3
      (iv) both ends pinned,   „   „   „   = 3
  (c)   The maximum number of supports allowed is $i*$.
  (d)   The maximum number of loading cases allowed over all spans is $j*$, (including all live loads).
  (e)   The total number of points for which the bending moments are calculated should not exceed $k*$.

Illustration
Consider the beam in Fig. 3.18. Find the shearing forces at the supports and the bending moments at the supports and midspan due to the dead load, and also their maximum and minimum values due to the given live loading.

The input data are shown on standard sheets in Fig. 3.16. All forces or loads are expressed in either kN or kN/m, and all lengths in m, so that all output shearing forces are in kN and bending moments in kNm. For simplicity the moments of inertia have been expressed in ratio form. In punching the data only the letters, numbers, and symbols contained in the rectangular boxes should be included.

The output is given in Fig. 3.19, in accordance with the above details and is virtually self-explanatory.

Notes on Example 6

1.   Bending moments and shearing forces can readily be obtained for other points in any span merely by altering the input data given for item (g) above.
2.   This program is particularly useful in preparing influence lines for the bending moment at any point on a continuous beam, and to check a proposed solution to a continuous beam design problem.

* The actual values of i, j, and k, depend on the type and size of computer used and can be determined either by trial or by establishing what storage remains in any particular computer once the program has been inserted. For example, in the Elliott 803 computer with 4096 words of store $i = 10$; $j = 40$; and $k = 60$.

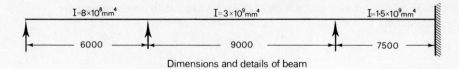

Dimensions and details of beam

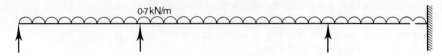

Dead load case

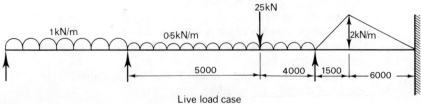

Live load case

[*note: The loadings on the centre span, if applied, act together as a single loading*]

FIG. 3.18

FIG. 3.19 – SHEET 1
PROGRAM NO. DCTCES 6:– NUMERICAL EXAMPLE
PRINT-OUT OF INPUT DATA
(As punched fr   m Input Data Sheet – see Fig. 3.16)

```
4,1,1,0,
6,9,7.5,
8,30,15,
2,2,2,
1,1,1,
0.7,0,0,0,2
0.7,0,0,0,2,
0.7,0,0,0,2,
1,2,1,
1,0,0,0,2,
0.5,0,0,0,2,
25,5,4,0,1,
2,1.5,6,0,4,
777,
```

FIG. 3.19 – SHEET 2
PROGRAM NO. DCTCES 6:– NUMERICAL EXAMPLE
OUTPUT FROM COMPUTER

NUMBER OF SUPPORTS   4

LEFT END PIN-JOINTED, RIGHT END FIXED

LOADING DATA

DEAD LOAD

		W	A	B	C	Y	L	J
SPAN	1						6.0	8.000
		0.70	0.00	0.00	0.00	2		
SPAN	2						9.0	30.00
		0.70	0.00	0.00	0.00	2		
SPAN	3						7.5	15.00
		0.70	0.00	0.00	0.00	2		

LIVE LOAD

		W	A	B	C	Y
SPAN	1					
		1.00	0.00	0.00	0.00	2
SPAN	2					
		0.50	0.00	0.00	0.00	2
		25.00	5.00	4.00	0.00	1
SPAN	3					
		2.00	1.50	6.00	0.00	4

SHEAR FORCE AT THE SUPPORTS

SPAN		DEAD LOAD	LIVE LOAD	
		Q	MAX Q	MIN Q
1	L	1.49	2.53	-1.69
1	R	-2.71	0.09	-5.25
2	L	3.10	12.81	-0.46
2	R	-3.20	0.45	-17.60
3	L	2.80	8.03	-0.15
3	R	-2.45	3.84	-3.46

BENDING MOMENTS DISTRIBUTION

SPAN	X	DEAD LOAD	LIVE LOAD	
		M	MAX M	MIN M
1	0	0.00	0.00	0.00
1	3.00	1.31	3.10	-5.08
1	6.00	-3.68	0.51	-13.49
2	4.50	3.17	40.38	-2.83
2	9.00	-4.15	0.74	-22.79
3	3.75	1.42	3.74	-4.80
3	7.50	-2.84	9.60	-6.31

General Comments on the Purpose and Use of the Examples

The reader should prepare the programs for use on whatever computer is available to him and check the various numerical examples given: he can then extend the process to his own problems. In addition he should test the various special cases and features in each program by selecting appropriate numerical data for processing. The validity and accuracy of all the output obtained can be appraised by comparing these results with solutions derived by other means.

An experienced computer operator can give the necessary assistance in punching the tapes or cards and in operating the computer, though the engineer should then learn how to do such things for himself. By such means the engineer can:

learn how to use an existing library program to solve a numerical problem in structural analysis;

compare this process with those already familiar to him, e.g. using manual calculations, charts, tables, etc.;

obtain an understanding of how computers and ancillary equipment are used; assess what is involved in the execution of the work, with regard to staffing, equipment and development work;

decide whether his own particular problems can be solved in this way; and learn what a program print-out looks like and appreciate something of the nature of the work involved in writing a program.

The reader has now become a computer user, capable of making use of any existing program available for his computer. (The preparation of numerical data for, and the use of, any library program is generally similar.) He should also be able to specify what program he would like to have written for him by an experienced programmer, based on a traditional method.

In this context it may be of interest that Programs DCTCES 1 and 2 were written by the author who is an engineer, DCTCES 3 and 4 were written by the author in collaboration with K. Lowe (a mathematician/programmer) and DCTCES 5 and 6 were written for the author by J. S. Roper (a computer expert).

These examples of the use of a computer to solve problems in structural analysis by traditional methods present the engineer with an opportunity to participate in useful practical work at this early stage before any attempt is made to study the various technologies necessary for a fuller understanding of the subject.

References

1. *Steel Designers' Manual.* 4th ed. Crosby Lockwood. 1972
2. Kleinlogel, A. *Rigid Frame Formulas.* 2nd ed. Ungar. 1958
   Kleinlogel, A. *Beam Formulas.* 7th ed. Ungar. 1953
   Kleinlogel, A. and Haselbach, A. *Multibay Frames.* 7th ed. Ungar. 1963
3. *Handbook for Constructional Engineers.* Dorman Long.

4. Program Number LC2—Analysis of a framework by Kani's Iterative Method. Elliott 803 Computer Application Program. (All enquiries now to International Computers Ltd., London.)
5. Program DCTCES 5: Analysis of a framework by Kani's Method. Department of Civil Engineering, Dundee College of Technology. 1971
6. Kani, G. *Analysis of Multistorey Frames*. Ungar. 1957
7. Program Number LC3—Calculation of Bending Moments and Shearing Forces in a Continuous Beam. Elliott 803 Computer Application Program. (All enquiries now to International Computers, Ltd., London.)
8. Program DCTCES 6: Calculation of Bending Moments and Shearing Forces in a Continuous Beam. Department of Civil Engineering, Dundee College of Technology. 1971

4
Structural Analysis— Applications of Automatic Techniques Using Methods Based on Modern Computational Processes

Outline

The structural engineer now has some experience in using the computer to solve problems involving traditional methods of analysis which are already familiar to him. Moreover, he should now also be capable of using any existing program written for the computer which is most readily available to him, and is therefore now equipped to use programs based on modern methods, even although he does not yet possess a working knowledge of such methods. These methods present a fresh approach to the various problems in structural analysis, and are devised to use the computer most efficiently and effectively.

This chapter, then, extends the process started in the previous one, giving numerical illustrations of problems based on, or to be solved by, modern methods and enabling the engineer to gain some insight into the value of these techniques.

Elastic Analysis of Plane Frames and Beams

This problem has aroused considerable attention for over a hundred years and a suitable background, describing the approach adopted in both traditional and modern methods, is given in Chapter 1.

In this particular context the object of modern methods is to find a means of structural analysis which can be used on all types of frames and beams—no matter how complicated or large—giving deflections and rotations as well as forces and moments, to a high degree of accuracy in the minimum possible time. Since all the resulting calculations are carried out by computer, no method need be rejected at the outset simply because of the nature and magnitude of the computational difficulties. Nevertheless there still will be some limitation on the calculations that can be effected by any computer, though this restriction is obviously considerably less than hitherto.

Three main methods have been devised to meet this specification requirement. They are as follows:

the Stiffness Method

74

the Flexibility Method

Energy Methods.

An outline of these methods, together with a suitable bibliography, is given in Chapter 7. (It is not necessary for the reader to have any knowledge of these methods at this stage.) Several library programs have been written utilizing such methods. Those based on the Stiffness Method are the most popular and a bibliography is given at the end of the chapter. In general, the function of such programs is to calculate the deflections and rotations of the joints together with the axial and shearing forces and the bending moments in the members of a frame or beam subjected to various loading conditions.

It is usually possible to analyse a wide variety of types of frames or beams supported by any combination of roller, pinned, fixed and sprung supports using a single program. The loading is generally applied at the joints and can consist of two perpendicular forces and a moment. However some programs permit various types of loads on the members (e.g. uniformly distributed, concentrated or trapezoidal loads, etc.) to be specified. With programs where this is not directly applicable, it is still possible to obtain a solution by partitioning the loading and applying the Principle of Superposition (an illustration of which is given later in the chapter). While the individual members must be of uniform section, the joints can be either pinned or fixed. A method of dealing with non-uniform members is described later. A selection of frames and beams differing substantially in appearance and constructional detail yet which can be analysed by one such program is shown in Fig. 4.1.

The features of this particular modern method can therefore be summarised as follows:

the elastic analysis of a wide variety of different types of frames and beams can be accomplished using the one basic method;

the structures involved can have many irregularities or special details, e.g. different types of support, pinned or fixed joints, etc;

large structures subjected to different loading conditions can generally be analysed;

the actual numerical calculations can be carried out to a high degree of accuracy using techniques which the computer is eminently suitable for effecting;

the output can be quite comprehensive, giving deflections and rotations as well as all internal and external forces and moments;

the basic concept of the method is simple, and the entire process can be readily modified, adapted or extended to deal with special problems or refinements;

the method allows for account to be taken of the effect of rib shortening of the individual members of the frame.

Examples of Problems

A series of numerical problems is given in order to illustrate both the use of the method and some of the above features.

SELECTION OF FRAMES AND BEAMS TO BE ANALYSED BY THE MODERN METHODS

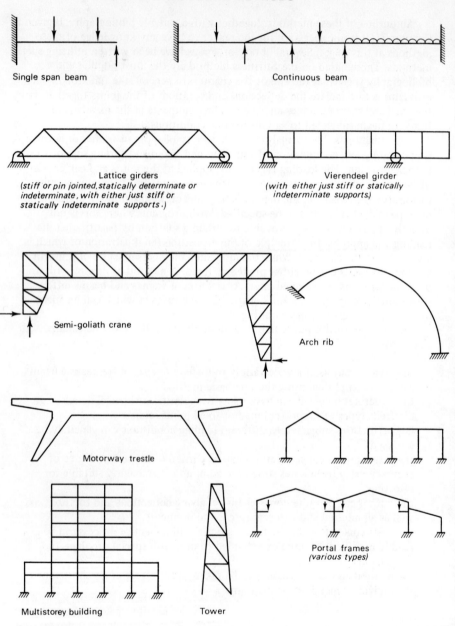

Single span beam

Continuous beam

Lattice girders
(stiff or pin jointed, statically determinate or indeterminate, with either just stiff or statically indeterminate supports.)

Vierendeel girder
(with either just stiff or statically indeterminate supports.)

Semi-goliath crane

Arch rib

Motorway trestle

Portal frames
(various types)

Multistorey building

Tower

FIG. 4.1

Problem 1:   Elastic analysis of multistorey frame

Consider the frame shown in Fig. 4.2. Since all dimensions, loads and member scantlings are given, the problem is essentially a design check calculation to find the deflection and rotation of the joints, the forces and moments in the members, and the reactions at the points of support when the frame is subjected to the given loading.

This particular numerical example deals with an irregular multistorey frame, illustrating the treatment of the following features:

(a)   The various types of end conditions for the members of the frame involving members which are either fixed or pinned at both ends, fixed

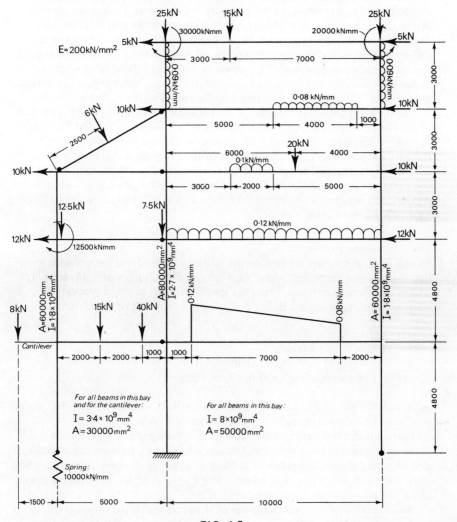

FIG. 4.2

at one end and pinned at the other end, or fixed at one end and free at the other end. At least one member is in each of these categories.
(b)   A sloping member.
(c)   Different types of frame reactions, involving fixed, pinned and elastic supports.
(d)   Applied loads at the joints of the frame, involving either horizontal or vertical forces with or without an applied external moment.
(e)   Applied loads on the members. These are: a concentrated load; a uniformly distributed load over the entire length of the member; a uniformly distributed load on part of a member; a trapezoidal load; or any combination of these.

A selection of existing library programs which can effect the elastic analysis of any plane frame is given in the bibliography at the end of the chapter, and any one of these programs, or others, can be used to solve this particular problem, depending on the computer available to the engineer. While each program has different features and various refinements, all do essentially the same basic job and are used in the same way. This generally consists of drawing up a list of numerical values for the input data in a specified order, according to the detailed instructions. This process is now illustrated by using such a program.

The program chosen is Program A27[1]: Elastic Analysis of Plane Framework, of the Glen Computing Centre Ltd. (associated with G. Maunsell and Partners, Consulting Engineers, London). (The author is indebted to G. Maunsell and Partners for permission to include details of the program in this book.)

The program is used to solve most of the numerical problems in this chapter, and therefore some descriptive details of it must first be given. It is written in ALGOL and is primarily intended for use on an Elliott 4100 computer having at least 24K words of store.

Use and features of Program A27
The method of solution is based on the Stiffness Method and the program is intended for the elastic analysis of rigid-jointed plane frameworks, loaded only within the plane of bending. From the specification of physical properties, restraint conditions and loading, the following results will be calculated:

    joint deflections
    axial forces in all members
    shearing forces at each end of all members
    bending moments at each end of all members.

Although the program is primarily intended for rigid-jointed frameworks, pinned joints can be incorporated at one or both ends of a member.

Certain restrictions are imposed by the theory adopted. These are that the frame is assumed to be perfectly elastic and subject only to small deformations, and that only frames composed of straight members of uniform section may be analysed. (This restriction may be overcome by the insertion of fictitious joints along curved or non-uniform members.)

An indication of the maximum size of structure which the program will analyse is given in the following condition. (This actually refers to Program

EAPF which is similar to A27. See Bibliography—Item 4.)

$$25\,m + 6j\,(3d + 4) \leqslant Q \qquad\qquad —(A)$$

where m is the number of members, j the number of joints, d the maximum joint number difference, and Q the amount of storage space available.

The value of Q depends on the computer configuration being used and may vary with different issues of the Algol Compiler. For example, the approximate values of Q using the ENL 15 ALGOL Compiler, and compiling and running separately were:

Storage Size	24K	32K
Value of Q	12000	20200

For further details of Program A27, which can handle much larger problems, reference should be made to the program brochure[1].

Input: The engineer should first prepare a sketch on which the member and joint numbering systems can be written. Such a diagram is shown in Fig. 4.3. Joints should be numbered in a manner which minimizes the greatest difference between the numbers defining member ends, since good numbering enables the program to be completed faster and the capacity of the program also depends on the maximum joint number difference (see equation (A) above). Joints must be introduced where two or more members meet, where a member changes direction or section, and where it is connected to an external support. Members of non-uniform section or curved members may be simulated by introducing joints as shown in Fig. 4.4 The frame members should be numbered and all lists of member properties given in ascending order of these numbers (see Fig. 4.3).

Finally a suitable origin should be chosen, with rectangular x and y axes that are clearly directed (see Fig. 4.3).

This completes the preliminary details and it is now possible to draw up the input data. As in the previous examples, it is normally advantageous to write this input on special standard data sheets, this time of the type shown in Fig. 4.5. The input data are entered in the appropriate enclosed blocks on these sheets, and consist of the following items in the order given.

Data Sheet 1   *General details*

The following information should be inserted.
- 1.1   Title block. This is for the user's reference. Anything written in this block will be output as a heading in the results. The title must be enclosed between the symbols / and \.
- 1.2   The total number of joints.
- 1.3   The total number of members.
- 1.4   The total number of joints with any specified zero deformations.
- 1.5   The total number of joints with elastic supports.

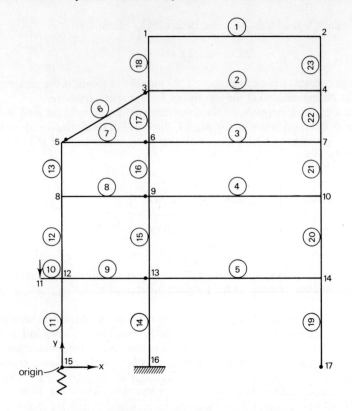

MEMBER & JOINT NUMBERING DIAGRAM

FIG. 4.3

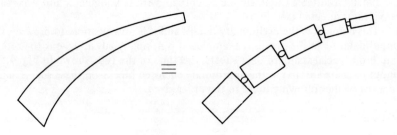

SIMULATION OF A NON-UNIFORM MEMBER

FIG. 4.4

1.6   The total number of cases of loading.

1.7   The total number of member loadings (see Data Sheet 7).

1.8   Young's Modulus for the material used.

1.9   The density of the material. If the self-weight of the members is to be taken into account, insert the appropriate value of density in compatible units; otherwise enter zero.

1.10 The program steering symbol. If a check output of member properties and joint loading is required, insert 1; otherwise insert 0.

Program No A27

Elastic Analysis of Plane Framework

Input data

Sheet 1. General details

Job title	/ CHECK ON FRAME 4·2 BY A27
Number of joints	17
Number of members	23
Number of joints with any zero deformations	3
Number of spring joints	I
Number of loadcases	I
Number of member loads	II
Young's Modulus	2×10^8
Density	O
Steering symbol (Output check required insert 1, otherwise 0)	I

FIG. 4.5

Data Sheet 2   *Joint co-ordinates*

The following information should be inserted.

2.1   Data sheet number. Insert 2.

2.2   A list of the x and y co-ordinates of all joints in ascending order of the joint numbers.

Sheet 2. Joint co-ordinates

Sheet number
(PUNCH THIS!)   2

JOINT NUMBER	x	y	JOINT NUMBER	x	y
1 →	5	18.6	→ 2 →	15	18.6
3	5	15.6	4	15	15.6
5	0	12.6	6	5	12.6
7	15	12.6	8	0	9.6
9	5	9.6	10	15	9.6
11	-1.5	4.8		0	4.8
	5	4.8		15	4.8
	0	0		5	0
	15	0			

FIG. 4.5 *(continued)*

Data Sheet 3   *Member properties*

Members must be of uniform cross-section between joints. The following information should be inserted.

3.1   Data sheet number. Insert 3.

3.2   Member properties. Five numbers should normally be entered here for each member (all of which are listed in turn in member number order). These are:

(a)   the joint number at end 1 of the member;

(b)   the joint number at end 2 of the member;

(c)   the hinge marker. This is:

0 if the member is rigidly fixed at both ends;

1 if the member is pinned at each end; and

2 if the member is rigidly jointed at end 1 and pinned at end 2;

(d)   the cross-sectional area of the member; and

(e)   the second moment of area of the member.

(Note: −1 may be inserted if the hinge condition, area and inertia are the same as for the previously entered member, though in that case the hinge marker, area and inertia should be omitted.) End 1 should normally be taken at the left-hand end of the member and end 2 at the right-hand end. If a pinned base is induced by the use of zero deformations, then a hinge should not be specified at that end of the member. Similarly a cantilever member should have a hinge marker of zero.

Sheet 3   Member properties

Sheet number
(PUNCH THIS!)

3

	END 1	END 2	HINGE	A	I
1	1	2	0	0·05	0·008
2	3	4	−1		
3	6	7	−1		
4	9	10	−1		
5	13	14	−1		
6	5	3	1	0·03	0·0034
7	5	6	2	0·03	0·0034
8	8	9	−1		
9	12	13	−1		
10	11	12	0	0·03	0·0034
11	15	12	0	0·06	0·0018
	12	8	−1		
	8	5	−1		
	16	13	0	0·08	0·0027
	13	9	−1		
	9	6	−1		
	6	3	−1		
	3	1	−1		
	17	14	0	0·06	0·0018
	14	10	−1		
	10	7	−1		
	7	4	−1		
	4	2	−1		

FIG. 4.5 (*continued*)

Data Sheet 4   *Zero deformations*

This data sheet may be omitted if there are no zero deformations, though there must either be zero deformations or elastic constraints on the structure.

  4.1   Data sheet number. Insert 4.

For each joint which has zero deformations the following information must be given.

  4.2   The joint number at which there is any zero deformation.
  4.3   If the joint is restrained in the x direction, insert 1; otherwise insert 0 (i.e. M1 = 1 or 0) (see Fig. 4.5).
  4.4   If the joint is restrained in the y direction, insert 1; otherwise 0 (i.e. M2 = 1 or 0).
  4.5   If the joint is restrained against rotation, insert 1; otherwise insert 0 (i.e. M3 = 1 or 0).

Sheet 4. Zero deformations

Sheet number
(PUNCH THIS!)   4

JOINT No.	M1	M2	M3
15	1	0	0
16	1	1	1
17	1	1	0

FIG. 4.5 (*continued*)

Data Sheet 5   *Joints with elastic supports*

This data sheet may be omitted if there are no joints with elastic supports.

   5.1   Data sheet number. Insert 5.

For each joint with an elastic support the following information must be given.

   5.2   The joint number at which there is an elastic support.
   5.3   The stiffness of the spring in the x direction.
   5.4   The stiffness of the spring in the y direction.
   5.5   The stiffness of the rotational spring.

If there is no elastic reaction in a particular direction, then zero should be
inserted in the appropriate position. The stiffness is the load required to
produce a unit displacement or, in the case of rotation, the moment required to
produce a rotation of one radian. Note that the elastic reactions may not be
resolved into component directions, as this does not result in the same structural
effect. The units adopted must be the same as those used elsewhere.

Description of loadings

The following two data sheets are required to describe the loadings for each load
case and both must be completed even if only one is used to specify the loading.
Any number of load cases may be applied.

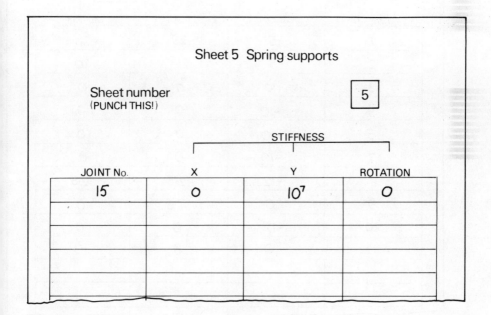

FIG. 4.5 *(continued)*

Data Sheet 6   *Joint loadings*

Loads should not be applied to restrained joints (see Data Sheet 4) where the restraint is in the direction of the applied loading, as these have no effect on the structure and are ignored in the program.

6.1   Data sheet number. Insert 6.
6.2   Loading case title block. The description of the loading case should be inserted here between the symbols / and \.
6.3   The total number of joints to be loaded. If there are no joint loadings, this data sheet must be completed with a title and zero for the number of loaded joints.

Sheet 6. Joint loading

Sheet number   6
(PUNCH THIS!)

Loadcase   / COMBINED LOADING
title

Number of joints loading   10

JOINT No.	Load X	Load Y	Moment
1	−5	25	−30
2	−5	25	20
3	−10	0	0
4	−10	0	0
5	−10	0	0
7	−10	0	0
8	−12	12·5	−12·5
9	0	7·5	0
10	−12	0	0
11	0	8	0

FIG. 4.5 (*continued*)

6.4 Joint loading block. For each loaded joint the following four numbers must be provided.
    (a)   The number of the joint loaded.
    (b)   The value of the force applied in the positive *x*-direction. If no load exists in this direction, insert zero.
    (c)   The value of the force in the negative *y*-direction (i.e. downward loads are positive). If no load exists in this direction, insert zero.
    (d)   The value of the moment applied to the joint. If no moment is imposed, insert zero. Due account must be taken of the sign convention (see below).

Data Sheet 7   *Member loadings*

7.1   Data Sheet Number. Insert 7.
7.2   The total number of member loadings in this load case. (Even if there

Sheet 7. Member loadings

Sheet number (PUNCH THIS!)   7

Number of member loading   11

LOADING TYPE	END 1	END 2	DETAILS OF LOADING				
1	1	2	1	15	3		
3	3	4	1	80	5	4	
1	6	7	1	20	6		
3	6	7	1	100	3	2	
2	9	10	1	120			
4	13	14	1	120	80	1	7
1	5	3	3	6	2·5		
1	12	13	1	15	2		
1	12	13	1	40	4		
2	3	1	2	−90			
2	4	2	2	−90			

FIG. 4.5 *(continued)*

are no loaded members, Data Sheet 7 must be submitted with the number of member loads as zero.) (For types of member loadings see Fig. 4.6.)

7.3   Member loading blocks. For each loaded member the following information must be supplied.

(a)   Type of loading. Insert:
1 if an internal concentrated load;
2 if a full uniformly distributed load;
3 if a partial uniformly distributed load; or
4 if a trapezoidal load.
(See Fig. 4.6.)

(b)   The joint number at end 1 of the loaded member.

(c)   The joint number at end 2 of the loaded member.

(d)   The direction in which the load acts. Insert:
1 if vertical, i.e. in a y-direction;
2 if horizontal, i.e. in an x-direction; or
3 if perpendicular to member where end 1 is on the left and end 2 is on the right. Positive is downwards. (See Fig. 4.6.)
Note that distances are always measured along the member, which is not necessarily normal to the direction of loading.

(e)   Details of the load. These vary according to the type of loading and are as follows.
Type 1: Specify P and a.
Type 2: Specify w.
Type 3: Specify w, a and b.
Type 4: Specify p, q, a and b.
(See Fig. 4.6.)

This completes the input for Program No. A27.

Output: The results are output in the following order.

1.   The user's reference or heading as on the data heading sheet.
2.   The table of member properties if required. This table repeats the area and inertia values given on the input data sheets. In addition, calculated values of member lengths and the sine and cosine of member inclinations to the x-axis are output.
3.   A table of total joint loading if required. This includes values for the x-force, y-force and moment at each joint, including the end forces and moments due to member loading.
4.   For every loading case a table of joint deflections containing details of deformation in the x, y and rotational directions. The units here are as for data input. (A simple visual check may be made to ensure that a zero value occurs at all joint positions where zero deformation has been specified.)
5.   For each loading case, a table of forces and moments. For each member there are six columns of results. The columns from left to right refer to the axial forces, the shearing forces and the internal moments at end 1 and end 2, respectively. The sign convention (see

DETAILS OF MEMBER LOADINGS

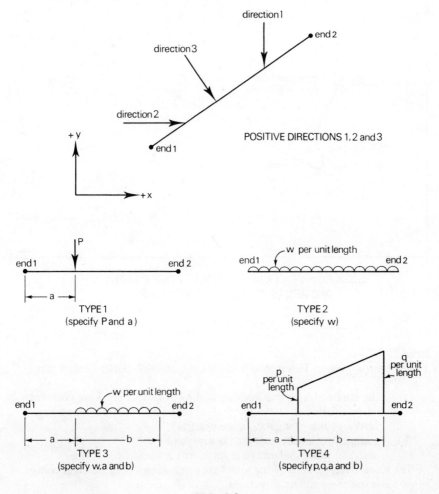

FIG. 4.6

below) is essentially the same as that for input data, though the results are now referred to member axes.

6. Finally, the total applied horizontal and vertical forces are given for each loading case. These values provide a quick check on the loading data, as the total force is the sum of all applied joint loads except those acting directly against zero deformations (which are ignored, as they have no effect on the structure), plus all applied member loads. The total force is actually computed by summing the external forces on the structure (and not by summing the input loading), which means it is also a check on the accuracy of the solution.

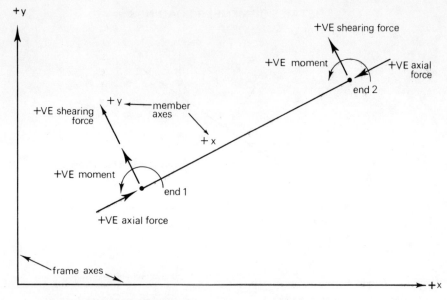

SIGN CONVENTION – OUTPUT FORCES AND
MOMENTS AT ENDS OF MEMBER

FIG. 4.7

Notes

1. *Sign Convention.*   This is shown in Figs. 4.6 and 4.7. It can be seen that:

(a)   the applied forces and loadings in directions 1, 2 and 3 are positive
when directed as shown (so that, in particular, vertically downward
forces, in the $-y$ direction, are positive);

(b)   compressive axial forces are positive; and

(c)   anticlockwise moments (and rotations) are positive.

The linear deformations of the joints are given as two perpendicular components in the $+x$ and $+y$ directions.

2. *Units.*   Compatible units must be used throughout, but there is no restriction as to what these might be.

3. *Self-weight.*   If a non-zero value of density is inserted on the input data
sheet, the self-weight, determined by the density x cross-sectional area of the
member, is added to each loading case as a uniformly distributed load.

4. *Error indications.*   In the case of incorrect data various error indications are
output where appropriate, full details being given in the program brochure.

5. *Punching instructions.*   Data should be punched in the form dictated by the
layout of the data blocks and tables on the standard input sheets. Successive
numbers should be separated either by two or more spaces, by a comma, or by
carriage return line feed (see also Chapter 3). Only numbers, etc., contained
within the rectangular 'boxes' should be punched (all other words, numbers,
symbols, etc., on the standard format sheets being ignored at this stage).

Numerical Example

The problem in Fig. 4.2 is now considered. The input data is given on the standard sheets in Fig. 4.5, and the output is shown in Fig. 4.8. All forces are expressed in kN and all lengths in m, so that moments are expressed in kNm, etc. The signs of the various quantities in the output have been derived with respect to local axes, there being one such set for each member. (These are chosen with the origin at end 1 of the member, the x-axis along the axis of the member being positive from end 1 to end 2, and with that y-axis which completes a right-hand

FIG. 4.8 (IN 5 SHEETS)
PROGRAM NO. A27 – PROBLEM 1: NUMERICAL EXAMPLE
(of multistorey frame shown in Fig. 4.2)

SHEETS 1 & 2 – PRINT-OUT OF INPUT DATA
(as punched from Input Data Sheets – see Fig. 4.5)

SHEET 1

```
'CHECK ON FRAME FIG 4.2 BY A27'
17,23,3,1,1,11,2,8,0,1,
2,
5,18.6,15,18.6,
5,15.6,15,15.6,
0,12.6,5,12.6,15,12.6,
0,9.6,5,9.6,15,9.6,
-1.5,4.8,0,4.8,5,4.8,15,4.8,
0,0,5,0,15,0,
3,
1,2,0,0.05,0.008,
3,4,-1,
6,7,-1,
9,10,-1,
13,14,-1,
5,3,1,0.03,0.0034,
5,6,2,0.03,0.0034,
8,9,-1,
12,13,-1,
11,12,0,0.03,0.0034,
15,12,0,0.06,0.0018,
12,8,-1,
8,5,-1,
16,13,0,0.08,0.0027,
13,9,-1,
9,6,-1,
6,3,-1,
3,1,-1,
17,14,0,0.06,0.0018,
14,10,-1,
10,7,-1,
7,4,-1,
4,2,-1,
4,
15,1,0,0,
16,1,1,1,
17,1,1,0,
5,
15,0,.7,0,
6,
```

SHEET 2

```
'COMBINED LOADING',
10,
1,-5,25,-30,
2,-5,25,20,
3,-10,0,0,
4,-10,0,0,
5,-10,0,0,
7,-10,0,0,
8,-12,12.5,-12.5,
9,0,7.5,0,
10,-12,0,0,
11,0,8,0,
7,
11,
1,1,2,1,15,3,
3,3,4,1,80,5,4,
1,6,7,1,20,6,
3,6,7,1,100,3,2,
2,9,10,1,120,
4,13,14,1,120,80,1,7,
1,5,3,3,6,2.5,
1,12,13,1,15,2,
1,12,13,1,40,4,
2,3,1,2,-90,
2,4,2,2,-90,
```

FIG. 4.8 (*continued*)

SHEETS 3, 4 & 5 — OUTPUT FROM COMPUTER

SHEET 3

CHECK ON FRAME FIG 4.2 BY A27 REVISED DEC 1970

YOUNGS MODULUS=   200000000

MAX JOINT DIFFERENCE=   4

MEMBER PROPERTIES

N C.	END1	END2	LENGTH	SIN	COS	X-AREA	M OF I.
1	1	2	10.00	0.0000	1.0000	5.000 ₁₀-02	8.000 ₁₀-03
2	3	4	10.00	0.0000	1.0000	5.000 ₁₀-02	8.000 ₁₀-03
3	6	7	10.00	0.0000	1.0000	5.000 ₁₀-02	8.000 ₁₀-03
4	9	10	10.00	0.0000	1.0000	5.000 ₁₀-02	8.000 ₁₀-03
5	13	14	10.00	0.0000	1.0000	5.000 ₁₀-02	8.000 ₁₀-03
6	5	3	5.83	0.5145	0.8575	3.000 ₁₀-02	0.000 ₁₀+00
7	5	6	5.00	0.0000	1.0000	3.000 ₁₀-02	3.400 ₁₀-03
8	8	9	5.00	0.0000	1.0000	3.000 ₁₀-02	3.400 ₁₀-03
9	12	13	5.00	0.0000	1.0000	3.000 ₁₀-02	3.400 ₁₀-03
10	11	12	1.50	0.0000	1.0000	3.000 ₁₀-02	3.400 ₁₀-03
11	15	12	4.80	1.0000	0.0000	6.000 ₁₀-02	1.800 ₁₀-03
12	12	8	4.80	1.0000	0.0000	6.000 ₁₀-02	1.800 ₁₀-03
13	8	5	3.00	1.0000	0.0000	6.000 ₁₀-02	1.800 ₁₀-03
14	16	13	4.80	1.0000	0.0000	8.000 ₁₀-02	2.700 ₁₀-03
15	13	9	4.80	1.0000	0.0000	8.000 ₁₀-02	2.700 ₁₀-03
16	9	6	3.00	1.0000	0.0000	8.000 ₁₀-02	2.700 ₁₀-03
17	6	3	3.00	1.0000	0.0000	8.000 ₁₀-02	2.700 ₁₀-03
18	3	1	3.00	1.0000	0.0000	8.000 ₁₀-02	2.700 ₁₀-03
19	17	14	4.80	1.0000	0.0000	6.000 ₁₀-02	1.800 ₁₀-03
20	14	10	4.80	1.0000	0.0000	6.000 ₁₀-02	1.800 ₁₀-03
21	10	7	3.00	1.0000	0.0000	6.000 ₁₀-02	1.800 ₁₀-03
22	7	4	3.00	1.0000	0.0000	6.000 ₁₀-02	1.800 ₁₀-03
23	4	2	3.00	1.0000	0.0000	6.000 ₁₀-02	1.800 ₁₀-03

SHEET 4

JOINT LOADINGS   COMBINED LOADING

JOINT NO.	X-FORCE	Y-FORCE	MOMENT
1	-1.4000000 ₁₀+02	3.6760000 ₁₀+01	-1.1955000 ₁₀+02
2	-1.4000000 ₁₀+02	2.8240000 ₁₀+01	-3.8050000 ₁₀+01
3	-1.4367647 ₁₀+02	7.6445882 ₁₀+01	-1.3836667 ₁₀+02
4	-1.4500000 ₁₀+02	2.4576000 ₁₀+02	4.9096667 ₁₀+02
5	-8.2365549 ₁₀+00	2.9390752 ₁₀+00	0.0000000 ₁₀+00
6	0.0000000 ₁₀+00	1.3624000 ₁₀+02	-3.0186667 ₁₀+02
7	-1.0000000 ₁₀+01	8.3760000 ₁₀+01	2.1946667 ₁₀+02
8	-1.2000000 ₁₀+01	1.2500000 ₁₀+01	-1.2500000 ₁₀+01
9	0.0000000 ₁₀+00	6.0750000 ₁₀+02	-1.0000000 ₁₀+03
10	-1.2000000 ₁₀+01	6.0000000 ₁₀+02	1.0000000 ₁₀+03

FIG. 4.8 SHEET 4 (*continued*)

11	$0.0000000_{10}+00$	$8.0000000_{10}+00$	$0.0000000_{10}+00$
12	$0.0000000_{10}+00$	$2.3720000_{10}+01$	$-3.3600000_{10}+01$
13	$0.0000000_{10}+00$	$4.4688400_{10}+02$	$-7.8652000_{10}+02$
14	$0.0000000_{10}+00$	$2.8439600_{10}+02$	$6.4381333_{10}+02$
15	$0.0000000_{10}+00$	$0.0000000_{10}+00$	$0.0000000_{10}+00$
16	$0.0000000_{10}+00$	$0.0000000_{10}+00$	$0.0000000_{10}+00$
17	$0.0000000_{10}+00$	$0.0000000_{10}+00$	$0.0000000_{10}+00$

LOADCASE   1   COMBINED LOADING

DEFORMATIONS

JOINT NO.	X-DIRECTION	Y-DIRECTION	ROTATION(DEGS.)
1	-0.0265511	-0.0006194	0.0154320
2	-0.0266278	-0.0005859	0.0036506
3	-0.0249726	-0.0006066	0.0103839
4	-0.0252060	-0.0005867	0.0355318
5	-0.0242491	-0.0006035	0.0340762
6	-0.0238859	-0.0006269	0.0239760
7	-0.0238605	-0.0005452	0.0218440
8	-0.0209236	-0.0005082	0.0639994
9	-0.0209315	-0.0006167	0.0164114
10	-0.0208524	-0.0005015	0.0905854
11	-0.0114216	-0.0023053	0.0763283
12	-0.0114216	-0.0003136	0.0755699
13	-0.0113441	-0.0003929	0.0631175
14	-0.0113942	-0.0002625	0.0731555
15	0.0000000	-0.0000627	0.1667187
16	0.0000000	0.0000000	0.0000000
17	0.0000000	0.0000000	0.1674344

SHEET 5

LOADCASE   1   COMBINED LOADING

FORCES AND MOMENTS ON MEMBERS

NO.	MEMBER		AXIAL FORCES		SHEAR FORCES		MOMENTS	
	END1	END2	END1	END2	END1	END2	END1	END2
1	1	2	76.64	76.64	43.09	-28.09	211.61	114.31
2	3	4	233.41	233.41	150.79	169.21	518.39	29.51
3	6	7	-25.42	-25.42	211.44	8.56	683.84	150.60
4	9	10	-79.18	-79.18	777.06	422.94	1678.18	92.45
5	13	14	50.12	50.12	641.43	58.57	1887.60	513.33
6	5	3	640.03	640.03	3.43	2.57	0.00	0.00
7	5	6	-435.80	-435.80	48.91	-48.91	244.56	0.00
8	8	9	9.54	9.54	92.92	-92.92	464.59	0.00
9	12	13	-93.09	-93.09	132.64	-77.64	578.20	0.00
10	11	12	0.00	0.00	-8.00	8.00	-0.00	-12.00

FIG. 4.8 SHEET 5 *(continued)*

11	15	12	627.20	627.20	-49.71	49.71	0.00	-238.63
12	12	8	486.56	486.56	-142.80	142.80	-327.58	-357.87
13	8	5	381.14	381.14	-121.26	121.26	-119.22	-244.56
14	16	13	1309.76	1309.76	-509.78	509.78	-1347.40	-1099.54
15	13	9	745.97	745.97	-366.57	366.57	-788.07	-971.48
16	9	6	54.33	54.33	-455.29	455.29	-706.70	-659.17
17	6	3	-108.21	-108.21	-44.91	44.91	-24.67	-110.07
18	3	1	68.09	68.09	-351.64	81.64	-408.32	-241.61
19	17	14	656.18	656.18	-51.42	51.42	0.00	-246.82
20	14	10	597.61	597.61	-101.54	101.54	-266.51	-220.88
21	10	7	174.68	174.68	-10.36	10.36	128.43	-159.51
22	7	4	166.12	166.12	25.06	-25.06	8.92	66.25
23	4	2	-3.09	-3.09	-198.36	-71.64	-95.76	-94.31

TOTAL HORIZONTAL FORCE=   -610.91

TOTAL VERTICAL FORCE=   2593.14

system (Fig. 4.7). Thus anticlockwise moments on the ends of a member are positive. See also Chapter 7.)

The output is reasonably self-explanatory and the engineer may round off the numerical results to practical values.

Problem 2:   Simply-supported beam carrying a uniformly distributed load

This is Example 1 from Chapter 3 and the problem will be solved once more, this time using Program A27. While it might be fair criticism to regard this as similar to using a sledgehammer to crack a walnut, it is important to realise that this program can analyse complicated frames and continuous beams on the one hand, and a single span beam on the other. It is precisely to illustrate this variation in the type of problem that can be solved by a single program that such a simple example has been chosen.

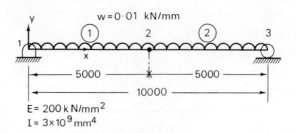

FIG. 4.9

The beam is shown in Fig. 4.9 and it can be seen that the span has been divided into two equal parts, to obtain the deflection of joint number 2 (i.e. the mid-point) in the output. Since this particular problem is exceptionally simple, the input is given as follows (p. 96) without the use of standard data sheets.

FIG. 4.10
PROBLEM 2: SIMPLY SUPPORTED BEAM CARRYING
A UNIFORMLY DISTRIBUTED LOAD

OUTPUT FROM COMPUTER

EXAMPLE 1 FROM CHAPTER 3 UDL SI UNITS

YOUNGS MODULUS= 200

MAX JOINT DIFFERENCE= 1

MEMBER PROPERTIES

NO.	END1	END2	LENGTH	SIN	COS	X-AREA	M OF I
1	1	2	5000.00	0.0000	1.0000	1.000ₑ+06	3.000ₑ+09
2	2	3	5000.00	0.0000	1.0000	1.000ₑ+06	3.000ₑ+09

JOINT LOADINGS   UNIFORMLY DISTRIBUTED LOAD

JOINT NO.	X-FORCE	Y-FORCE	MOMENT
1	0.0000000ₑ+00	2.5000000ₑ+01	-2.0833333ₑ+04
2	0.0000000ₑ+00	5.0000000ₑ+01	0.0000000ₑ+00
3	0.0000000ₑ+00	2.5000000ₑ+01	2.0833333ₑ+04

LOADCASE 1 UNIFORMLY DISTRIBUTED LOAD

DEFORMATIONS

JOINT NC.	X-DIRECTION	Y-DIRECTION	ROTATION(DEGS.)
1	0.0000000	0.0000000	-0.0397887
2	0.0000000	-2.1701389	0.0000000
3	0.0000000	0.0000000	0.0397887

LOADCASE 1 UNIFORMLY DISTRIBUTED LOAD

FORCES AND MOMENTS ON MEMBERS

MEMBER			AXIAL FORCES		SHEAR FORCES		MOMENTS	
NC.	END1	END2	END 1	END 2	END 1	END 2	END 1	END 2
1	1	2	0.00	0.00	50.00	-0.00	0.00	125000.00
2	2	3	0.00	0.00	0.00	50.00	-125000.00	0.00

TOTAL HORIZONTAL FORCE= 0.00

TOTAL VERTICAL FORCE= 100.00

```
/ EXAMPLE 1 FROM CHAPTER 3 UDL SI UNITS \
3    2    2    0    1    2    200    0    1
2
0    0    5000        0   10000        0
3
1    2    0    10⁶  3 x 10⁹
2    3   -1
4
1    1    1    0
3    0    1    0
6
/ UNIFORMLY DISTRIBUTED LOAD \
0
7
2
2    1    2    1    0.01
2    2    3    1    0.01
```

The output from the computer is given in Fig. 4.10. All the forces are in kN and the lengths are in mm. These results can be compared with those for Example 1 in Chapter 3 and it can be seen that they are in close agreement.

Note that in this method it is necessary to specify a cross sectional area for the beam. As this effect was neglected in the former solution, a fair comparison can best be made by specifying a large value—taken here as 10^6 mm^2.

Problem 3:   Elastic analysis of a built-in beam carrying a concentrated load

This is the same as Example 2 in Chapter 3 and shows how a statically-indeterminate beam may also be analysed by this method.

Referring to the example already given, it is not possible here to locate the position of the maximum deflection except by a trial-and-adjustment process. However by inserting imaginary joints on the beam, the output will include the deflection of these points. Thus, by using the result previously obtained that the maximum deflection occurs where $x = 8571.4286$ mm (Fig. 3.3), an imaginary joint can be chosen at this point and a check made on the deflection obtained.

The analysis is carried out by using Program A27 once more. The beam is divided into four parts (Fig. 4.11) with imaginary joints at mid-span, at the point of maximum deflection on the span, and at the load point, so that the deflections and bending moments at these locations can be obtained.

The input for Program A27 is brief and is given without the use of the standard data sheets.

```
/ EXAMPLE 2 CHAPTER 3 \
5       4    2      0    1    0    200    0    1
2
0       0    7500   0    8571.4286        0
10000   0           15000  0
3
1       2    0           10⁶  4 x 10⁹
2       3   -1
3       4   -1
4       5   -1
4
1       1    1      1
5       1    1      1
6
/ CONCENTRATED LOAD \
1
4       0    80     0
7
0
```

This data can then be used with Program A27 in the usual way. and the output obtained from the computer is given in Fig. 4.12. While this data is rather more comprehensive than that given by the analysis in Chapter 3 using Program DCTCES 2, a direct comparison between the two solutions shows that the results are in agreement.

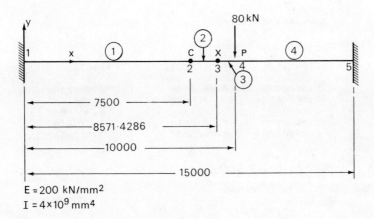

$E = 200 \ kN/mm^2$
$I = 4 \times 10^9 \ mm^4$

FIG. 4.11

FIG. 4.12
PROBLEM 3: ELASTIC ANALYSIS OF A BUILT-IN BEAM
CARRYING A CONCENTRATED LOAD

OUTPUT FROM COMPUTER

EXAMPLE 2 CHAPTER 3

YOUNGS MODULUS= 200

MAX JOINT DIFFERENCE= 1

MEMBER PROPERTIES

NO.	END1	END2	LENGTH	SIN	COS	X-AREA	M OF I
1	1	2	7500.00	0.0000	1.0000	1.000ₘ+06	4.000ₘ+09
2	2	3	1071.43	0.0000	1.0000	1.000ₘ+06	4.000ₘ+09
3	3	4	1428.57	0.0000	1.0000	1.000ₘ+06	4.000ₘ+09
4	4	5	5000.00	0.0000	1.0000	1.000ₘ+06	4.000ₘ+09

JOINT LOADINGS   CONCENTRATED LOAD

JOINT NO.	X-FORCE	Y-FORCE	MOMENT
1	0.0000000ₘ+00	0.0000000ₘ+00	0.0000000ₘ+00
2	0.0000000ₘ+00	0.0000000ₘ+00	0.0000000ₘ+00
3	0.0000000ₘ+00	0.0000000ₘ+00	0.0000000ₘ+00
4	0.0000000ₘ+00	8.0000000ₘ+01	0.0000000ₘ+00
5	0.0000000ₘ+00	0.0000000ₘ+00	0.0000000ₘ+00

LOADCASE   1   CONCENTRATED LOAD

DEFORMATIONS

JOINT NO.	X-DIRECTION	Y-DIRECTION	ROTATION(DEGS.)
1	0.0000000	0.0000000	0.0000000
2	0.0000000	-1.3020833	-0.0059683
3	0.0000000	-1.3605442	0.0000000
4	0.0000000	-1.2345679	0.0106103
5	0.0000000	0.0000000	0.0000000

LOADCASE   1   CONCENTRATED LOAD

FORCES AND MOMENTS ON MEMBERS

MEMBER			AXIAL FORCES		SHEAR FORCES		MOMENTS	
NO.	END1	END2	END 1	END 2	END 1	END 2	END 1	END 2
1	1	2	0.00	0.00	20.74	-20.74	88888.89	66666.67
2	2	3	0.00	0.00	20.74	-20.74	-66666.67	88888.89
3	3	4	0.00	0.00	20.74	-20.74	-88888.89	118518.52
4	4	5	0.00	0.00	-59.26	59.26	-118518.52	-177777.78

TOTAL HORIZONTAL FORCE=   0.00

TOTAL VERTICAL FORCE=   80.00

Problem 4:   Propped cantilever with concentrated load

This is the same as Example 3 in Chapter 3 (Fig. 3.5).

As in the previous problem, the location of the maximum deflection can be obtained from the results already given and an imaginary joint can be inserted at this point. The beam has, in fact, been divided into three parts (Fig. 4.13) to include also the load point.

The analysis can now be carried out using Program A27 again. The input is as follows.

```
/ PROPPED CANTILEVER WITH CONCENTRATED LOAD \
4    3    2    0    1    0       200      0      1
2
0    0    55384.615       0    60000    0    90000    0
3
1    2    0    10⁶   1.2 x 10¹⁰
2    3   -1
3    4   -1
4
1    1    1    1
4    0    1    0
6
/ CONCENTRATED LOAD \
1
3    0    100  0
7
0
```

The computer output is shown in Fig. 4.14 and it can be seen that the results are in agreement with those already quoted.

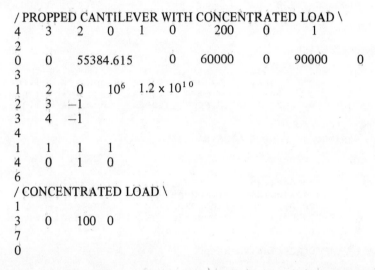

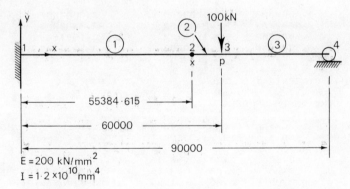

FIG. 4.13

FIG. 4.14
PROBLEM 4: PROPPED CANTILEVER WITH CONCENTRATED LOAD

OUTPUT FROM COMPUTER

PROPPED CANTILEVER WITH CONCENTRATED LOAD

YOUNGS MODULUS=   200

MAX JOINT DIFFERENCE=   1

MEMBER PROPERTIES

NO.	END1	END2	LENGTH	SIN	COS	X-AREA	M OF I
1	1	2	55384.62	0.0000	1.0000	1.000 ₒ+06	1.200 ₒ+10
2	2	3	4615.38	0.0000	1.0000	1.000 ₒ+06	1.200 ₒ+10
3	3	4	30000.00	0.0000	1.0000	1.000 ₒ+06	1.200 ₒ+10

JOINT LOADINGS   CONCENTRATED LOAD

JOINT NO.	X-FORCE	Y-FORCE	MOMENT
1	0.0000000 ₒ+00	0.0000000 ₒ+00	0.0000000 ₒ+00
2	0.0000000 ₒ+00	0.0000000 ₒ+00	0.0000000 ₒ+00
3	0.0000000 ₒ+00	1.0000000 ₒ+02	0.0000000 ₒ+00
4	0.0000000 ₒ+00	0.0000000 ₒ+00	0.0000000 ₒ+00

LOADCASE   1   CONCENTRATED LOAD

DEFORMATIONS

JOINT NO.	X-DIRECTION	Y-DIRECTION	ROTATION(DEGS.)
1	0.0000000	0.0000000	0.0000000
2	0.0000000	-2.840 ₒ+02	-0.0000000
3	0.0000000	-2.778 ₒ+02	0.1591549
4	0.0000000	0.0000000	0.7161972

LOADCASE   1   CONCENTRATED LOAD

FORCES AND MOMENTS ON MEMBERS

MEMBER			AXIAL FORCES		SHEAR FORCES		MOMENTS	
NO.	END1	END2	END 1	END 2	END 1	END 2	END 1	END 2
1	1	2	0.00	0.00	48.15	-48.15	1.333 ₒ+06	1.333 ₒ+06
2	2	3	0.00	0.00	48.15	-48.15	-1.333 ₒ+06	1.556 ₒ+06
3	3	4	0.00	0.00	-51.85	51.85	-1.556 ₒ+06	0.00

TOTAL HORIZONTAL FORCE=   0.00

TOTAL VERTICAL FORCE=   100.00

Problem 5:   Single bay, pitched roof, portal frame with pinned bases

A check on the solution given to Example 4 of Chapter 3 can be effected using Program A27.

The frame is shown in Fig. 4.15. The joints and members have been numbered and a suitable set of co-ordinate axes chosen. The input to the computer is given on the special format sheets in Fig. 4.16 and the resulting output in Fig. 4.17. The results can be compared with those already obtained and are seen to be in agreement. The output here is more comprehensive, giving the deflections of the joints as well as the axial and shearing forces and bending moments in the members. Note that on this occasion it was necessary to specify values for Young's Modulus and the cross-sectional areas of the members.

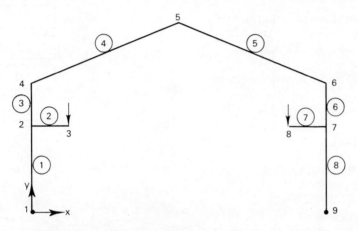

Numbering of joints and members.
(For dimensions, loading etc. see fig.3.7)

FIG. 4.15

Program No A27
Elastic Analysis of Plane Framework
Input data
Sheet 1. General details

Job title	/ PROBLEM 5. SINGLE BAY PITCHED ROOF PORTAL FRAME

Number of joints	9
Number of members	8
Number of joints with any zero deformations	2
Number of spring joints	0
Number of loadcases	1
Number of member loads	5
Young's Modulus	200
Density	0
Steering symbol (Output check required insert 1, otherwise 0)	1

Sheet 2. Joint co-ordinates

Sheet number
(PUNCH THIS!)

2

JOINT NUMBER	x	y		JOINT NUMBER	x	y
1 →	0	0	→ 2 →		0	6000
3	1500	6000		4	0	8000
5	6000	10400		6	12000	8000
7	12000	6000		8	10500	6000
9	12000	0		10		
11						

FIG. 4.16

Sheet 3 Member properties

Sheet number
(PUNCH THIS!)

$\boxed{3}$

	END 1	END 2	HINGE	A	I
1	1	2	0	30000	2×10^9
2	2	3	-1		
3	2	4	-1		
4	4	5	0	30000	8×10^8
5	5	6	-1		
6	7	6	0	30000	2×10^9
7	8	7	-1		
8	9	7	-1		
9					
10					
11					

Sheet 4. Zero deformations

Sheet number
(PUNCH THIS!)

$\boxed{4}$

JOINT No.	M1	M2	M3
1	1	1	0
9	1	1	0

FIG. 4.16 (*continued*)

Sheet 6. Joint loading

Sheet number | 6 |
(PUNCH THIS!)

Loadcase title | /   LOADING AS ON FIG.3·7 |

Number of joints loading | 7 |

JOINT No.	Load X	Load Y	Moment
2	10	0	0
3	0	60	0
4	6	0	0
5	8	0	0
6	4	0	0
7	10	0	0
8	0	20	0

Sheet 7. Member loadings

Sheet number | 7 |
(PUNCH THIS!)

Number of member loading | 5 |

LOADING TYPE	END 1	END 2		DETAILS OF LOADING			
2	1	2	2	0·0014			
2	2	4	2	0·0014			
2	5	6	1	0·005571			
2	7	6	2	0·0006			
2	9	7	2	0·0006			

FIG. 4.16 (*continued*)

FIG. 4.17
PROBLEM 5: SINGLE BAY, PITCHED ROOF, PORTAL
FRAME WITH PINNED BASES.

PRINT-OUT OF INPUT DATA
(as punched from the Special Format Sheets in Fig. 4.16)

```
*MARCH 1971 PITCHED ROOF FRAME BY A27*

9,8,2,0,1,5,200,0,1,
2,
0,0,0,6000,1500,6000,0,8000,
6000,10400,12000,8000,12000,6000,
10500,6000,12000,0,
3,
1,2,0,30000,2.9,
2,3,-1,
2,4,-1,
4,5,0,30000,8.8,
5,6,-1,
7,6,0,30000,2.9,
8,7,-1,
9,7,-1,
4,
1,1,1,0,
9,1,1,0,
6,

*LOADING AS ON FIG. 3.7*

7,
2,10,0,0,
3,0,60,0,
4,6,0,0,
5,8,0,0,
6,4,0,0,
7,10,0,0,
8,0,20,0,
7,
5,
2,1,2,2,0.0014,
2,2,4,2,0.0014,
2,5,6,1,0.005571,
2,7,6,2,0.0006,
2,9,7,2,0.0006,
```

FIG. 4.17 (*continued*)

OUTPUT FROM COMPUTER

MARCH 1971 PITCHED ROOF FRAME BY A27

YOUNGS MODULUS=   200

MAX JOINT DIFFERENCE=   2

MEMBER PROPERTIES

NC.	END1	END2	LENGTH	SIN	COS	X-AREA	M OF I
1	1	2	6000.00	1.0000	0.0000	3.000 ₒ+04	2.000 ₒ+09
2	2	3	1500.00	0.0000	1.0000	3.000 ₒ+04	2.000 ₒ+09
3	2	4	2000.00	1.0000	0.0000	3.000 ₒ+04	2.000 ₒ+09
4	4	5	6462.20	0.3714	0.9285	3.000 ₒ+04	8.000 ₒ+08
5	5	6	6462.20	-0.3714	0.9285	3.000 ₒ+04	8.000 ₒ+08
6	7	6	2000.00	1.0000	0.0000	3.000 ₒ+04	2.000 ₒ+09
7	8	7	1500.00	0.0000	1.0000	3.000 ₒ+04	2.000 ₒ+09
8	9	7	6000.00	1.0000	0.0000	3.000 ₒ+04	2.000 ₒ+09

JOINT LOADINGS   LOADING AS ON FIG. 3.7

JOINT NC.	X-FORCE	Y-FORCE	MOMENT
1	4.2000000 ₒ+00	0.0000000 ₒ+00	-4.2000000 ₒ+03
2	1.5600000 ₒ+01	0.0000000 ₒ+00	3.7333333 ₒ+03
3	0.0000000 ₒ+00	6.0000000 ₒ+01	0.0000000 ₒ+00
4	7.4000000 ₒ+00	0.0000000 ₒ+00	4.6666667 ₒ+02
5	8.0000000 ₒ+00	1.8000452 ₒ+01	-1.8000452 ₒ+04
6	4.6000000 ₒ+00	1.8000452 ₒ+01	1.8200452 ₒ+04
7	1.2400000 ₒ+01	0.0000000 ₒ+00	1.6000000 ₒ+03
8	0.0000000 ₒ+00	2.0000000 ₒ+01	0.0000000 ₒ+00
9	1.8000000 ₒ+00	0.0000000 ₒ+00	-1.8000000 ₒ+03

LOADCASE   1   LOADING AS ON FIG. 3.7

DEFORMATIONS

JOINT NC.	X-DIRECTION	Y-DIRECTION	ROTATION(DEGS.)
1	0.0000000	0.0000000	-0.2427354
2	2.390 ₒ+01	-0.0350669	-0.2010031
3	2.390 ₒ+01	-5.4660652	-0.2106717
4	3.002 ₒ+01	-0.0267559	-0.1499044
5	3.046 ₒ+01	-1.1053593	0.0639910
6	3.082 ₒ+01	-0.1012453	-0.1293949
7	2.532 ₒ+01	-0.0809340	-0.1839115
8	2.532 ₒ+01	4.6776068	-0.1806886
9	0.0000000	0.0000000	-0.2710946

FIG. 4.17 *(continued)*

LOADCASE   1   LOADING AS ON FIG. 3.7

FORCES AND MOMENTS ON MEMBERS

MEMBER			AXIAL FORCES		SHEAR FORCES		MOMENTS	
NO.	END1	END2	END 1	END 2	END 1	END 2	END 1	END 2
1	1	2	35.07	35.07	18.99	-10.59	0.00	88715.55
2	2	3	0.00	0.00	60.00	-60.00	90000.00	-0.00
3	2	4	-24.93	-24.93	0.59	2.21	-178715.54	177087.39
4	4	5	-1.63	-1.63	-26.20	26.20	-177087.39	7774.96
5	5	6	24.31	37.68	-17.13	59.55	-7774.96	-210912.62
6	7	6	60.93	60.93	21.41	-20.21	-169284.46	210912.62
7	8	7	-0.00	-0.00	-20.00	20.00	0.00	-30000.46
8	9	7	80.93	80.93	35.01	-31.41	0.00	199284.46

TOTAL HORIZONTAL FORCE=   54.00

TOTAL VERTICAL FORCE=   116.00

Problem 6:   Multistorey frames

An illustration of the use of Program A27 to analyse a multistorey frame has
already been given in Problem 1 above. However, it is also possible to check the
solution given for the multistorey frame in Example 5 of Chapter 3.

This frame is numbered as shown in Fig. 4.18, the input being given in
Fig. 4.19, and the output in Fig. 4.20. It can be seen that the results are in fair
agreement, considering that these have been obtained by two different methods
of structural analysis.

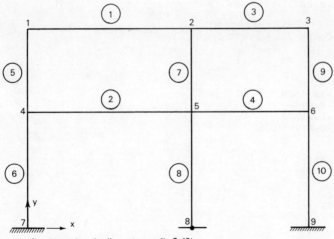

(for dimensions loadings etc; see fig. 3.12)

FIG. 4.18

FIG. 4.19
PROBLEM 6: MULTISTOREY FRAMES
INPUT TO COMPUTER

```
'EXAMPLE 5 CHAPTER 3 BY MAUNSELLS A27 REVISED DEC 1970',

9,10,3,0,3,9,200,0,1,
2,
0,11000,12000,11000,21000,11000,
0,6000,12000,6000,21000,6000,
0,0,12000,0,21000,0,
3,
1,2,0, ,12,3 ,9,
4,5,-1,
2,3,0, ,12,2 ,9,
5,6,-1,
1,4,-1,
4,7,-1,
2,5,-1,
5,8,0, ,12,3 ,9,
3,6,0, ,12, ,9,
6,9,0, ,12,2 ,9,
4,
7,1,1,1,8,1,1,0,9,1,1,1,
6,

'DEAD LCAD',

0,
7,
5,
4,1,2,1,0,0.01,0,12000,
2,4,5,1,0.009,
3,4,5,1,0.02,6000,4000,
4,2,3,1,0.01,0,0,9000,
2,5,6,1,0.008,
6,

'SUPER LOADS',

0,
7,
4,
1,1,2,1,30,6000,
2,4,5,1,0.015,
1,5,6,1,20,5000,
2,5,6,1,0.01,
6,

'WIND LOADS',

2,
1,10,0,0,
4,20,0,0,
7,
0,
```

FIG. 4.20
PROBLEM 6: MULTISTOREY FRAMES
OUTPUT FROM COMPUTER

EXAMPLE 5 CHAPTER 3 BY MAUNSELLS A27 REVISED DEC 1970

YCUNGS MODULUS= 200

MAX JCINT DIFFERENCE= 3

MEMBER PROPERTIES

NC.	END1	END2	LENGTH	SIN	COS	X-AREA	M OF I
1	1	2	12000·00	0·0000	1·0000	1·000 ₀+12	3·000 ₀+09
2	4	5	12000·00	0·0000	1·0000	1·000 ₀+12	3·000 ₀+09
3	2	3	9000·00	0·0000	1·0000	1·000 ₀+12	2·000 ₀+09
4	5	6	9000·00	0·0000	1·0000	1·000 ₀+12	2·000 ₀+09
5	1	4	5000·00	-1·0000	0·0000	1·000 ₀+12	2·000 ₀+09
6	4	7	6000·00	-1·0000	0·0000	1·000 ₀+12	2·000 ₀+09
7	2	5	5000·00	-1·0000	0·0000	1·000 ₀+12	2·000 ₀+09
8	5	8	6000·00	-1·0000	0·0000	1·000 ₀+12	3·000 ₀+09
9	3	6	5000·00	-1·0000	0·0000	1·000 ₀+12	1·000 ₀+09
10	6	9	6000·00	-1·0000	0·0000	1·000 ₀+12	2·000 ₀+09

JCINT LCADINGS   DEAD LCAD

JCINT NC.	X-FCRCE	Y-FCRCE	MOMENT
1	0·0000000 ₀+00	1·8000000 ₀+01	-4·8000000 ₀+04
2	0·0000000 ₀+00	7·3500000 ₀+01	3·1500000 ₀+04
3	0·0000000 ₀+00	1·3500000 ₀+01	2·7000000 ₀+04
4	0·0000000 ₀+00	7·5481481 ₀+01	-1·7911111 ₀+05
5	0·0000000 ₀+00	1·4851852 ₀+02	1·8733333 ₀+05
6	0·0000000 ₀+00	3·6000000 ₀+01	5·4000000 ₀+04
7	0·0000000 ₀+00	0·0000000 ₀+00	0·0000000 ₀+00
8	0·0000000 ₀+00	0·0000000 ₀+00	0·0000000 ₀+00
9	0·0000000 ₀+00	0·0000000 ₀+00	0·0000000 ₀+00

JCINT LCADINGS   SUPER LCADS

JOINT NC.	X-FCRCE	Y-FCRCE	MOMENT
1	0·0000000 ₀+00	1·5000000 ₀+01	-4·5000000 ₀+04
2	0·0000000 ₀+00	1·5000000 ₀+01	4·5000000 ₀+04
3	0·0000000 ₀+00	0·0000000 ₀+00	0·0000000 ₀+00
4	0·0000000 ₀+00	9·0000000 ₀+01	-1·8000000 ₀+05
5	0·0000000 ₀+00	1·4334019 ₀+02	9·2746914 ₀+04
6	0·0000000 ₀+00	5·6659808 ₀+01	9·2191358 ₀+04
7	0·0000000 ₀+00	0·0000000 ₀+00	0·0000000 ₀+00
8	0·0000000 ₀+00	0·0000000 ₀+00	0·0000000 ₀+00
9	0·0000000 ₀+00	0·0000000 ₀+00	0·0000000 ₀+00

FIG. 4.20 (*continued*)

JOINT LOADINGS   WIND LOADS

JOINT NO.	X-FORCE	Y-FORCE	MOMENT
1	$1.0000000_{\text{m}}+01$	$0.0000000_{\text{m}}+00$	$0.0000000_{\text{m}}+00$
2	$0.0000000_{\text{m}}+00$	$0.0000000_{\text{m}}+00$	$0.0000000_{\text{m}}+00$
3	$0.0000000_{\text{m}}+00$	$0.0000000_{\text{m}}+00$	$0.0000000_{\text{m}}+00$
4	$2.0000000_{\text{m}}+01$	$0.0000000_{\text{m}}+00$	$0.0000000_{\text{m}}+00$
5	$0.0000000_{\text{m}}+00$	$0.0000000_{\text{m}}+00$	$0.0000000_{\text{m}}+00$
6	$0.0000000_{\text{m}}+00$	$0.0000000_{\text{m}}+00$	$0.0000000_{\text{m}}+00$
7	$0.0000000_{\text{m}}+00$	$0.0000000_{\text{m}}+00$	$0.0000000_{\text{m}}+00$
8	$0.0000000_{\text{m}}+00$	$0.0000000_{\text{m}}+00$	$0.0000000_{\text{m}}+00$
9	$0.0000000_{\text{m}}+00$	$0.0000000_{\text{m}}+00$	$0.0000000_{\text{m}}+00$

LOADCASE   1   DEAD LOAD

DEFORMATIONS

JOINT NO.	X-DIRECTION	Y-DIRECTION	ROTATION(DEGS.)
1	0.1177512	-0.0000000	-0.0008517
2	0.1177512	-0.0000000	-0.0007414
3	0.1177512	-0.0000000	0.0040138
4	0.0872817	-0.0000000	-0.0150049
5	0.0872817	-0.0000000	0.0117685
6	0.0872817	-0.0000000	0.0021666
7	0.0000000	0.0000000	0.0000000
8	0.0000000	0.0000000	-0.0071344
9	0.0000000	0.0000000	0.0000000

LOADCASE   1   DEAD LOAD

FORCES AND MOMENTS ON MEMBERS

MEMBER			AXIAL FORCES		SHEAR FORCES		MOMENTS	
NO.	END1	END2	END 1	END 2	END 1	END 2	END 1	END 2
1	1	2	25.41	25.41	17.30	42.70	43733.14	-76074.46
2	4	5	-9.90	-9.90	74.07	113.93	147273.86	-226442.22
3	2	3	5.76	5.76	33.19	11.81	44426.51	-15696.27
4	5	6	-1.25	-1.25	43.21	28.79	93876.54	-29019.79
5	1	4	17.30	17.30	-25.40	25.40	-43733.14	-83256.53
6	4	7	91.37	91.37	-15.52	15.52	-64017.33	-29099.28
7	2	5	75.89	75.89	19.65	-19.65	31647.95	66582.10
8	5	8	233.02	233.02	11.00	-11.00	65983.57	0.00
9	3	6	11.81	11.81	5.76	-5.76	15696.27	13117.15
10	6	9	40.60	40.60	4.46	-4.46	15902.65	10860.71

TOTAL HORIZONTAL FORCE=   -0.06

TOTAL VERTICAL FORCE=   365.00

FIG. 4.20 (*continued*)

LOADCASE   2   SUPER LOADS

DEFORMATIONS

JOINT NO.	X-DIRECTION	Y-DIRECTION	ROTATION(DEGS.)
1	0.2193898	-0.0000000	-0.0023837
2	0.2193898	-0.0000000	0.0022866
3	0.2193898	-0.0000000	-0.0034811
4	0.0798134	-0.0000000	-0.0145948
5	0.0798134	-0.0000000	0.0047677
6	0.0798134	-0.0000000	0.0073591
7	0.0000000	0.0000000	0.0000000
8	0.0000000	0.0000000	-0.0035271
9	0.0000000	0.0000000	0.0000000

LOADCASE   2   SUPER LOADS

FORCES AND MOMENTS ON MEMBERS

	MEMBER		AXIAL FORCES		SHEAR FORCES		MOMENTS	
NO.	END1	END2	END 1	END 2	END 1	END 2	END 1	END 2
1	1	2	23.10	23.10	14.96	15.04	40670.14	-41178.56
2	4	5	-7.87	-7.87	85.71	94.29	137375.92	-188830.22
3	2	3	5.90	5.90	-0.62	0.62	1694.35	-7253.75
4	5	6	4.45	4.45	59.61	50.39	113463.22	-61960.98
5	1	4	14.96	14.96	-23.09	23.09	-40670.14	-74769.75
6	4	7	100.67	100.67	-15.21	15.21	-62606.17	-28642.64
7	2	5	14.42	14.42	17.18	-17.18	39484.21	46412.69
8	5	8	168.32	168.32	4.83	-4.83	28954.31	0.00
9	3	6	0.62	0.62	5.93	-5.93	7253.75	22389.48
10	6	9	51.01	51.01	10.34	-10.34	39571.50	22446.20

TOTAL HORIZONTAL FORCE=   -0.05

TOTAL VERTICAL FORCE=   320.00

LOADCASE   3   WIND LOADS

DEFORMATIONS

JOINT NO.	X-DIRECTION	Y-DIRECTION	ROTATION(DEGS.)
1	1.3319716	0.0000000	-0.0020460
2	1.3319716	0.0000000	-0.0021165
3	1.3319716	-0.0000000	-0.0012149
4	0.9100949	0.0000000	-0.0065307
5	0.9100949	0.0000000	-0.0033192
6	0.9100949	-0.0000000	-0.0070218
7	0.0000000	0.0000000	0.0000000
8	0.0000000	0.0000000	-0.0113765
9	0.0000000	0.0000000	0.0000000

FIG. 4.20 (*continued*)

L CADCASE   3   WIND LOADS

FORCES AND MOMENTS ON MEMBERS

MEMBER			AXIAL FORCES		SHEAR FORCES		MOMENTS	
NC.	END1	END2	END 1	END 2	END 1	END 2	END 1	END 2
1	1	2	8.49	8.49	-1.82	1.82	-10835.84	-10958.81
2	4	5	9.22	9.22	-4.30	4.30	-28589.70	-22984.55
3	2	3	0.97	0.97	-1.72	1.72	-8451.78	-7053.09
4	5	6	11.16	11.16	-5.35	5.35	-21192.63	-26936.90
5	1	4	-1.82	-1.82	1.83	-1.83	10835.84	-1687.88
6	4	7	-6.11	-6.11	12.63	-12.63	30277.58	45475.29
7	2	5	0.09	0.09	7.09	-7.09	19410.59	16051.85
8	5	8	-0.96	-0.96	4.69	-4.69	28125.34	-0.00
9	3	6	1.72	1.72	1.20	-1.20	7053.09	-1054.94
10	6	9	7.07	7.07	12.05	-12.05	27991.83	44332.41

TCTAL HCRIZONTAL FORCE=   29.37

TCTAL VERTICAL FORCE=   -0.00

Problem 7:   Continuous beams

Continuous beams can also be analysed using Program A27. Consider the continuous beam solved in Example 6 of Chapter 3. This is shown in Fig. 4.21 and the input to, and the output from, the computer are given in Figs. 4.22 and 4.23 respectively. Once more it can be seen that the two sets of results are in agreement.

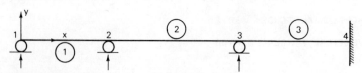

(*for dimensions loading etc; see fig. 3.18*)

FIG. 4.21

FIG. 4.22
PROBLEM 7: CONTINUOUS BEAMS
INPUT TO COMPUTER

```
*EXAMPLE 6 CHAPTER 3 CONTINUOUS BEAM*

4,3,4,0,4,8,2.8,0,1,
2,
0,0,6,0,15,0,22.5,0,
3,
1,2,0,100,0.0008,
2,3,0,100,0.003,
3,4,0,100,0.0015,
4,
1,0,1,0,
2,0,1,0,
3,0,1,0,
4,1,1,1,
6,

*DEAD LOAD*

0,
7,
3,
2,1,2,1,0.7,
2,2,3,1,0.7,
2,3,4,1,0.7,
6,

*LIVE LOAD ON LEFT SPAN*

0,
7,
1,
2,1,2,1,1,
6,

*LIVE LOAD ON CENTRE SPAN*

0,
7,
2,
1,2,3,1,25,5,
2,2,3,1,0.5,
6,

*LIVE LOAD ON RIGHT SPAN*

0,
7,
2,
4,3,4,1,0,2,0,1.5,
4,3,4,1,2,0,1.5,6,
```

FIG. 4.23
PROBLEM 7: CONTINUOUS BEAMS
OUTPUT FROM COMPUTER

EXAMPLE 6 CHAPTER 3 CONTINUOUS BEAM

YOUNGS MODULUS= 200000000

MAX JOINT DIFFERENCE= 1

MEMBER PROPERTIES

N C.	END1	END2	LENGTH	SIN	COS	X-AREA	M OF I
1	1	2	6·00	0·0000	1·0000	1·000 ₙ+02	8·000 ₙ-04
2	2	3	9·00	0·0000	1·0000	1·000 ₙ+02	3·000 ₙ-03
3	3	4	7·50	0·0000	1·0000	1·000 ₙ+02	1·500 ₙ-03

JOINT LOADINGS   DEAD LOAD

JOINT NO.	X-FORCE	Y-FORCE	MOMENT
1	0·0000000 ₙ+00	2·1000000 ₙ+00	-2·1000000 ₙ+00
2	0·0000000 ₙ+00	5·2500000 ₙ+00	-2·6250000 ₙ+00
3	0·0000000 ₙ+00	5·7750000 ₙ+00	1·4437500 ₙ+00
4	0·0000000 ₙ+00	2·6250000 ₙ+00	3·2812500 ₙ+00

JOINT LOADINGS   LIVE LOAD ON LEFT SPAN

JOINT NO.	X-FORCE	Y-FORCE	MOMENT
1	0·0000000 ₙ+00	3·0000000 ₙ+00	-3·0000000 ₙ+00
2	0·0000000 ₙ+00	3·0000000 ₙ+00	3·0000000 ₙ+00
3	0·0000000 ₙ+00	0·0000000 ₙ+00	0·0000000 ₙ+00
4	0·0000000 ₙ+00	0·0000000 ₙ+00	0·0000000 ₙ+00

JOINT LOADINGS   LIVE LOAD ON CENTRE SPAN

JOINT NO.	X-FORCE	Y-FORCE	MOMENT
1	0·0000000 ₙ+00	0·0000000 ₙ+00	0·0000000 ₙ+00
2	0·0000000 ₙ+00	1·2675240 ₙ+01	-2·8066358 ₙ+01
3	0·0000000 ₙ+00	1·6824760 ₙ+01	3·4239198 ₙ+01
4	0·0000000 ₙ+00	0·0000000 ₙ+00	0·0000000 ₙ+00

JOINT LOADINGS   LIVE LOAD ON RIGHT SPAN

JOINT NO.	X-FORCE	Y-FORCE	MOMENT
1	0·0000000 ₙ+00	0·0000000 ₙ+00	0·0000000 ₙ+00
2	0·0000000 ₙ+00	0·0000000 ₙ+00	0·0000000 ₙ+00
3	0·0000000 ₙ+00	4·7220000 ₙ+00	-6·2700000 ₙ+00
4	0·0000000 ₙ+00	2·7780000 ₙ+00	4·6050000 ₙ+00

FIG. 4.23 (*continued*)

LOADCASE   1   DEAD LOAD

DEFORMATIONS

JOINT NO.	X-DIRECTION	Y-DIRECTION	ROTATION(DEGS.)
1	0.0000000	0.0000000	-0.0009377
2	0.0000000	0.0000000	-0.0003806
3	0.0000000	0.0000000	0.0003128
4	0.0000000	0.0000000	0.0000000

LOADCASE   1   DEAD LOAD

FORCES AND MOMENTS ON MEMBERS

MEMBER			AXIAL FORCES		SHEAR FORCES		MOMENTS	
NO.	END1	END2	END 1	END 2	END 1	END 2	END 1	END 2
1	1	2	0.00	0.00	1.49	2.71	0.00	-3.68
2	2	3	0.00	0.00	3.10	3.20	3.68	-4.15
3	3	4	0.00	0.00	2.80	2.45	4.15	-2.84

TOTAL HORIZONTAL FORCE=   0.00

TOTAL VERTICAL FORCE=   15.75

LOADCASE   2   LIVE LOAD ON LEFT SPAN

DEFORMATIONS

JOINT NO.	X-DIRECTION	Y-DIRECTION	ROTATION(DEGS.)
1	0.0000000	0.0000000	-0.0020341
2	0.0000000	0.0000000	0.0008453
3	0.0000000	0.0000000	-0.0002642
4	0.0000000	0.0000000	0.0000000

LOADCASE   2   LIVE LOAD ON LEFT SPAN

FORCES AND MOMENTS ON MEMBERS

MEMBER			AXIAL FORCES		SHEAR FORCES		MOMENTS	
NO.	END1	END2	END 1	END 2	END 1	END 2	END 1	END 2
1	1	2	0.00	0.00	2.45	3.55	0.00	-3.32
2	2	3	0.00	0.00	0.45	-0.45	3.32	0.74
3	3	4	0.00	0.00	-0.15	0.15	-0.74	-0.37

TOTAL HORIZONTAL FORCE=   0.00

TOTAL VERTICAL FORCE=   6.00

FIG. 4.23 (*continued*)

LOADCASE   3   LIVE LOAD ON CENTRE SPAN

DEFORMATIONS

JOINT NO.	X-DIRECTION	Y-DIRECTION	ROTATION(DEGS.)
1	0.0000000	0.0000000	0.0036412
2	0.0000000	0.0000000	-0.0072824
3	0.0000000	0.0000000	0.0068736
4	0.0000000	0.0000000	0.0000000

LOADCASE   3   LIVE LOAD ON CENTRE SPAN

FORCES AND MOMENTS ON MEMBERS

MEMBER			AXIAL FORCES		SHEAR FORCES		MOMENTS	
NO.	END1	END2	END 1	END 2	END 1	END 2	END 1	END 2
1	1	2	0.00	0.00	-1.69	1.69	0.00	-10.17
2	2	3	0.00	0.00	12.36	17.14	10.17	-19.19
3	3	4	0.00	0.00	3.84	-3.84	19.19	9.60

TOTAL HORIZONTAL FORCE=   0.00

TOTAL VERTICAL FORCE=   29.50

LOADCASE   4   LIVE LOAD ON RIGHT SPAN

DEFORMATIONS

JOINT NO.	X-DIRECTION	Y-DIRECTION	ROTATION(DEGS.)
1	0.0000000	0.0000000	-0.0001840
2	0.0000000	0.0000000	0.0003681
3	0.0000000	0.0000000	-0.0009570
4	0.0000000	0.0000000	0.0000000

LOADCASE   4   LIVE LOAD ON RIGHT SPAN

FORCES AND MOMENTS ON MEMBERS

MEMBER			AXIAL FORCES		SHEAR FORCES		MOMENTS	
NO.	END1	END2	END 1	END 2	END 1	END 2	END 1	END 2
1	1	2	0.00	0.00	0.09	-0.09	0.00	0.51
2	2	3	0.00	0.00	-0.46	0.46	-0.51	-3.60
3	3	4	0.00	0.00	4.19	3.31	3.60	-5.94

TOTAL HORIZONTAL FORCES=   0.00

TOTAL VERTICAL FORCES=   7.50

Problem 8: Lattice frames

It is also possible to analyse lattice frames using Program A27. While this program is particularly intended for frames with rigid joints, frames with pinned joints can also be successfully analysed by this means.

As a numerical illustration, the Warren girder shown in Fig. 4.24 is analysed using Program A27, first with pinned joints throughout and then with all joints fully rigid.

In the case of the pin-jointed frame, the input is shown in Fig. 4.25 and the output in Fig. 4.26.

For the rigid-jointed frame the input is given in Fig. 4.27 and the output in Fig. 4.28. This latter case is, in fact, the traditional 'secondary stresses' analysis and gives the moments induced in the members due to end fixity.

It can thus be seen that this modern method can replace traditional methods such as Bow's Notation, Resolution, Method of Sections, Williot-Mohr Diagrams and the use of Castigliano's Theorems, etc. This method can equally well be applied to redundantly braced lattice girders with multiple supports. Various loading cases can be considered. For example, the analysis can be carried out as illustrated above for a series of positions of a unit load, thus obtaining a comprehensive set of influence lines for the reactions, internal forces, deflections, etc. Finally, it should also be realised that structures such as Vierendeel girders, braced arches, roof trusses, lattice girders in buildings, bridges and cranes, etc., can all be analysed by this one method, and typical examples have already been indicated in Fig. 4.1.

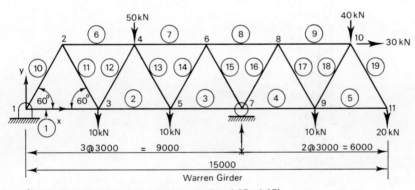

(For sectional properties of members etc. see figs. 4.25 , 4.27)

FIG. 4.24

FIG. 4.25
PROBLEM 8: ANALYSIS OF LATTICE FRAMES
PIN JOINTED FRAME

INPUT TO COMPUTER

```
'PIN JOINTED FRAME (WARREN GIRDER)'

11,19,2,0,1,0,200,0,1,
2,
0,0,1500,2600,3000,0,
4500,2600,6000,0,7500,2600,
9000,0,10500,2600,12000,0,
13500,2600,15000,0,
3,
1,3,1,900,300000,
3,5,-1,
5,7,-1,
7,9,-1,
9,11,-1,
2,4,1,1500,400000,
4,6,-1,
6,8,-1,
8,10,-1,
1,2,1,250,200000,
2,3,-1,
3,4,-1,
4,5,1,900,300000,
5,6,-1,
6,7,-1,
7,8,-1,
8,9,-1,
9,10,-1,
10,11,1,250,200000,
4,
1,1,1,1,0,
7,0,1,0,
6,

'SUPER LOAD'

6,
3,0,10,0,
4,0,50,0,
5,0,10,0,
9,0,10,0,
10,30,40,0,
11,0,20,0,
7,
0,
```

FIG. 4.26
PROBLEM 8: ANALYSIS OF LATTICE FRAMES
PIN JOINTED FRAME

OUTPUT FROM COMPUTER

PIN JOINTED FRAME (WARREN GIRDER)

YOUNGS MODULUS= 200

MAX JOINT DIFFERENCE= 2

MEMBER PROPERTIES

NC.	END1	END2	LENGTH	SIN	COS	X-AREA	M OF I
1	1	3	3000.00	0.0000	1.0000	$9.000_{10}+02$	$0.000_{10}+00$
2	3	5	3000.00	0.0000	1.0000	$9.000_{10}+02$	$0.000_{10}+00$
3	5	7	3000.00	0.0000	1.0000	$9.000_{10}+02$	$0.000_{10}+00$
4	7	9	3000.00	0.0000	1.0000	$9.000_{10}+02$	$0.000_{10}+00$
5	9	11	3000.00	0.0000	1.0000	$9.000_{10}+02$	$0.000_{10}+00$
6	2	4	3000.00	0.0000	1.0000	$1.500_{10}+03$	$0.000_{10}+00$
7	4	6	3000.00	0.0000	1.0000	$1.500_{10}+03$	$0.000_{10}+00$
8	6	8	3000.00	0.0000	1.0000	$1.500_{10}+03$	$0.000_{10}+00$
9	8	10	3000.00	0.0000	1.0000	$1.500_{10}+03$	$0.000_{10}+00$
10	1	2	3001.67	0.8662	0.4997	$2.500_{10}+02$	$0.000_{10}+00$
11	2	3	3001.67	-0.8662	0.4997	$2.500_{10}+02$	$0.000_{10}+00$
12	3	4	3001.67	0.8662	0.4997	$2.500_{10}+02$	$0.000_{10}+00$
13	4	5	3001.67	-0.8662	0.4997	$9.000_{10}+02$	$0.000_{10}+00$
14	5	6	3001.67	0.8662	0.4997	$9.000_{10}+02$	$0.000_{10}+00$
15	6	7	3001.67	-0.8662	0.4997	$9.000_{10}+02$	$0.000_{10}+00$
16	7	8	3001.67	0.8662	0.4997	$9.000_{10}+02$	$0.000_{10}+00$
17	8	9	3001.67	-0.8662	0.4997	$9.000_{10}+02$	$0.000_{10}+00$
18	9	10	3001.67	0.8662	0.4997	$9.000_{10}+02$	$0.000_{10}+00$
19	10	11	3001.67	-0.8662	0.4997	$2.500_{10}+02$	$0.000_{10}+00$

JOINT LOADINGS   SUPER LOAD

JOINT NC.	X-FORCE	Y-FORCE	MOMENT
1	$0.0000000_{10}+00$	$0.0000000_{10}+00$	$0.0000000_{10}+00$
2	$0.0000000_{10}+00$	$0.0000000_{10}+00$	$0.0000000_{10}+00$
3	$0.0000000_{10}+00$	$1.0000000_{10}+01$	$0.0000000_{10}+00$
4	$0.0000000_{10}+00$	$5.0000000_{10}+01$	$0.0000000_{10}+00$
5	$0.0000000_{10}+00$	$1.0000000_{10}+01$	$0.0000000_{10}+00$
6	$0.0000000_{10}+00$	$0.0000000_{10}+00$	$0.0000000_{10}+00$
7	$0.0000000_{10}+00$	$0.0000000_{10}+00$	$0.0000000_{10}+00$
8	$0.0000000_{10}+00$	$0.0000000_{10}+00$	$0.0000000_{10}+00$
9	$0.0000000_{10}+00$	$1.0000000_{10}+01$	$0.0000000_{10}+00$
10	$3.0000000_{10}+01$	$4.0000000_{10}+01$	$0.0000000_{10}+00$
11	$0.0000000_{10}+00$	$2.0000000_{10}+01$	$0.0000000_{10}+00$

FIG. 4.26 *(continued)*

LOADCASE   1   SUPER LOAD

DEFORMATIONS

JOINT NO.	X-DIRECTION	Y-DIRECTION	ROTATION(DEGS.)
1	0.0000000	0.0000000	0.0000000
2	2.3003067	-0.5002801	0.0000000
3	0.4006410	-0.7694211	0.0000000
4	2.4195374	-0.3072009	0.0000000
5	0.5064103	0.1523246	0.0000000
6	3.0618451	0.4635514	0.0000000
7	-0.8365385	0.0000000	0.0000000
8	4.6310759	-4.7102368	0.0000000
9	-2.2788462	$-1.025_{10}+01$	0.0000000
10	5.3926144	$-1.601_{10}+01$	0.0000000
11	-2.4711538	$-2.215_{10}+01$	0.0000000

LOADCASE   1   SUPER LOAD

FORCES AND MOMENTS ON MEMBERS

MEMBER			AXIAL FORCES		SHEAR FORCES		MOMENTS	
NO.	END1	END2	END 1	END 2	END 1	END 2	END 1	END 2
1	1	3	-24.04	-24.04	0.00	0.00	0.00	0.00
2	3	5	-6.35	-6.35	0.00	0.00	0.00	0.00
3	5	7	80.58	80.58	0.00	0.00	0.00	0.00
4	7	9	86.54	86.54	0.00	0.00	0.00	0.00
5	9	11	11.54	11.54	0.00	0.00	0.00	0.00
6	2	4	-11.92	-11.92	0.00	0.00	0.00	0.00
7	4	6	-64.23	-64.23	0.00	0.00	0.00	0.00
8	6	8	-156.92	-156.92	0.00	0.00	0.00	0.00
9	8	10	-76.15	-76.15	0.00	0.00	0.00	0.00
10	1	2	-11.93	-11.93	-0.00	0.00	0.00	0.00
11	2	3	11.93	11.93	-0.00	0.00	0.00	0.00
12	3	4	-23.47	-23.47	-0.00	0.00	0.00	0.00
13	4	5	81.20	81.20	-0.00	0.00	0.00	0.00
14	5	6	-92.74	-92.74	-0.00	0.00	0.00	0.00
15	6	7	92.74	92.74	-0.00	0.00	0.00	0.00
16	7	8	80.81	80.81	-0.00	0.00	0.00	0.00
17	8	9	-80.81	-80.81	-0.00	0.00	0.00	0.00
18	9	10	69.27	69.27	-0.00	0.00	0.00	0.00
19	10	11	-23.09	-23.09	-0.00	0.00	0.00	0.00

TOTAL HORIZONTAL FORCE=   30.00

TOTAL VERTICAL FORCE=   140.00

FIG. 4.27
PROBLEM 8: ANALYSIS OF LATTICE FRAMES
STIFF JOINTED FRAME

INPUT TO COMPUTER

```
'STIFF JOINTED FRAME (WARREN GIRDER)'

11,19,2,0,1,0,200,0,1,
2,
0,0,1500,2600,3000,0,
4500,2600,6000,0,7500,2600,
9000,0,10500,2600,12000,0,
13500,2600,15000,0,
3,
1,3,1,900,300000,
3,5,-1,
5,7,-1,
7,9,-1,
9,11,-1,
2,4,0,1500,400000,
4,6,-1,
6,8,-1,
8,10,-1,
1,2,0,250,200000,
2,3,-1,
3,4,-1,
4,5,0,900,300000,
5,6,-1,
6,7,-1,
7,8,-1,
8,9,-1,
9,10,-1,
10,11,0,250,200000,
4,
1,1,1,0,
7,0,1,0,
6,

'SUPER LOAD'

6,
3,0,10,0,
4,0,50,0,
5,0,10,0,
9,0,10,0,
10,30,40,0,
11,0,20,0,
7,
0,
```

FIG. 4.28
PROBLEM 8: ANALYSIS OF LATTICE FRAMES
STIFF JOINTED FRAME

OUTPUT FROM COMPUTER

STIFF JOINTED FRAME (WARREN GIRDER)

YOUNGS MODULUS=   200

MAX JOINT DIFFERENCE=   2

MEMBER PROPERTIES

NO.	END1	END2	LENGTH	SIN	COS	X-AREA	M OF I
1	1	3	3000.00	0.0000	1.0000	9.000₁₀+02	3.000₁₀+05
2	3	5	3000.00	0.0000	1.0000	9.000₁₀+02	3.000₁₀+05
3	5	7	3000.00	0.0000	1.0000	9.000₁₀+02	3.000₁₀+05
4	7	9	3000.00	0.0000	1.0000	9.000₁₀+02	3.000₁₀+05
5	9	11	3000.00	0.0000	1.0000	9.000₁₀+02	3.000₁₀+05
6	2	4	3000.00	0.0000	1.0000	1.500₁₀+03	4.000₁₀+05
7	4	6	3000.00	0.0000	1.0000	1.500₁₀+03	4.000₁₀+05
8	6	8	3000.00	0.0000	1.0000	1.500₁₀+03	4.000₁₀+05
9	8	10	3000.00	0.0000	1.0000	1.500₁₀+03	4.000₁₀+05
10	1	2	3001.67	0.8662	0.4997	2.500₁₀+02	2.000₁₀+05
11	2	3	3001.67	-0.8662	0.4997	2.500₁₀+02	2.000₁₀+05
12	3	4	3001.67	0.8662	0.4997	2.500₁₀+02	2.000₁₀+05
13	4	5	3001.67	-0.8662	0.4997	9.000₁₀+02	3.000₁₀+05
14	5	6	3001.67	0.8662	0.4997	9.000₁₀+02	3.000₁₀+05
15	6	7	3001.67	-0.8662	0.4997	9.000₁₀+02	3.000₁₀+05
16	7	8	3001.67	0.8662	0.4997	9.000₁₀+02	3.000₁₀+05
17	8	9	3001.67	-0.8662	0.4997	9.000₁₀+02	3.000₁₀+05
18	9	10	3001.67	0.8662	0.4997	9.000₁₀+02	3.000₁₀+05
19	10	11	3001.67	-0.8662	0.4997	2.500₁₀+02	2.000₁₀+05

JOINT LOADINGS   SUPER LOAD

JOINT NO.	X-FORCE	Y-FORCE	MOMENT
1	0.0000000₁₀+00	0.0000000₁₀+00	0.0000000₁₀+00
2	0.0000000₁₀+00	0.0000000₁₀+00	0.0000000₁₀+00
3	0.0000000₁₀+00	1.0000000₁₀+01	0.0000000₁₀+00
4	0.0000000₁₀+00	5.0000000₁₀+01	0.0000000₁₀+00
5	0.0000000₁₀+00	1.0000000₁₀+01	0.0000000₁₀+00
6	0.0000000₁₀+00	0.0000000₁₀+00	0.0000000₁₀+00
7	0.0000000₁₀+00	0.0000000₁₀+00	0.0000000₁₀+00
8	0.0000000₁₀+00	0.0000000₁₀+00	0.0000000₁₀+00
9	0.0000000₁₀+00	1.0000000₁₀+01	0.0000000₁₀+00
10	3.0000000₁₀+01	4.0000000₁₀+01	0.0000000₁₀+00
11	0.0000000₁₀+00	2.0000000₁₀+01	0.0000000₁₀+00

FIG. 4.28 (*continued*)

LOADCASE   1   SUPER LOAD

DEFORMATIONS

JOINT NO.	X-DIRECTION	Y-DIRECTION	ROTATION(DEGS.)
1	0.0000000	0.0000000	-0.0313412
2	2.2968201	-0.4983720	-0.0213067
3	0.4004332	-0.7676307	-0.0107482
4	2.4161748	-0.3072671	0.0022721
5	0.5061062	0.1524016	-0.0012538
6	3.0584541	0.4629223	-0.0366070
7	-0.8361281	0.0000000	-0.1001171
8	4.6268556	-4.7062576	-0.1651093
9	-2.2776718	-1.024.+01	-0.2037735
10	5.3883468	-1.600.+01	-0.2027879
11	-2.4701198	-2.213.+01	-0.2157091

LOADCASE   1   SUPER LOAD

FORCES AND MOMENTS ON MEMBERS

MEMBER			AXIAL FORCES		SHEAR FORCES		MOMENTS	
NO.	END1	END2	END 1	END 2	END 1	END 2	END 1	END 2
1	1	3	-24.03	-24.03	-0.01	0.01	-20.56	-6.18
2	3	5	-6.34	-6.34	-0.03	0.03	-52.68	-46.06
3	5	7	80.53	80.53	-0.07	0.07	-65.55	-134.57
4	7	9	86.49	86.49	0.06	-0.06	127.67	55.31
5	9	11	11.55	11.55	0.02	-0.02	40.39	32.06
6	2	4	-11.94	-11.94	-0.02	0.02	-47.74	-25.80
7	4	6	-64.23	-64.23	-0.06	0.06	-70.92	-107.11
8	6	8	-156.84	-156.84	-0.00	0.00	53.85	-65.77
9	8	10	-76.15	-76.15	0.06	-0.06	106.07	71.00
10	1	2	-11.93	-11.93	0.02	-0.02	20.56	25.23
11	2	3	11.90	11.90	0.02	-0.02	22.52	27.43
12	3	4	-23.42	-23.42	0.02	-0.02	31.44	37.49
13	4	5	81.11	81.11	0.04	-0.04	59.22	56.76
14	5	6	-92.61	-92.61	0.03	-0.03	54.84	30.17
15	6	7	92.66	92.66	0.00	-0.00	23.09	-21.22
16	7	8	80.75	80.75	0.00	-0.00	28.12	-17.23
17	8	9	-80.69	-80.69	-0.02	0.02	-23.08	-50.06
18	9	10	69.19	69.19	-0.03	0.03	-45.64	-44.95
19	10	11	-23.07	-23.07	-0.02	0.02	-26.05	-32.06

TOTAL HORIZONTAL FORCE=   30.00

TOTAL VERTICAL FORCE=   140.00

Problem 9:   Approximate analysis of a fixed arch rib

It has previously been briefly shown how it is possible to analyse a structure consisting of curved or tapering members using Program A27 (in spite of the restrictions specified earlier) by simulating the actual structure by an idealised form consisting of a set of members of uniform section. The success of this approximation naturally depends on how close the simulation is to the actual form, and a valuable subsidiary issue would be to compare the behaviour of the actual and the simulated structures under a load-deformation-strain test, involving either full-scale or model specimens. Research along these lines could well become important, if not essential, should this type of approach in structural analysis become popular.

This method of analysis is rather akin to the treatment of an integral as a limit of a sum—as more elements are introduced, the approximation becomes better until in the limit when the structure consists of an infinite number of uniform members, the simulation is ideal. This is, in fact, the basic concept of the Finite Element Method illustrated later in the chapter.

Consider now the fixed arch rib shown in Fig. 4.29. The actual arch has been simulated by a set of 21 straight members of uniform section, numbered in the usual way. This simulated arch rib can now be analysed using Program A27 and the computer input and output are given in Figs. 4.30 and 4.31 respectively.

This method can clearly be extended to structures such as motorway trestles (Fig. 4.1), shell roofs, and plate and slab structures. The approximate analysis obtained in all cases should be compared with the results from an approved, established, method, so that a critical appraisal of the new process can be made. There is obviously much scope here for development work in the form of both design studies and loading tests, involving investigations in which the actual structure is simulated by different numbers of elements, so establishing, among other things, the minimum number of equivalent uniform members required to give a satisfactory solution.

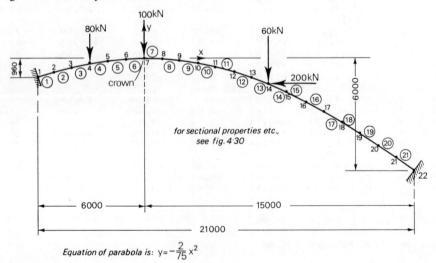

Equation of parabola is: $y = -\frac{2}{75}x^2$

FIG. 4.29

FIG. 4.30
PROBLEM 9: APPROXIMATE ANALYSIS OF A FIXED
ARCH RIB

INPUT TO COMPUTER

```
'FIXED ARCH (BY SIMULATION)'
22,21,2,0,1,0,13.333,0,1,
2,
-6000,-960,-5000,-667,-4000,-426,
-3000,-240,-2000,-107,-1000,-27,
0,0,1000,-27,2000,-107,
3000,-240,4000,-426,5000,-667,
6000,-960,7000,-1308,8000,-1708,
9000,-2160,10000,-2667,11000,-3230,
12000,-3840,13000,-4510,14000,-5230,
15000,-6000,
3,
1,2,0,2.5,6,1.4,12,
2,3,-1,
3,4,-1,
4,5,-1,
5,6,-1,
6,7,-1,
7,8,-1,
8,9,-1,
9,10,-1,
10,11,-1,
11,12,-1,
12,13,-1,
13,14,-1,
14,15,-1,
15,16,-1,
16,17,-1,
17,18,-1,
18,19,-1,
19,20,-1,
20,21,-1,
21,22,-1,
4,
1,1,1,1,
22,1,1,1,
6,

'HORIZONTAL AND VERTICAL LOADING'

3,
3,0,80,0,
7,0,100,0,
14,-200,60,0,
7,
0,
```

FIG. 4.31
PROBLEM 9: APPROXIMATE ANALYSIS OF A FIXED
ARCH RIB

OUTPUT FROM COMPUTER

FIXED ARCH (BY SIMULATION)

YOUNGS MODULUS=   13

MAX JOINT DIFFERENCE=   1

—MEMBER PROPERTIES

N O.	END1	END2	LENGTH	SIN	COS	X-AREA	M OF I
1	1	2	1042.04	0.2812	0.9597	2.500 ₒ+06	1.400 ₒ+12
2	2	3	1028.63	0.2343	0.9722	2.500 ₒ+06	1.400 ₒ+12
3	3	4	1017.15	0.1829	0.9831	2.500 ₒ+06	1.400 ₒ+12
4	4	5	1008.81	0.1318	0.9913	2.500 ₒ+06	1.400 ₒ+12
5	5	6	1003.19	0.0797	0.9968	2.500 ₒ+06	1.400 ₒ+12
6	6	7	1000.36	0.0270	0.9996	2.500 ₒ+06	1.400 ₒ+12
7	7	8	1000.36	-0.0270	0.9996	2.500 ₒ+06	1.400 ₒ+12
8	8	9	1003.19	-0.0797	0.9968	2.500 ₒ+06	1.400 ₒ+12
9	9	10	1008.81	-0.1318	0.9913	2.500 ₒ+06	1.400 ₒ+12
10	10	11	1017.15	-0.1829	0.9831	2.500 ₒ+06	1.400 ₒ+12
11	11	12	1028.63	-0.2343	0.9722	2.500 ₒ+06	1.400 ₒ+12
12	12	13	1042.04	-0.2812	0.9597	2.500 ₒ+06	1.400 ₒ+12
13	13	14	1058.82	-0.3287	0.9444	2.500 ₒ+06	1.400 ₒ+12
14	14	15	1077.03	-0.3714	0.9285	2.500 ₒ+06	1.400 ₒ+12
15	15	16	1097.41	-0.4119	0.9112	2.500 ₒ+06	1.400 ₒ+12
16	16	17	1121.18	-0.4522	0.8919	2.500 ₒ+06	1.400 ₒ+12
17	17	18	1147.18	-0.4906	0.8714	2.500 ₒ+06	1.400 ₒ+12
18	18	19	1171.37	-0.5208	0.8537	2.500 ₒ+06	1.400 ₒ+12
19	19	20	1203.70	-0.5566	0.8308	2.500 ₒ+06	1.400 ₒ+12
20	20	21	1232.23	-0.5843	0.8115	2.500 ₒ+06	1.400 ₒ+12
21	21	22	1262.10	-0.6101	0.7923	2.500 ₒ+06	1.400 ₒ+12

JOINT LOADINGS   HORIZONTAL AND VERTICAL LOADING

JOINT NO.	X-FORCE	Y-FORCE	MOMENT
1	0.0000000 ₒ+00	0.0000000 ₒ+00	0.0000000 ₒ+00
2	0.0000000 ₒ+00	0.0000000 ₒ+00	0.0000000 ₒ+00
3	0.0000000 ₒ+00	8.0000000 ₒ+01	0.0000000 ₒ+00
4	0.0000000 ₒ+00	0.0000000 ₒ+00	0.0000000 ₒ+00
5	0.0000000 ₒ+00	0.0000000 ₒ+00	0.0000000 ₒ+00

FIG. 4.31 *(continued)*

6	0.0000000ₑ+00	0.0000000ₑ+00	0.0000000ₑ+00
7	0.0000000ₑ+00	1.0000000ₑ+02	0.0000000ₑ+00
8	0.0000000ₑ+00	0.0000000ₑ+00	0.0000000ₑ+00
9	0.0000000ₑ+00	0.0000000ₑ+00	0.0000000ₑ+00
10	0.0000000ₑ+00	0.0000000ₑ+00	0.0000000ₑ+00
11	0.0000000ₑ+00	0.0000000ₑ+00	0.0000000ₑ+00
12	0.0000000ₑ+00	0.0000000ₑ+00	0.0000000ₑ+00
13	0.0000000ₑ+00	0.0000000ₑ+00	0.0000000ₑ+00
14	-2.0000000ₑ+02	6.0000000ₑ+01	0.0000000ₑ+00
15	0.0000000ₑ+00	0.0000000ₑ+00	0.0000000ₑ+00
16	0.0000000ₑ+00	0.0000000ₑ+00	0.0000000ₑ+00
17	0.0000000ₑ+00	0.0000000ₑ+00	0.0000000ₑ+00
18	0.0000000ₑ+00	0.0000000ₑ+00	0.0000000ₑ+00
19	0.0000000ₑ+00	0.0000000ₑ+00	0.0000000ₑ+00
20	0.0000000ₑ+00	0.0000000ₑ+00	0.0000000ₑ+00
21	0.0000000ₑ+00	0.0000000ₑ+00	0.0000000ₑ+00
22	0.0000000ₑ+00	0.0000000ₑ+00	0.0000000ₑ+00

LOADCASE   1   HORIZONTAL AND VERTICAL LOADING

DEFORMATIONS

JOINT NO.	X-DIRECTION	Y-DIRECTION	ROTATION(DEGS.)
1	0.0000000	0.0000000	0.0000000
2	-0.0065794	-0.0093012	-0.0007363
3	-0.0108668	-0.0286068	-0.0011944
4	-0.0143878	-0.0533866	-0.0014644
5	-0.0185792	-0.0816229	-0.0016321
6	-0.0240005	-0.1111910	-0.0016583
7	-0.0308915	-0.1392781	-0.0015029
8	-0.0391513	-0.1632497	-0.0012781
9	-0.0484347	-0.1833184	-0.0010952
10	-0.0583987	-0.1998545	-0.0009127
11	-0.0686237	-0.2124990	-0.0006882
12	-0.0784703	-0.2201259	-0.0003771
13	-0.0868408	-0.2208530	0.0000665
14	-0.0920743	-0.2119360	0.0006917
15	-0.0873779	-0.1930973	0.0013184
16	-0.0777053	-0.1649666	0.0017386
17	-0.0638126	-0.1311152	0.0019546
18	-0.0470459	-0.0950901	0.0019676
19	-0.0296507	-0.0604542	0.0017782
20	-0.0137408	-0.0307008	0.0013860
21	-0.0026742	-0.0093835	0.0007931
22	0.0000000	0.0000000	0.0000000

FIG. 4.31 *(continued)*

LOADCASE   1   HORIZONTAL AND VERTICAL LOADING

FORCES AND MOMENTS ON MEMBERS

MEMBER			AXIAL FORCES		SHEAR FORCES		MOMENTS	
NO.	END1	END2	END 1	END 2	END 1	END 2	END 1	END 2
1	1	2	285.63	285.63	75.35	-75.35	269451.78	-190931.35
2	2	3	281.64	281.64	89.12	-89.12	190931.35	-99259.27
3	3	4	261.93	261.93	25.15	-25.15	99259.27	-73676.79
4	4	5	260.28	260.28	38.65	-38.65	73676.79	-34689.73
5	5	6	257.90	257.90	52.22	-52.22	34689.73	17701.89
6	6	7	254.78	254.78	65.77	-65.77	-17701.89	83498.09
7	7	8	253.56	253.56	-20.54	20.54	-83498.09	62951.78
8	8	9	254.29	254.29	-7.12	7.12	-62951.78	55810.04
9	9	10	254.32	254.32	6.21	-6.21	-55810.04	62072.87
10	10	11	253.66	253.66	19.34	-19.34	-62072.87	81740.27
11	11	12	252.29	252.29	32.64	-32.64	-81740.27	115318.08
12	12	13	250.41	250.41	44.84	-44.84	-115318.08	162047.54
13	13	14	247.86	247.86	57.27	-57.27	-162047.54	222687.40
14	14	15	81.58	81.58	-61.47	61.47	-222687.40	156478.92
15	15	16	84.21	84.21	-57.82	57.82	-156478.92	93022.10
16	16	17	86.71	86.71	-54.00	54.00	-93022.10	32475.68
17	17	18	88.98	88.98	-50.18	50.18	-32475.68	-25107.42
18	18	19	90.68	90.68	-47.04	47.04	25107.42	-80203.44
19	19	20	92.60	92.60	-43.13	43.13	80203.44	-132124.48
20	20	21	94.00	94.00	-39.99	39.99	132124.48	-181399.70
21	21	22	95.23	95.23	-36.95	36.95	181399.70	-228029.09

TOTAL HORIZONTAL FORCE=   -200.00

TOTAL VERTICAL FORCE=   240.00

Miscellaneous Comments on the Analysis of Plane Frames and Beams

1.   When a program which does not allow for member loading, or which does not cater for a particular case of member loading, is being used, an analysis is still possible by partitioning the loading and applying the Principle of Super-position. Consider the simple rectangular portal frame shown in Fig. 4.32. The actual problem can be replaced by a sum of the two separate loading conditions shown in the diagram. These are:

(A)   All joints rigidly clamped
Here only the beam is subjected to any bending and shearing, and the bending moment diagram for the beam can be obtained from the values of the fixed-end moments, here equal to $\pm WL/12$. In other more-complicated cases reference can be made to the suitable tables (see the list of references in Chapter 3).

The computer is not required for this analysis. There are no moments on the columns (as the ends are fully fixed) and there is no transverse loading.

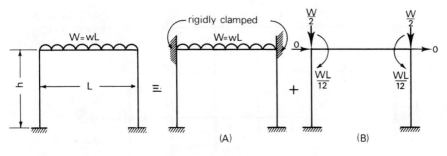

FIG. 4.32

(B)   Forces and moments applied only at the joints of the frame
These applied forces and moments are equal and opposite to the fixed-end
shearing forces and bending moments acting on the joints in condition (A)
above, and their values are therefore already known. The structural analysis can
now be carried out by computer using any of the relevant library programs
referred to in the chapter, since here all external forces and moments are applied
solely at the joints.

The total bending moments and shearing forces in the members can now be
obtained by adding together the corresponding results for conditions (A) and
(B). Since there are no deflections nor rotations of the joints in condition (A),
the computer output in condition (B) gives the total deflections and rotations in
the actual problem.

2.   None of the illustrations given above is particularly large. However, it is
possible to analyse larger frames and beams, provided always that the size of the
structure is within the limits laid down in the program specification for the given
computer.

For example, for an Elliott 4100 computer with 24K storage size, if the
number of joints j = 80, and the maximum joint number difference d = 5, then
the maximum number of members m = 115 (see equation (A) above). This data
actually refers to Program EAPF which is similar to A27. Obviously larger
computers can deal with larger frames. Moreover, if additional backing storage is
provided—such as by magnetic tape, etc.—it is possible to modify existing library
programs so that larger structures can be analysed. Some notes on the treatment
of large frames by this and other methods are given in Chapter 9.

3.   It is not generally possible to carry out the analysis of grids, grillages and
plane frames in which the loading is applied perpendicular to the plane of the
frame using the range of library programs suitable for problems 1 to 9 inclusive.
Nevertheless, the Stiffness Method especially can still be used to deal with grids,
and other library programs have been written to solve such problems e.g.
Programs LC8[2] and A28[3].

A typical grid is shown in Fig. 4.33. Loads are applied at the joints
perpendicular to the plane of the frame, i.e. in the z-direction. Moments can also
be applied to the joints about axes parallel to the x- and y-directions.

The input to the computer for use with either of the above programs should
be drawn up according to the instructions given. This is essentially similar to that
for Program A27, but also includes the torsional properties of the members.

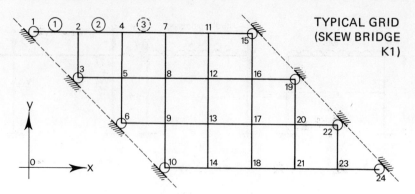

FIG. 4.33

In the case of Program LC8 the output is as shown in Fig. 4.34. This provides deflections and rotations of the joints as well as forces, moments and torques on the members.

Finally it should be noted that space frames can be effectively analysed by the Stiffness Method, though library programs for such structures are neither readily nor widely available at the present time.

Elastic Analysis of Other Types of Structures

All the previous applications have dealt primarily with structures comprising frames and beams, and all the library programs referred to were specially written to deal with such structural systems, though a brief indication has also been given as to how other types of structures (e.g. arch ribs, motorway trestles, etc.) might also be approximately analysed using the same methods and programs.

However, a completely fresh approach can also be adopted for other types of structures such as non-uniform beams and frames; monolithic structures, such as beam-column-slab buildings, retaining walls, dams, etc.; shell roofs and support-ing buildings; bent plate construction; cellular construction (e.g. lightweight reinforced concrete flooring, box girders, aircraft wings and fuselage, etc.); and so on. In such cases the entire system is regarded as a single structural continuum rather than as a collection of readily identifiable, basic components of beams, struts and ties.

The method of analysis consists of replacing or simulating the continuum by a large number of appropriately interconnected elements of fairly standard geometrical form in either two or three dimensions (though some elements such as uniform beams, struts or ties might even be one-dimensional) and then, by using various properties which can be established or assumed for the elements and taking account of the conditions for static equilibrium and compatibility of deformations between adjacent elements, of calculating the various structural quantities (e.g. forces, reactions, bending moments, deflections, etc.) at specified locations throughout the structure.

This approach obviously gives a realistic portrayal of the structural system and the resulting analysis can be expected to be very accurate. The calculations

FORM OF OUTPUT FROM COMPUTER USING PROGRAM NO. LC8

STRUCTURAL ANALYSIS OF A GRID STRUCTURE. LC8.
SKEW BRIDGE K1.

MEMBER PROPERTIES

NO. OF MEMBER	NO. OF JOINT AT END 1	NO. OF JOINT AT END 2	L	SIN	COS	$\frac{GJ}{E}$	I
1	1	2	etc	etc			
2	2	4					
.	.	.					
.	.	.					
.	.	.					

LOAD CASE 1

DEFORMATIONS

JOINT NO.	X ROTATION	Y ROTATION	VERTICAL DEFLECTION
1	etc	etc	
2			
.			
.			
.			

FORCES, MOMENTS & TORQUES IN MEMBERS

NO. OF MEMBER	JOINT NO. AT END 1	JOINT NO. AT END 2	AXIAL TORSION	BENDING MOMENTS END 1 END 2	VERTICAL SHEARING FORCE
1	1	2	etc	etc	
2	2	4			
.	.	.			
.	.	.			
.	.	.			

FIG. 4.34

can be quite complex and extensive, however, and consequently the methods evolved have generally been devised with the computer very much in mind. Further details of this approach, known as the Finite Element Method, are given in Chapter 7. Analysis by this method can be based on elastic, plastic, nonlinear or dynamic theory, and a range of library programs have been, or will be prepared to deal with a wide variety of problems.

Elastic Analysis:
To illustrate the use of the Finite Element Method, an elastic analysis of a plate of uniform thickness with a hole, and subjected to a one-dimensional tension field in the plane of the plate, is now effected (Fig. 4.35). There is a concentration of stress around the hole and the main purpose of the elastic analysis is to obtain the stress distribution throughout the specimen due to the

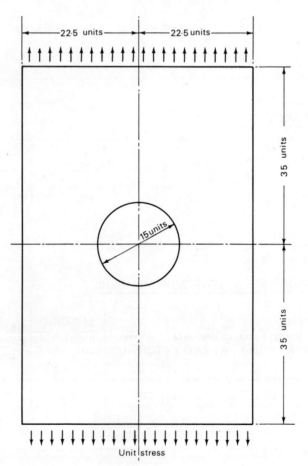

FIG. 4.35

given loading (though the distributions of the other structural quantities are obtained as well). The plate is symmetrical about both axes in its plane and consequently it is only necessary to consider a quadrant, as shown in Fig. 4.36. By this arrangement it is possible to either reduce the storage and machine-time requirements of whatever computer might be used, or, alternatively, to provide greater accuracy by increasing the number of elements which can be specified with a given computer.

The next stage in the process is to divide the quadrant of the plate into a series of elements as shown. It can be seen that in this particular case all the elements have been chosen to be triangular and that those in the proximity of the hole are small and closely banded together—thereby enabling many values of stress and other structural quantities to be obtained in this region of stress concentration and steep stress gradients. (One set of values is obtained from each element.) Away from this area, where steady conditions are more likely to prevail, the elements have been made much larger in size. (The layout of elements can be altered as a result of a pilot analysis, and a trial and adjustment approach adopted to obtain the best final arrangement.)

It is now necessary to number the elements and their joints, as shown in Fig. 4.36 and it can be seen that there are 390 elements and 224 joints in the quadrant. It is assumed that the elements are connected (structurally) together only at the joints, which are therefore termed nodes or nodal points. Thus the quadrant is replaced by 390 triangular elements connected together at 224 nodal points to form a simulation of the actual specimen, and the structural analysis is now carried out on this simulation.

A large library program was available for the computer at the National Engineering Laboratory, East Kilbride, Scotland, and the elastic analysis of the simulated specimen was effected on that machine and with that program.

Details of the Computer:
The UNIVAC 1108 at the National Engineering Laboratory has a working access of 132K and a backing store consisting of: 6 tape units, 2 fast random drum sets (of 22 millions words apiece), and 3 sets of 2 flying head drums (of ½ million words each).

Details of the Program:
The program used is part of a large, finite element package (which includes a cathode ray tube display, etc.), and is usually referred to as the 'Plane/Stress/ Strain Finite Element Solution Package'[4].

The program was originated by Professor Zienkiewicz of the University of Wales (Swansea) (see Chapter 7) but has been modified by the Laboratory. It is written in FORTRAN and in its present form is limited to 600 nodal points and 1,000 elements (though this is potentially capable of being extended to approximately 3,000 elements). (The "use" of this package is available to all at the Laboratory and suitable problems can be analysed there either as a bureau service or as a consultant service, though the program itself cannot be divulged.)

Elastic Analysis by Computer:
In this case no attempt is made to give a complete verbatim list of computer input and output. Instead the type of information required in compiling the

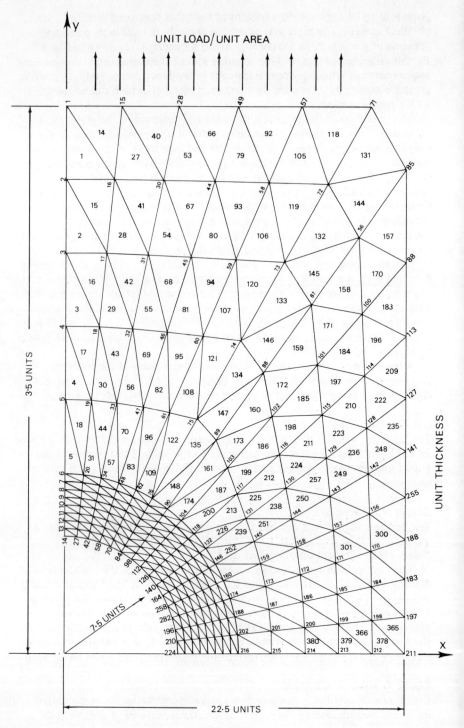

FIG. 4.36

input is indicated and some details of the results obtained as output are given.
Some of the input data are shown in Fig. 4.37. These include:
Fig. 4.37a the co-ordinates of the nodal points;
Fig. 4.37b the details of connections between elements;
Fig. 4.37c the boundary conditions; and
Fig. 4.37d the loading.

The x and y axes for the node co-ordinates are indicated in Fig. 4.36. In Fig.
4.37B, a list is given defining the 390 elements, indicating which nodes are
interconnected and specifying the 'type' of material. The figure quoted for
'material type' acts as a steering symbol and the properties of the material are
declared in the following way:

Material Type	E	v	t
1	1	0.3	1
2	1	0.3	1.5
3	1	0.3	2
4	1.5	0.28	1

so that, when the 'material type' is called, the program uses the relevant values
of E, v and t.
    In Fig. 4.37C the boundary conditions are tabulated for all the nodes. In the
case of the steering symbols, 0 indicates 'free' and 1 indicates 'restrained', with
reference to either the x or y direction. The 'specified displacements' are defined
by two numbers—one with respect to x and the other to y. When the steering
symbol is unity (either for x or y), this must be accompanied by a finite
numerical value in the appropriate column. In all other cases this value should be
given as zero.
    The loading is given in Fig. 4.37D. The values in the table are the externally
applied loads at the nodes and have been obtained here by the appropriate
substitution of a concentrated load for a uniformly distributed load, as for
example:

$$\begin{aligned} \text{force on node 29} &= \text{stress} \times \text{area} \\ &= 1 \times 3.845 \times 1 \\ &= 3.845 \text{ units.} \end{aligned}$$

Some of the results obtained in the output are shown in Fig. 4.38. In Fig. 4.38E,
the displacements in the x and y directions of all 224 nodes are given. The
stresses at the centroids of the elements and at the nodes are given in Fig. 4.38F,
where θ is the angle between the maximum principle stress σ_{max} and the y-axis.
    Finally a series of graphical outputs of results obtained direct from the
computer is shown in Figs. 4.38G to L inclusive in respect of the distribution of
displacements and stresses. The stresses shown in Figs. 4.38H to L inclusive are
the average stresses at the nodes. It has been assumed that the stresses in any

FIG. 4.37

SAMPLE DETAILS OF INPUT DATA

(A)
CO-ORDINATES

Nodal Points	Co-ordinates	
	x	y
1	0.000	35.000
2	0.000	30.300
3	0.000	25.600
14	0.000	7.500
15	3.680	35.000
16	3.186	30.293
28	0.790	7.460
29	7.440	35.000
30	6.429	30.244
42	1.560	7.340
43	11.370	35.000
44	9.805	30.182
223	8.000	0.000
224	7.500	0.000

(B)
CONNECTIONS BETWEEN ELEMENTS

Element Number	Nodal Connection Numbers			Material Type
1	1	2	16	1
2	2	3	17	1
3	3	4	18	1
14	1	16	15	1
15	2	17	16	1
300	156	170	169	1
301	157	171	170	1
365	197	198	211	1
378	198	212	211	1
390	210	224	223	1

FIG. 4.37 *(continued)*

(C)
BOUNDARY CONDITIONS

Node	Steering Symbol		Specified Displacements	
	x	y	x	y
1	1	0	0.000	0.000
2	1	0	0.000	0.000
3	1	0	0.000	0.000
4	1	0	0.000	0.000
12	1	0	0.000	0.000
13	1	0	0.000	0.000
14	1	0	0.000	0.000
211	0	1	0.000	0.000
212	0	1	0.000	0.000
213	0	1	0.000	0.000
223	0	1	0.000	0.000
224	0	1	0.000	0.000

(D)
LOADING
NON ZERO LOADS

Node	Load in X direction	Load in Y direction
1	0	1.85
15	0	3.70
29	0	3.85
43	0	4.10
57	0	4.40
71	0	4.60

element are constant and equal to the values at the centroid, and the stresses at any node are obtained by simply taking the average of the appropriate values in all elements meeting at that particular node.

Compatible unspecified units have been used throughout the example.

Conclusions

The main purpose of this chapter is to show both the scope and usefulness of modern methods of structural analysis by various numerical examples. The engineer is invited to try the various problems himself on his own computer to become familiar with the techniques propounded. This is the main reason for giving details of the library **Program A27**.

There are, however, several implications which arise if the engineer decides to use the techniques described in this book on 'production' work, and the author considers these to be sufficiently important to devote Chapter 11 to this issue. The engineer should therefore study that chapter very carefully before attempting to make any serious use of these techniques professionally.

A matter of particular importance is the question of responsibility—who is responsible for the validity of the results? The author considers that this must rest fairly and squarely with the engineer, and it is therefore vital that the engineer should be completely satisfied with the program method used and have carried out his own various checks and trial runs, etc., before attempting to make serious use of any program or technique thereon. For example, there could well be an error in a program (perhaps even in A27!), or it might be that the originator of the program did not intend it to be used in a certain way, so it is clearly advisable for the engineer to realise his position of responsibility and act accordingly.

Finally, while it appears that **Program A27** can totally replace programs such as those given in Chapter 3, it must be remembered that these other programs always provide effective, independent checks on any problem by another method of analysis. Moreover, since these have all been written for specific tasks, it could well be that, on certain occasions, the solutions obtained by these alternative means might require less computer time (i.e. 'the sledgehammer and the walnut'!). It should also be realised that short, simple, programs are less likely to contain errors and, even if they do, can be much more easily checked and rectified.

References

1. Program A27 Elastic Analysis of a Plane Frame. Glen Computing Centre, Ltd., Stag Place, London, S.W.1 (associated with G. Maunsell & Partners, Consulting Engineers, London). (For Elliott 4100 Computer).
2. Program LC8 Structural Analysis of a Grid. Sept. 1962. Elliott 803 Computer Application Program. (All enquiries now to International Computers, Ltd., London).

3. Program A28 Elastic Analysis of a Grid. Glen Computing Centre, Ltd. (For Elliott 4100 Computer). (See 1 above).
4. "Plane/Stress/Strain Finite Element Solution Package". National Engineering Laboratory, East Kilbride (For UNIVAC 1108 Computer)

Bibliography

In addition to the library programs given above, a wide selection of further programs based on modern methods exist, which have been written in various languages for different computers. The selection includes:

Program Number LC7 Structural Analysis of a Plane Framework. Sept. 1962. Elliott 803 Computer Application Program. (All enquiries now to International Computers, Ltd., London.)

Program Number LC29 Structural Analysis of a Plane Framework. Elliott 4100 Applied Program. (See note above.)

Program Number DHCVO1 FINSTRUCTPROG. Elastic Analysis of a Plane Frame. University of Newcastle upon Tyne, 1965. (For English Electric KDF9 Computer.)

Program Number EAPF Elastic Analysis of Plane Framework. Elliott 4100 Applied Program, June 1968 (See note above.)

Program Numbers DCx/AU/JSR where x = 1, 2, −5. Range of programs dealing with the elastic analysis of plane frames. University of Durham. (See Chapter 9.) (For Elliott 803 Computer.)

FIG. 4.38

SAMPLE DETAILS OF OUTPUT DATA

(E)

NODAL DISPLACEMENTS

Node	x-direction	y-direction
1	0.000000	0.404131 +02
2	0.000000	0.365456 +02
3	0.000000	0.328296 +02
15	−0.400157 +00	0.400491 +02
16	−0.665906 +00	0.363472 +02
197	−0.160491 +02	0.181339 +01
198	−0.153203 +02	0.195664 +01
211	−0.164144 +02	0.000000
212	−0.155609 +02	0.000000
213	−0.147316 +02	0.000000
222	−0.109994 +02	0.000000
223	−0.105095 +02	0.000000
224	−0.984393 +01	0.000000

FIG. 4.38 (continued)

(F)

STRESSES AT ELEMENT CENTROIDS

Element	σ_x	σ_y	τ_{xy}	σ_{max}	σ_{min}	θ
1	.193379 +00	.987140 +00	-.232543 -01	.987821 +00	.192698 +00	-1.677
2	.132397 +00	.925577 +00	-.243396 -01	.926324 +00	.131651 +00	-1.756
14	.313340 +00	.100994 +01	-.119441 -01	.101015 +01	.313135 +00	-0.982
15	.175740 +00	.945881 +00	-.163563 -01	.946229 +00	.175393 +00	-1.216
250	-.526459 -01	.133019 +01	.790975 -00	.133470 +01	-.571556 -01	3.263
390	.445077 -01	.351713 +01	-.642948 -01	.351832 +01	.433178 -01	-1.060

STRESSES AT NODES

Node	σ_x	σ_y	σ_{xy}	σ_{max}	σ_{min}	θ
1	.253359 +00	.998541 +00	-.175992 -01	.998957 +00	.252944 +00	-1.352
2	.167172 +00	.952866 +00	-.213168 -01	.953444 +00	.166594 +00	-1.553
14	-.138281 +01	.693885 -02	-.101490 -01	.701296 -02	-.138289 +01	-.418
15	.250877 +00	.991895 +00	-.261614 -01	.992817 +00	.249954 +00	-2.019
100	-.116115 -01	.112031 +01	-.650359 -01	.112404 +01	-.153359 -01	-3.278
223	.946308 -01	.315883 +01	-.126462 +00	.316404 +01	.894205 -01	-2.359
224	.445077 -01	.351713 +01	-.642948 -01	.351832 +01	.433178 -01	-1.060

FIG. 4.38 (*continued*)
(G)
DISPLACEMENT PATTERN

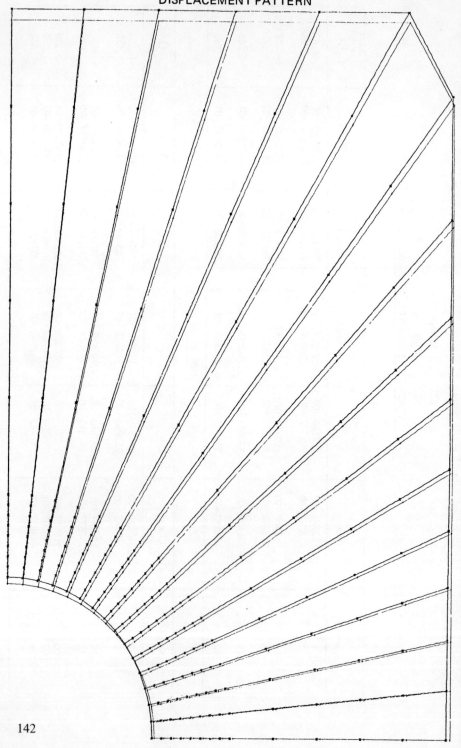

142

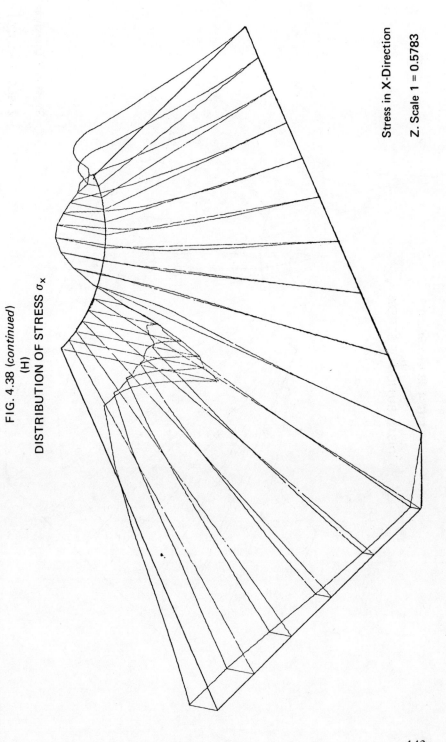

FIG. 4.38 (*continued*)
(H)
DISTRIBUTION OF STRESS σ_x

Stress in X-Direction

Z. Scale 1 = 0.5783

FIG. 4.38 (continued)
(I)
DISTRIBUTION OF STRESS σ_y

Stress in Y-Direction

Z. Scale 1 = 1.1201

144

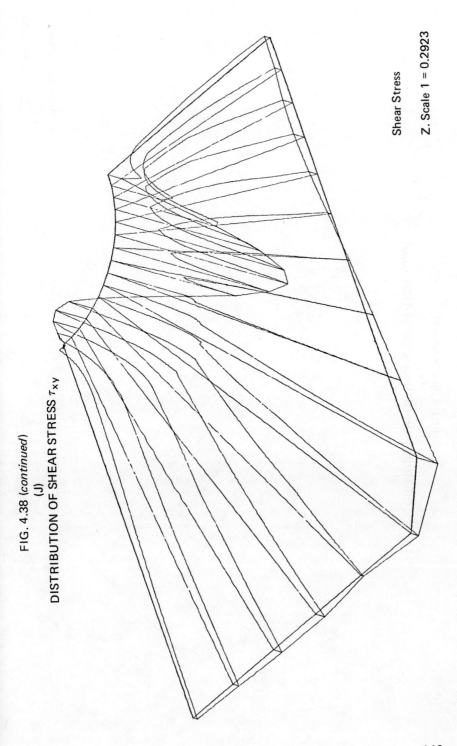

FIG. 4.38 *(continued)*
(J)
DISTRIBUTION OF SHEAR STRESS τ_{xy}

Shear Stress

Z. Scale 1 = 0.2923

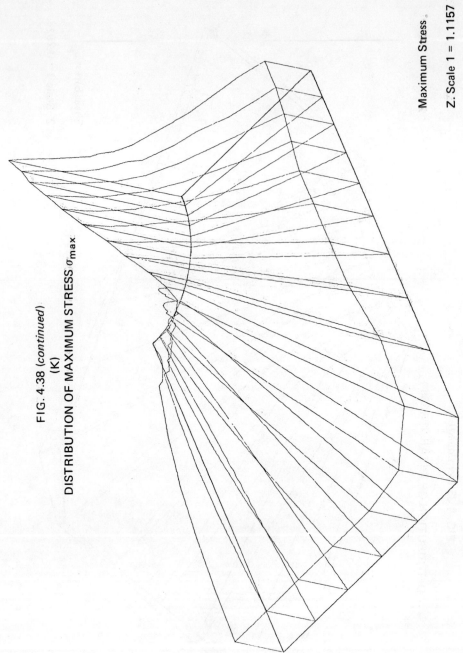

Maximum Stress

Z. Scale 1 = 1.1157

FIG. 4.38 (*continued*)
(K)
DISTRIBUTION OF MAXIMUM STRESS σ_{max}

146

FIG. 4.38 (continued)
(L)
DISTRIBUTION OF MINIMUM STRESS σ_{min}

Minimum Stress

Z. Scale 1 = 0.5594

147

5
Matrix Algebra

Outline

Before the theory behind modern analytical methods can be studied it is first necessary to acquire a basic knowledge of matrix algebra, since this is extensively employed in deriving and manipulating the formulae and expressions used in such methods. The following notes do not provide a comprehensive coverage of the subject, but are intended to explain the reasons for using matrix algebra in this context and to illustrate the more common processes used in manipulating matrices. Finally a bibliography is given to enable those readers, who so wish, to study this subject in greater detail.

Types of Matrices

A matrix is a rectangular array of elements drawn up in a series of rows and columns. The number of elements in any row of a given matrix must be the same, and likewise for columns, although the number of rows need not equal the number of columns. By this means a record of numerical data, etc., can be provided in an orderly manner. This is clearly particularly advantageous when the data is extensive.

The engineer is already familiar with the use of tables of data. Such tables can be regarded as matrices and do become so when the conventional format described below is used. If, for example,

$$A = \begin{bmatrix} a_{11} & a_{12} & a_{13} \\ a_{21} & a_{22} & a_{23} \end{bmatrix}$$

then A is a matrix consisting of two rows and three columns. In this example there are six elements, $a_{11}, a_{12}, \ldots a_{23}$.

These elements may be numbers, algebraic functions, differential operators, etc., or might even themselves be matrices. Consequently a matrix as such has no single numerical value but is rather a neat method of presenting a collection or series of mathematical quantities. The elements of a matrix are contained between square or rounded brackets, thus:

$$\begin{bmatrix} a_{11} & a_{12} & a_{13} \\ a_{21} & a_{22} & a_{23} \end{bmatrix} \text{ or } \begin{pmatrix} a_{11} & a_{12} & a_{13} \\ a_{21} & a_{22} & a_{23} \end{pmatrix}$$

(Care must be taken to distinguish between matrices and determinants, because of the similarity in notation. The terms in a determinant are contained between two vertical lines. Moreover, a determinant has a specific mathematical value and the number of rows must always be equal to the number of columns. For example, if determinant

$$C = \begin{vmatrix} 5 & 2 \\ 4 & 3 \end{vmatrix},$$

then $C = (5 \times 3) - (4 \times 2) = 7$.

On the other hand if C is a matrix, then

$$C = \begin{bmatrix} 5 & 2 \\ 4 & 3 \end{bmatrix} \text{ or } \begin{pmatrix} 5 & 2 \\ 4 & 3 \end{pmatrix},$$

and matrix C does not have a single numerical value. In this case matrix C is a means of recording four numbers which could, for example, be the co-ordinates of points P (5,2) and Q (4,3) and clearly it would be wrong and meaningless to give matrix C a single, derived, numerical value.)

The order or size of a matrix is defined by the number of rows and the number of columns it contains. Thus the matrix A above is of order 2 x 3. Generally, if m represents the number of rows and n represents the number of columns in a matrix, then the matrix is of order m x n. Where $m = n$, the matrix is known as a *square matrix*. A square matrix of order 1 x 1 is termed a *scalar* when the element is a number. A matrix with one row, i.e. $m = 1$, is known as a *row vector* and a matrix with one column, i.e. $n = 1$, is known as a *column vector*. Finally a typical element in a matrix of order m x n can be denoted by a_{ij}, where i is the row number and j the column number.

Examples

(i)   The general form of a matrix

$$A = \begin{bmatrix} a_{11} & a_{12} & a_{13} & \cdots & \cdots & \cdots & a_{1j} & \cdots & \cdots & a_{1n} \\ a_{21} & a_{22} & a_{23} & \cdots & \cdots & \cdots & a_{2j} & \cdots & \cdots & a_{2n} \\ a_{31} & a_{32} & a_{33} & & & & a_{3j} & & & a_{3n} \\ \cdots & \cdots & \cdots & \cdots & \cdots & \cdots & \cdots & \cdots & \cdots & \cdots \\ \cdots & \cdots & \cdots & \cdots & \cdots & \cdots & \cdots & \cdots & \cdots & \cdots \\ a_{i1} & a_{i2} & a_{i3} & \cdots & \cdots & \cdots & a_{ij} & \cdots & \cdots & a_{in} \\ \cdots & \cdots & \cdots & \cdots & \cdots & \cdots & \cdots & \cdots & \cdots & \cdots \\ \cdots & \cdots & \cdots & \cdots & \cdots & \cdots & \cdots & \cdots & \cdots & \cdots \\ a_{m1} & a_{m2} & a_{m3} & \cdots & \cdots & \cdots & a_{mj} & \cdots & \cdots & a_{mn} \end{bmatrix}$$

Matrix A here is of order m x n. There are m rows, n columns, and m x n elements. The general term for an element is a_{ij} which is in the ith row and jth column.

(ii)   Miscellaneous matrices

$$\begin{bmatrix} 1 & 9 & -3 & 2 & 6 \\ 2 & 15 & 25 & -11 & 7 \end{bmatrix}$$

This matrix is of order 2 x 5 and consists entirely of positive and negative integers.

$$\begin{bmatrix} 1 & 3 & -2.7 \\ 7 & -8 & 4 \\ -2.5 & 0 & 5 \\ 9 & -7.9 & -1 \\ 0 & 0 & 0 \\ -1 & -1 & -1 \\ 4.8 & 5.3 & -8.2 \end{bmatrix}$$

This is a 7 x 3 matrix in which the elements are all positive, negative or zero numbers.

$$\begin{bmatrix} \dfrac{dy}{dx} & x^3 & xy & -y^2 \\ x^2y & 2y^3 & \dfrac{-d^2y}{dx^2} & 3x\dfrac{dy}{dx} \\ -xy^2 & \dfrac{d^3y}{dx^3} & 4xy & \sqrt{x^2+y^2} \end{bmatrix}$$

This is a 3 x 4 matrix in which the elements consist of algebraic functions, differential coefficients and numbers.

$$\begin{bmatrix} \sin\theta & \cos\theta & \tan\theta \\ \sec\theta & \operatorname{cosec}\theta & \cotan\theta \end{bmatrix}$$

This is a 2 x 3 matrix involving trigonometric functions.

$$K = \begin{bmatrix} A & B \\ C & D \\ E & F \end{bmatrix}$$

where $A = \begin{bmatrix} a_{11} & a_{12} & a_{13} \\ a_{21} & a_{22} & a_{23} \end{bmatrix}$, $B = \begin{bmatrix} b_{11} & b_{12} & b_{13} \\ b_{21} & b_{22} & b_{23} \end{bmatrix}$,   etc.

In this example, K is a matrix of order 3 x 2 in which the elements themselves are all matrices of order 2 x 3.

(iii)   The transpose of a matrix

The transpose of a given matrix T of any order is denoted by the matrix T' (or sometimes T^* or T^T) and is obtained by making the rows of matrix T the columns of matrix T' and vice versa. This means that if T is of order $m \times n$, then T' is of order $n \times m$. For example if

$$T = \begin{bmatrix} 7 & 8 & 9 \\ 4 & 5 & 6 \end{bmatrix}, \text{ then } T' = \begin{bmatrix} 7 & 4 \\ 8 & 5 \\ 9 & 6 \end{bmatrix}$$

order 2 x 3   order 3 x 2

T' is sometimes referred to as the transpose matrix.

This process can be expressed in general terms thus:

$a'_{ij} = a_{ji}$ for all values of i and j, where a_{ji} is the element in the jth row and ith column of the given matrix, and a'_{ij} is the element in the ith row and jth column of the transpose matrix. It is important to realise that, by this rule, the sequence in which the elements in the columns of T' are written is the same as that for the rows in T, and vice versa.

(iv)   The square matrix
A square matrix has the same number of rows as columns. For example

$$\begin{array}{c} \text{Leading} \\ \text{Diagonal} \end{array} \begin{bmatrix} 6 & 3 & \sqrt{2} & -5 \\ -8 & \sqrt{3} & 1 & 0 \\ 7 & -8.6 & -4.5 & 9 \\ 0 & 0 & -2 & 3.14 \end{bmatrix}$$

This is a square matrix of order 4 x 4. The elements on the sloping line are known as those on the leading diagonal.

There are several types of square matrices, each having special features and consequently an individual name. The more-common types are given below.

(v)   Symmetrical matrix
This is a square matrix which is equal to its own transpose. For example

$$\begin{bmatrix} \dfrac{EA}{L} & 0 & 0 \\[2ex] 0 & \dfrac{12EI}{L^3} & \dfrac{6EI}{L^2} \\[2ex] 0 & \dfrac{6EI}{L^2} & \dfrac{4EI}{L} \end{bmatrix}$$

This is a symmetrical matrix of order 3 x 3 in which the elements are algebraic expressions. It can be seen that the elements in the first row are equal to the elements in the first column, and so on, or in general terms $a_{ij} = a_{ji}$, for all values of i and j. (This is the same formula as is given above for the transpose of a matrix.) Thus in the above example, $a_{12} = a_{21} = 0$, $a_{23} = a_{32} = 6EI/L^2$, etc.

It is clear that symmetrical matrices must be square, and are also equal to their transpose matrices.

Symmetrical matrices are used extensively in modern methods of structural analysis, and the example given is typical (being in fact a stiffness matrix relating the displacements and rotations to the forces and moments at the ends of a structural member. See Chapter 7).

(vi)   Skew or anti-symmetric matrix
This is a square matrix the sign of which is changed by transposing. Thus, if a given matrix S is such that $S = -S'$, then S is a skew matrix. Moreover if the general element is s_{ij}, then $s_{ij} = -s_{ji}$ for all values of i and j. Hence all elements on the leading diagonal of a skew matrix must be zero. For example

$$S = \begin{bmatrix} 0 & 5 & -3 & 2 \\ -5 & 0 & 7 & -6 \\ 3 & -7 & 0 & -4 \\ -2 & 6 & 4 & 0 \end{bmatrix}$$

is a skew matrix of order 4 x 4.

(vii)   Diagonal matrix
This is a square matrix in which all the elements are zero except those on the leading diagonal. For example

$$D = \begin{bmatrix} d_{11} & 0 & 0 & 0 \\ 0 & d_{22} & 0 & 0 \\ 0 & 0 & d_{33} & 0 \\ 0 & 0 & 0 & d_{44} \end{bmatrix}$$

is a diagonal matrix of order 4 x 4. Matrices of this type occur in problems involving scaling factors.

(viii)   Unit matrix
This is a special type of diagonal matrix in which all the elements on the leading diagonal are equal to unity. For example

$$I = \begin{bmatrix} 1 & 0 & 0 \\ 0 & 1 & 0 \\ 0 & 0 & 1 \end{bmatrix}$$

is the unit matrix of order 3 x 3. This type of matrix is used in the manipulation of matrices, especially in the case of matrix inversion.

(ix)   Triangular matrix
This is a square matrix in which all the elements either above or below the leading diagonal are zero. For example

$$L = \begin{bmatrix} 5 & 0 & 0 \\ -2 & 7 & 0 \\ 4 & 1 & 3 \end{bmatrix} \quad \text{is a lower triangular matrix and}$$

$$U = \begin{bmatrix} 4 & 2 & 3 \\ 0 & 5 & -1 \\ 0 & 0 & 8 \end{bmatrix} \quad \text{is an upper triangular matrix.}$$

Such matrices play an important role in the solving of linear simultaneous equations. Their manipulation can be simplified, especially in the case of multiplication.

(x) Banded matrix

This is a square matrix in which all the elements are zero except those on the leading diagonal and on some other diagonals adjacent to the leading diagonal. It is usual for the number of diagonals with non-zero elements above the leading diagonal to equal the number below the leading diagonal, so that the total number of such diagonals is odd. Moreover this number of such diagonals gives a measure of the band width; this means that diagonal and unit matrices are both special types of banded matrices which have a band width of one diagonal. For example

$$B = \begin{bmatrix} 5 & 2 & 0 & 0 & 0 & 0 & 0 \\ 6 & 7 & 3 & 0 & 0 & 0 & 0 \\ 0 & -1 & 8 & -2 & 0 & 0 & 0 \\ 0 & 0 & 2 & -4 & 1 & 0 & 0 \\ 0 & 0 & 0 & 8 & 9 & -5 & 0 \\ 0 & 0 & 0 & 0 & -3 & 1 & 6 \\ 0 & 0 & 0 & 0 & 0 & 4 & 3 \end{bmatrix}$$

is a banded matrix of order 7 x 7, having a band width of 3 diagonals. Such matrices need not be symmetrical.

(xi) Null matrix

This is a matrix of any order, not necessarily square, all of whose elements are zero. For example

$$\begin{bmatrix} 0 & 0 \\ 0 & 0 \end{bmatrix}, \quad \begin{bmatrix} 0 & 0 & 0 & 0 & 0 \\ 0 & 0 & 0 & 0 & 0 \\ 0 & 0 & 0 & 0 & 0 \end{bmatrix}$$

are null matrices.

(xii) The inverse of a matrix, or matrix inversion

This type of matrix is square and is obtained from a given matrix which clearly must also be square. The inverse of a given matrix A is denoted by A^{-1} where $A \times A^{-1} = I$, the unit matrix of the same order as A.

This introduces the concept of matrix multiplication and it is necessary therefore for the reader to have some knowledge of the manipulation of matrices before matrix inversion can be fully understood. (The inverse of a matrix is mentioned here merely to complete the list of the more common types of matrices encountered, a more detailed description being given later.)

This completes the list of the more common types of matrices encountered in solving civil and structural engineering problems. The next step is to show how these can be manipulated and used.

Operations with Matrices

Addition and Subtraction

If A and B are two matrices, then the operations $A \pm B$ can only be defined when A and B are of the same order. In all other cases these operations have no meaning and cannot be used. The resulting matrix $C = A \pm B$ is defined as that matrix of the same order as A and B in which the elements are obtained from the formula: $c_{ij} = a_{ij} \pm b_{ij}$, for all values of i and j, where c_{ij}, a_{ij}, and b_{ij} are elements in the matrices C, A and B respectively. For example, if

$$A = \begin{bmatrix} 3 & 2 & 1 \\ -5 & 7 & 2 \end{bmatrix} \text{ and } B = \begin{bmatrix} -2 & 5 & 3 \\ 8 & 1 & -9 \end{bmatrix}$$

then $A + B = \begin{bmatrix} 1 & 7 & 4 \\ 3 & 8 & -7 \end{bmatrix}$ and $A - B = \begin{bmatrix} 5 & -3 & -2 \\ -13 & 6 & 11 \end{bmatrix}$

Note that all matrices above are of the order 2 x 3. Now if $A - B = O$ where O is the null matrix of the same order as A and B (i.e. a matrix which consists completely of zero elements), then $A = B + O = B$. Thus two matrices A and B are said to be equal or identical if, and only if, they are of the same order and if $a_{ij} = b_{ij}$ for all i and j.

Multiplication of a matrix by a scalar

If A is a matrix and K is an ordinary number or scalar, then matrix $B = K. A$ is of the same order as matrix A and $b_{ij} = K a_{ij}$ for all values of i and j. For example, if

$$A = \begin{bmatrix} 7 & 8 & 3 & 1 \\ 2 & 9 & 4 & 5 \end{bmatrix} \text{ and } K = 3,$$

then $B = K \cdot A = \begin{bmatrix} 21 & 24 & 9 & 3 \\ 6 & 27 & 12 & 15 \end{bmatrix}.$

That is, all elements of A are multiplied by K (unlike the process for determinants where only one column is multiplied by the factor).

Scalar product of two vectors

This process applies only to the multiplication of a row vector by a column vector in which both vectors are of the same order.

Thus if   row vector $\mathbf{A} = [a_1 \ a_2 \ a_3 \ \dots \ \dots \ \dots \ a_n]$

and if   column vector $\mathbf{B} = \begin{bmatrix} b_1 \\ b_2 \\ b_3 \\ \cdots \\ \cdots \\ b_n \end{bmatrix}$

then both vectors $\mathbf{A}$ and $\mathbf{B}$ are of order n and the product $\mathbf{A} \cdot \mathbf{B}$ is given by

$$\mathbf{A} \cdot \mathbf{B} = [a_1 \quad a_2 \quad a_3 \cdots \cdots \cdots a_n] \begin{bmatrix} b_1 \\ b_2 \\ b_3 \\ \cdots \\ \cdots \\ \cdots \\ \cdots \\ \cdots \\ b_n \end{bmatrix}$$

$$= a_1 b_1 + a_2 b_2 + a_3 b_3 + \cdots \cdots \cdots + a_n b_n$$

$$= \sum_{i=1}^{n} a_i b_i$$

which is a scalar. Thus a definite single value is obtained.

This is the general case and defines the product of row vector $\mathbf{A}$ and column vector $\mathbf{B}$.

Example: If $\mathbf{A} = [9 \ 8 \ 4 \ 7]$

and $\mathbf{B} = \begin{bmatrix} 2 \\ 6 \\ 3 \\ 5 \end{bmatrix}$,

then both $\mathbf{A}$ and $\mathbf{B}$ are of order 4 and

$$\mathbf{A} \cdot \mathbf{B} = [9 \ 8 \ 4 \ 7] \begin{bmatrix} 2 \\ 6 \\ 3 \\ 5 \end{bmatrix}$$

$$= (9 \times 2) + (8 \times 6) + (4 \times 3) + (7 \times 5)$$

$$= 18 + 48 + 12 + 35 = 113$$

This multiplication rule enables vectors (and later matrices) to be used with great effect to solve various physical problems. In the above example, for instance, if the elements in $\mathbf{A}$ are the forces acting on a rigid body and the elements in **B** are the corresponding displacements of these forces (in the same directions as the forces), then the scalar product **A·B** gives the work done on the rigid body. The forces in **A** and the displacements in **B** are vectors, while the product **A·B** (which is the work done) is a scalar.

Multiplication of matrices

In this process a matrix is multiplied by a matrix giving a result which is yet another matrix. It is only possible to multiply one matrix by another when the order of each matrix is compatible. This occurs only when the number of columns in the first matrix is equal to the number of rows in the second matrix. For example, if A is a matrix of order $p \times q$, and B is a matrix of order $r \times s$, then $A \cdot B$ can only be evaluated if $q = r$. Moreover, if $A \cdot B = C$, then C must always be a matrix of order $p \times s$. Thus the pattern is

$$A \cdot B = C$$
$$(p \times q)(q \times s) \quad (p \times s)$$

The elements in matrix C are scalars which are obtained from the scalar products of the various row vectors in A and column vectors in B.

In all cases the appropriate row and column vectors can be obtained by drawing a horizontal line through the elements in a row vector of matrix A and extending this line into matrix B until it intersects with a vertical line drawn through the elements of a column vector in B. The scalar product of these vectors is evaluated as described above and the resulting scalar is inserted in matrix C at that location which is given by the intersection of the two lines. This process can best be illustrated by a numerical example.

If

$$A = \begin{bmatrix} 3 & 4 \\ 1 & 3 \\ 2 & 5 \\ 3 & 2 \end{bmatrix} \text{ and } B = \begin{bmatrix} 5 & 8 & 2 & 1 & 4 & 3 \\ 1 & 5 & 3 & 1 & 2 & 0 \end{bmatrix}$$

it is required to find $A \cdot B = C$. Matrix A is of order 4×2 and B is of order 2×6, so the matrices are compatible for matrix multiplication and matrix C is of order 4×6. Then

$$\begin{bmatrix} 3 & 4 \\ 1 & 3 \\ 2 & 5 \\ 3 & 2 \end{bmatrix} \begin{bmatrix} 5 & 8 & 2 & 1 & 4 & 3 \\ 1 & 5 & 3 & 1 & 2 & 0 \end{bmatrix} = \begin{bmatrix} c_{11} & c_{12} & c_{13} & c_{14} & c_{15} & c_{16} \\ c_{21} & c_{22} & c_{23} & c_{24} & c_{25} & c_{26} \\ c_{31} & c_{32} & c_{33} & c_{34} & c_{35} & c_{36} \\ c_{41} & c_{42} & c_{43} & c_{44} & c_{45} & c_{46} \end{bmatrix}$$

$$(4 \times 2) \quad\quad (2 \times 6) \quad\quad\quad\quad (4 \times 6)$$

where $c_{11} = (3 \times 5) + (4 \times 1) = 15 + 4 = 19$, i.e.,

$$\begin{bmatrix} 3 & 4 \\ \dots & \dots \\ \dots & \dots \\ \dots & \dots \end{bmatrix} \begin{bmatrix} 5 & \dots & \dots & \dots & \dots & \dots \\ 1 & \dots & \dots & \dots & \dots & \dots \end{bmatrix} = \begin{bmatrix} 19 & \dots & \dots & \dots & \dots & \dots \\ \dots & \dots & \dots & \dots & \dots & \dots \\ \dots & \dots & \dots & \dots & \dots & \dots \\ \dots & \dots & \dots & \dots & \dots & \dots \end{bmatrix}$$

(The rule for this process can be remembered as "row by column".)

$c_{12} = (3 \times 8) + (4 \times 5) = 24 + 20 = 44$, i.e.,

$$\begin{bmatrix} -3- & -4- \\ \cdots & \cdots \\ \cdots & \cdots \\ \cdots & \cdots \end{bmatrix} \begin{bmatrix} \cdots & -8 & \cdots & \cdots & \cdots & \cdots \\ \cdots & 5 & \cdots & \cdots & \cdots & \cdots \end{bmatrix} = \begin{bmatrix} \cdots & 44 & \cdots & \cdots & \cdots & \cdots \\ \cdots & \cdots & \cdots & \cdots & \cdots & \cdots \\ \cdots & \cdots & \cdots & \cdots & \cdots & \cdots \\ \cdots & \cdots & \cdots & \cdots & \cdots & \cdots \end{bmatrix}$$

$$c_{13} = (3 \times 2) + (4 \times 3) = 6 + 12 = 18$$
$$c_{14} = (3 \times 1) + (4 \times 1) = 3 + 4 = 7$$
$$c_{15} = (3 \times 4) + (4 \times 2) = 12 + 8 = 20$$
$$c_{16} = (3 \times 3) + (4 \times 0) = 9 + 0 = 9$$
$$c_{21} = (1 \times 5) + (3 \times 1) = 5 + 3 = 8, \text{i.e.}$$

$$\begin{bmatrix} \cdots & \cdots \\ -1- & -3- \\ \cdots & \cdots \\ \cdots & \cdots \end{bmatrix} \begin{bmatrix} 5 & \cdots & \cdots & \cdots & \cdots & \cdots \\ 1 & \cdots & \cdots & \cdots & \cdots & \cdots \end{bmatrix} = \begin{bmatrix} \cdots & \cdots & \cdots & \cdots & \cdots & \cdots \\ 8 & \cdots & \cdots & \cdots & \cdots & \cdots \\ \cdots & \cdots & \cdots & \cdots & \cdots & \cdots \\ \cdots & \cdots & \cdots & \cdots & \cdots & \cdots \end{bmatrix}$$

$$c_{22} = (1 \times 8) + (3 \times 5) = 8 + 15 = 23$$
$$c_{23} = (1 \times 2) + (3 \times 3) = 2 + 9 = 11$$
$$c_{24} = (1 \times 1) + (3 \times 1) = 1 + 3 = 4$$
$$c_{25} = (1 \times 4) + (3 \times 2) = 4 + 6 = 10$$
$$c_{26} = (1 \times 3) + (3 \times 0) = 3 + 0 = 3$$
$$c_{31} = (2 \times 5) + (5 \times 1) = 10 + 5 = 15$$
$$c_{32} = (2 \times 8) + (5 \times 5) = 16 + 25 = 41$$
$$c_{33} = (2 \times 2) + (5 \times 3) = 4 + 15 = 19$$
$$c_{34} = (2 \times 1) + (5 \times 1) = 2 + 5 = 7, \text{i.e.}$$

$$\begin{bmatrix} \cdots & \cdots \\ \cdots & \cdots \\ -2- & -5- \\ \cdots & \cdots \end{bmatrix} \begin{bmatrix} \cdots & \cdots & \cdots & 1 & \cdots & \cdots \\ \cdots & \cdots & \cdots & 1 & \cdots & \cdots \end{bmatrix} = \begin{bmatrix} \cdots & \cdots & \cdots & \cdots & \cdots & \cdots \\ \cdots & \cdots & \cdots & \cdots & \cdots & \cdots \\ \cdots & \cdots & \cdots & 7 & \cdots & \cdots \\ \cdots & \cdots & \cdots & \cdots & \cdots & \cdots \end{bmatrix}$$

$$c_{35} = (2 \times 4) + (5 \times 2) = 8 + 10 = 18$$
$$c_{36} = (2 \times 3) + (5 \times 0) = 6 + 0 = 6$$
$$c_{41} = (3 \times 5) + (2 \times 1) = 15 + 2 = 17$$
$$c_{42} = (3 \times 8) + (2 \times 5) = 24 + 10 = 34$$
$$c_{43} = (3 \times 2) + (2 \times 3) = 6 + 6 = 12$$
$$c_{44} = (3 \times 1) + (2 \times 1) = 3 + 2 = 5$$
$$c_{45} = (3 \times 4) + (2 \times 2) = 12 + 4 = 16$$
$$c_{46} = (3 \times 3) + (2 \times 0) = 9 + 0 = 9, \text{i.e.}$$

$$\begin{bmatrix} \cdots & \cdots \\ \cdots & \cdots \\ \cdots & \cdots \\ -3- & -2- \end{bmatrix} \begin{bmatrix} \cdots & \cdots & \cdots & \cdots & \cdots & 3 \\ \cdots & \cdots & \cdots & \cdots & \cdots & 0 \end{bmatrix} = \begin{bmatrix} \cdots & \cdots & \cdots & \cdots & \cdots & \cdots \\ \cdots & \cdots & \cdots & \cdots & \cdots & \cdots \\ \cdots & \cdots & \cdots & \cdots & \cdots & \cdots \\ \cdots & \cdots & \cdots & \cdots & \cdots & 9 \end{bmatrix}$$

All the elements in matrix C have now been obtained and the result of the matrix multiplication can now be expressed thus.

$$
\begin{bmatrix} 3 & 4 \\ 1 & 3 \\ 2 & 5 \\ 3 & 2 \end{bmatrix}
\begin{bmatrix} 5 & 8 & 2 & 1 & 4 & 3 \\ 1 & 5 & 3 & 1 & 2 & 0 \end{bmatrix}
=
\begin{bmatrix} 19 & 44 & 18 & 7 & 20 & 9 \\ 8 & 23 & 11 & 4 & 10 & 3 \\ 15 & 41 & 19 & 7 & 18 & 6 \\ 17 & 34 & 12 & 5 & 16 & 9 \end{bmatrix}
$$

Turning now to the general case, if $AB = C$ and matrix A is of order $i \times k$, and matrix B of order $k \times j$, then matrix C is of order $i \times j$, and

$$
c_{ij} = \sum_{k=1}^{n} a_{ik} b_{kj}
$$

where c_{ij} is the element in the ith row of the jth column of matrix C, a_{ik} is the element in the ith row of the kth column of matrix A and b_{kj} is the element in the kth row of the jth column of matrix B.

It is important to realise that the product $B \cdot A$ has generally no meaning since the order of the matrices are incompatible when taken this way round. It is only when A and B are both square matrices of the same order that it is possible to evaluate both AB and BA, though even then, in general, $AB \neq BA$.

Thus the order of matrices in a product is most important. In the product $A \cdot B$, B is pre-multiplied by A, or conversely, A is post-multiplied by B. An exception to this rule is the product $A \cdot I$ which is equal to $I \cdot A$ where I is the unit matrix. For example

$$
\underset{A}{\begin{bmatrix} 3 & 8 & -2 \\ -4 & 5 & 1 \\ 6 & -7 & -1 \end{bmatrix}}
\underset{I}{\begin{bmatrix} 1 & 0 & 0 \\ 0 & 1 & 0 \\ 0 & 0 & 1 \end{bmatrix}}
=
\underset{I}{\begin{bmatrix} 1 & 0 & 0 \\ 0 & 1 & 0 \\ 0 & 0 & 1 \end{bmatrix}}
\underset{A}{\begin{bmatrix} 3 & 8 & -2 \\ -4 & 5 & 1 \\ 6 & -7 & -1 \end{bmatrix}}
=
\underset{A}{\begin{bmatrix} 3 & 8 & -2 \\ -4 & 5 & 1 \\ 6 & -7 & -1 \end{bmatrix}}
$$

The engineer is invited to practise his own knowledge of matrix multiplication at this stage by carrying out the above numerical example for himself. Thus, in the special case of the unit matrix: $A \cdot I = I \cdot A = A$.

Consider now the following examples of matrix multiplication involving some of the types of matrices described above:

(i)   Diagonal matrices

If $A = \begin{bmatrix} a_{11} & a_{12} & a_{13} \\ a_{21} & a_{22} & a_{23} \\ a_{31} & a_{32} & a_{33} \end{bmatrix}$ and $D = \begin{bmatrix} d_{11} & 0 & 0 \\ 0 & d_{22} & 0 \\ 0 & 0 & d_{33} \end{bmatrix}$

then $D \cdot A = \begin{bmatrix} d_{11} & 0 & 0 \\ 0 & d_{22} & 0 \\ 0 & 0 & d_{33} \end{bmatrix} \begin{bmatrix} a_{11} & a_{12} & a_{13} \\ a_{21} & a_{22} & a_{23} \\ a_{31} & a_{32} & a_{33} \end{bmatrix} = \begin{bmatrix} d_{11}a_{11} & d_{11}a_{12} & d_{11}a_{13} \\ d_{22}a_{21} & d_{22}a_{22} & d_{22}a_{23} \\ d_{33}a_{31} & d_{33}a_{32} & d_{33}a_{33} \end{bmatrix}$

Here A is *pre-multiplied* by the diagonal matrix D, and it is seen that the rows in the product are each multiplied by the corresponding element on the leading diagonal of D.

$$\text{Also } A \cdot D = \begin{bmatrix} a_{11} & a_{12} & a_{13} \\ a_{21} & a_{22} & a_{23} \\ a_{31} & a_{32} & a_{33} \end{bmatrix} \begin{bmatrix} d_{11} & 0 & 0 \\ 0 & d_{22} & 0 \\ 0 & 0 & d_{33} \end{bmatrix} = \begin{bmatrix} d_{11}a_{11} & d_{22}a_{12} & d_{33}a_{13} \\ d_{11}a_{21} & d_{22}a_{22} & d_{33}a_{23} \\ d_{11}a_{31} & d_{22}a_{32} & d_{33}a_{33} \end{bmatrix}$$

This time A is *post-multiplied* by the diagonal matrix D, and now the columns in the product are each multiplied by the corresponding element on the leading diagonal of D.

Finally if E is another diagonal matrix $= \begin{bmatrix} e_{11} & 0 & 0 \\ 0 & e_{22} & 0 \\ 0 & 0 & e_{33} \end{bmatrix}$

$$\text{then } DE = \begin{bmatrix} d_{11} & 0 & 0 \\ 0 & d_{22} & 0 \\ 0 & 0 & d_{33} \end{bmatrix} \begin{bmatrix} e_{11} & 0 & 0 \\ 0 & e_{22} & 0 \\ 0 & 0 & e_{33} \end{bmatrix} = \begin{bmatrix} d_{11}e_{11} & 0 & 0 \\ 0 & d_{22}e_{22} & 0 \\ 0 & 0 & d_{33}e_{33} \end{bmatrix} = ED.$$

(ii) Triangular matrices

Matrix multiplication of two lower triangular matrices always produces a resultant matrix which is also of lower triangular form, and similarly for upper triangular matrices. The product of a lower with an upper triangular matrix gives a general matrix A (which is not triangular in form). For example,

$$\text{if} \qquad L_1 = \begin{bmatrix} 2 & 0 & 0 \\ -1 & 3 & 0 \\ 2 & 1 & 3 \end{bmatrix} \text{ and } L_2 = \begin{bmatrix} 1 & 0 & 0 \\ 2 & -1 & 0 \\ 3 & -2 & 4 \end{bmatrix}$$

$$\text{then} \qquad L_1 L_2 = \begin{bmatrix} 2 & 0 & 0 \\ -1 & 3 & 0 \\ 2 & 1 & 3 \end{bmatrix} \begin{bmatrix} 1 & 0 & 0 \\ 2 & -1 & 0 \\ 3 & -2 & 4 \end{bmatrix} = \begin{bmatrix} 2 & 0 & 0 \\ 5 & -3 & 0 \\ 13 & -7 & 12 \end{bmatrix} = L,$$

which is also of lower triangular form.

$$\text{If} \qquad U_1 = \begin{bmatrix} 2 & 1 & -2 \\ 0 & -3 & 1 \\ 0 & 0 & 3 \end{bmatrix} \text{ and } U_2 = \begin{bmatrix} 3 & 1 & -2 \\ 0 & 2 & 1 \\ 0 & 0 & -1 \end{bmatrix}$$

$$\text{then} \qquad U_1 U_2 = \begin{bmatrix} 2 & 1 & -2 \\ 0 & -3 & 1 \\ 0 & 0 & 3 \end{bmatrix} \begin{bmatrix} 3 & 1 & -2 \\ 0 & 2 & 1 \\ 0 & 0 & -1 \end{bmatrix} = \begin{bmatrix} 6 & 4 & -1 \\ 0 & -6 & -4 \\ 0 & 0 & -3 \end{bmatrix} = U$$

which is also of upper triangular form.

$$\text{Finally,} \qquad LU = \begin{bmatrix} 2 & 0 & 0 \\ 5 & -3 & 0 \\ 13 & -7 & 12 \end{bmatrix} \begin{bmatrix} 6 & 4 & -1 \\ 0 & -6 & -4 \\ 0 & 0 & -3 \end{bmatrix} = \begin{bmatrix} 12 & 8 & -2 \\ 30 & 38 & 7 \\ 78 & 94 & -21 \end{bmatrix} = A,$$

which is not triangular in form.

Thus, with both diagonal and triangular matrices, the process of multiplication is simplified. This also applies to banded matrices.

Powers of a matrix

As in the case of ordinary algebra the following notation is used to denote powers of a matrix.

$$A \cdot A = A^2, \quad A \cdot A \cdot A = A^2 \cdot A = A \cdot A^2 = A^3, \text{ etc.}$$

It is clear that, for compatibility in the resulting matrix multiplications, A must always be a square matrix. For example,

$$\text{if} \quad A = \begin{bmatrix} 3 & 1 & 2 \\ 1 & 5 & 3 \\ 4 & 2 & -1 \end{bmatrix} \text{ then } A^2 = \begin{bmatrix} 3 & 1 & 2 \\ 1 & 5 & 3 \\ 4 & 2 & -1 \end{bmatrix} \begin{bmatrix} 3 & 1 & 2 \\ 1 & 5 & 3 \\ 4 & 2 & -1 \end{bmatrix} = \begin{bmatrix} 18 & 12 & 7 \\ 20 & 32 & 14 \\ 10 & 12 & 15 \end{bmatrix}$$

Matrix inversion

This process has already been referred to earlier. The inverse of a given square matrix A is denoted by A^{-1} where $AA^{-1} = I$, the unit matrix of the same order as A. Once more, the notation is similar to that used in ordinary algebra so that the power to which the inverse matrix is raised is -1. (Note that AA^{-1} cannot possibly be equal to the scalar value of unity since it has already been shown that matrix multiplication produces a result which is also a matrix—the unit matrix I in this case.)

It is only possible to evaluate A^{-1} providing that matrix A is non-singular. A square matrix A is singular if the determinant of the elements of A, i.e. $|A|$, described as det. A, is zero. Conversely if det. $A \neq O$ then A is non-singular. The process of finding A^{-1} from a given square matrix A is known as inversion.

A matrix A and its inverse A^{-1} are commutative with respect to multiplication. Thus

$$AA^{-1} = A^{-1}A = I$$

The inverse matrix is used to perform that function in matrix algebra which is analogous to division in ordinary algebra. For example,
if $AB = C$ where A, B, and C are compatible matrices,
then post-multiplying both sides by B^{-1},
$ABB^{-1} = CB^{-1}$, and since $BB^{-1} = I$,
$AI = CB^{-1}$ and $\therefore A = CB^{-1}$.
Now pre-multiplying both sides by A^{-1},
$A^{-1}AB = A^{-1}C$ and hence $IB = A^{-1}C$ and $\therefore B = A^{-1}C$.
These results give the matrix equivalent of division, first by B and then by A.

The sequence or order in which the matrices are written in the above equations is most important, and the notation used in ordinary algebra, viz. $\dfrac{AB}{C}$

would be ambiguous in this context and must therefore never be used. (This expression could be interpreted as $C^{-1}AB$, $AC^{-1}B$, or ABC^{-1}, giving three different possible results which would obviously be quite unacceptable.) However, the associative law does apply to matrix algebra, so that for example

$$A \cdot B \cdot C = A(B \cdot C) = (A \cdot B)C$$

There are many different methods of matrix inversion. The basic process has been steadily improved to achieve greater accuracy and efficiency, and also to cater for the various exceptional cases which can arise and which cannot be dealt with satisfactorily by some of the simpler methods.

The following method is given to enable the reader to obtain manually (i.e. by using only a slide rule, logarithmic tables, or similar means) the inverse of any matrix, and so to gain some experience in manipulating such matrices, leading to a better understanding of this concept.

This method of matrix inversion makes use of the adjoint matrix and of the basic properties of determinants. The engineer should therefore revise his knowledge of determinants at this stage. (Suitable textbooks are listed in the bibliography at the end of the chapter.)

Let a_{ij} be the general element in a square matrix A, let $D = \det. A$, and let A_{ij} be the signed minor of a_{ij} in the determinant D.
If, for example,

$$A = \begin{bmatrix} a_{11} & a_{12} & a_{13} \\ a_{21} & a_{22} & a_{23} \\ a_{31} & a_{32} & a_{33} \end{bmatrix},$$

then
$$D = \det. A = \begin{vmatrix} a_{11} & a_{12} & a_{13} \\ a_{21} & a_{22} & a_{23} \\ a_{31} & a_{32} & a_{33} \end{vmatrix}$$

and the signed minors are as follows,

$$A_{11} = + \begin{vmatrix} a_{22} & a_{23} \\ a_{32} & a_{33} \end{vmatrix} = a_{22} \cdot a_{33} - a_{32} \cdot a_{23}$$

$$A_{12} = - \begin{vmatrix} a_{21} & a_{23} \\ a_{31} & a_{33} \end{vmatrix} = -(a_{21} \cdot a_{33} - a_{31} \cdot a_{23})$$

$$A_{13} = + \begin{vmatrix} a_{21} & a_{22} \\ a_{31} & a_{32} \end{vmatrix} = a_{21} \cdot a_{32} - a_{31} \cdot a_{22}$$

$$A_{21} = - \begin{vmatrix} a_{12} & a_{13} \\ a_{32} & a_{33} \end{vmatrix} = -(a_{12} \cdot a_{33} - a_{32} \cdot a_{13})$$

$$A_{22} = + \begin{vmatrix} a_{11} & a_{13} \\ a_{31} & a_{33} \end{vmatrix} = a_{11} \cdot a_{33} - a_{31} \cdot a_{13}$$

$$A_{23} = - \begin{vmatrix} a_{11} & a_{12} \\ a_{31} & a_{32} \end{vmatrix} = -(a_{11} \cdot a_{32} - a_{31} \cdot a_{12})$$

$$A_{31} = + \begin{vmatrix} a_{12} & a_{13} \\ a_{22} & a_{23} \end{vmatrix} = a_{12} \cdot a_{23} - a_{22} \cdot a_{13}$$

$$A_{32} = - \begin{vmatrix} a_{11} & a_{13} \\ a_{21} & a_{23} \end{vmatrix} = -(a_{11} \cdot a_{23} - a_{21} \cdot a_{13})$$

$$A_{33} = + \begin{vmatrix} a_{11} & a_{12} \\ a_{21} & a_{22} \end{vmatrix} = a_{11} \cdot a_{22} - a_{21} \cdot a_{12}$$

The signed minors $A_{11}, A_{12}, \ldots$ etc, are scalars, and numerical values can be obtained from the above formulae for any specified numerical values of the elements $a_{11}, a_{12}, \ldots$ etc. of the given square matrix A.

Note also that the signs of the signed minors are obtained according to the following pattern

$$\begin{bmatrix} + & - & + & - & + & \cdots & \cdots & \cdots & \cdots & \cdots & \cdots \\ - & + & - & + & - & \cdots & \cdots & \cdots & \cdots & \cdots & \cdots \\ + & - & + & - & + & \cdots & \cdots & \cdots & \cdots & \cdots & \cdots \\ - & + & - & + & - & \cdots & \cdots & \cdots & \cdots & \cdots & \cdots \\ \vdots & \vdots & \vdots & \vdots & \vdots & \vdots & \vdots & \vdots & \vdots & \vdots & \vdots \\ \vdots & \vdots & \vdots & \vdots & \vdots & \vdots & \vdots & \vdots & \vdots & \vdots & \vdots \\ \vdots & \vdots & \vdots & \vdots & \vdots & \vdots & \vdots & \vdots & \vdots & \vdots & \vdots \\ \vdots & \vdots & \vdots & \vdots & \vdots & \vdots & \vdots & \vdots & \vdots & \vdots & \vdots \end{bmatrix}$$

The adjoint matrix of A, written adj. A, is formed by making its ijth element α_{ij} equal to A_{ji}; that is, the adjoint matrix is the transpose of a matrix formed from the signed minors of the determinant of A. Thus in the above example:

$$\text{If } B = \begin{bmatrix} A_{11} & A_{12} & A_{13} \\ A_{21} & A_{22} & A_{23} \\ A_{31} & A_{32} & A_{33} \end{bmatrix}$$

Then the transpose of $B = B' = \begin{bmatrix} A_{11} & A_{21} & A_{31} \\ A_{12} & A_{22} & A_{32} \\ A_{13} & A_{23} & A_{33} \end{bmatrix} = \text{adj. } A$

Also det. $A = a_{11}(a_{22} \cdot a_{33} - a_{32} \cdot a_{23}) - a_{12}(a_{21} \cdot a_{33} - a_{31} \cdot a_{23}) + a_{13}(a_{21} \cdot a_{32} - a_{31} \cdot a_{22}) = a_{11}A_{11} + a_{12}A_{12} + a_{13}A_{13} = D$

It is now possible to state the formula which will give A^{-1}. This is:

$$\text{The inverse of } A = A^{-1} = \frac{\text{adj. } A}{\text{det. } A} = \frac{\text{adj. } A}{D}$$

Numerical example

Given

$$A = \begin{bmatrix} 3 & 2 & 1 \\ 1 & 3 & 2 \\ 4 & 1 & 3 \end{bmatrix}$$

find the value of the inverse matrix A^{-1}

$$\text{Det. } A = D = 3(9-2) - 2(3-8) + 1(1-12)$$
$$= 21 + 10 - 11 \quad = 20$$

Signed minors of D.

$$
\begin{aligned}
A_{11} &= \quad (3 \times 3) - (1 \times 2) &&= \quad 7 \\
A_{12} &= -[(1 \times 3) - (4 \times 2)] &&= \quad 5 \\
A_{13} &= \quad (1 \times 1) - (4 \times 3) &&= -11 \\
A_{21} &= -[(2 \times 3) - (1 \times 1)] &&= -5 \\
A_{22} &= \quad (3 \times 3) - (4 \times 1) &&= \quad 5 \\
A_{23} &= -[(3 \times 1) - (4 \times 2)] &&= \quad 5 \\
A_{31} &= \quad (2 \times 2) - (3 \times 1) &&= \quad 1 \\
A_{32} &= -[(3 \times 2) - (1 \times 1)] &&= -5 \\
A_{33} &= \quad (3 \times 3) - (1 \times 2) &&= \quad 7
\end{aligned}
$$

Therefore

$$B = \begin{bmatrix} A_{11} & A_{12} & A_{13} \\ A_{21} & A_{22} & A_{23} \\ A_{31} & A_{32} & A_{33} \end{bmatrix} = \begin{bmatrix} 7 & 5 & -11 \\ -5 & 5 & 5 \\ 1 & -5 & 7 \end{bmatrix}$$

$$\text{Adj. } A = B' = \begin{bmatrix} 7 & -5 & 1 \\ 5 & 5 & -5 \\ -11 & 5 & 7 \end{bmatrix}$$

and

$$A^{-1} = \frac{\text{adj} A}{\det A} = \frac{1}{20} \begin{bmatrix} 7 & -5 & 1 \\ 5 & 5 & -5 \\ -11 & 5 & 7 \end{bmatrix} - \begin{bmatrix} 0.35 & -0.25 & 0.05 \\ 0.25 & 0.25 & -0.25 \\ -0.55 & 0.25 & 0.35 \end{bmatrix}$$

Check (by matrix multiplication).

$$AA^{-1} = \begin{bmatrix} 3 & 2 & 1 \\ 1 & 3 & 2 \\ 4 & 1 & 3 \end{bmatrix} \begin{bmatrix} 0.35 & -0.25 & 0.05 \\ 0.25 & 0.25 & -0.25 \\ -0.55 & 0.25 & 0.35 \end{bmatrix}$$

$$= \begin{bmatrix} 1 & 0 & 0 \\ 0 & 1 & 0 \\ 0 & 0 & 1 \end{bmatrix} = I$$

Thus the value obtained is satisfactory.

The engineer now has a means of obtaining the inverse of a matrix. The method is perhaps cumbersome and lacking in elegance, but it does give a satisfactory solution within the scope of the knowledge which has been acquired by this stage.

Solution of Linear Simultaneous Equations

Consider the following set of equations:

$$a_{11}x_1 + a_{12}x_2 + \ldots \ldots \ldots \ldots + a_{1n}x_n = b_1$$
$$a_{21}x_1 + a_{22}x_2 + \cdots \cdots \cdots \cdots \cdots + a_{2n}x_n = b_2$$
$$a_{31}x_1 + a_{32}x_2 + \cdots \cdots \cdots \cdots \cdots + a_{3n}x_n = b_3$$
$$\cdots \cdots \cdots \cdots \cdots \qquad \cdots \cdots \cdots \cdots \cdots \cdots \cdots$$
$$\cdots \cdots \cdots \cdots \cdots \cdots \qquad \cdots \cdots \cdots \cdots \cdots \cdots \cdots$$
$$a_{n1}x_1 + a_{n2}x_2 + \cdots \cdots \cdots \cdots \cdots + a_{nn}x_n = b_n$$

There are n linear equations with n unknowns, namely $x_1, x_2, x_3, \ldots \ldots \ldots x_n$.

If
$$A = \begin{bmatrix} a_{11} & a_{12} & a_{13} & \cdots & \cdots & \cdots & \cdots & \cdots & a_{1n} \\ a_{21} & a_{22} & a_{23} & \cdots & \cdots & \cdots & \cdots & \cdots & a_{2n} \\ a_{31} & a_{32} & a_{33} & - & \cdots & \cdots & \cdots & \cdots & a_{3n} \\ \cdots & \cdots & \cdots & \cdots & \cdots & \cdots & \cdots & \cdots & \cdots \\ \cdots & \cdots & \cdots & \cdots & \cdots & \cdots & \cdots & \cdots & \cdots \\ a_{n1} & a_{n2} & a_{n3} & \cdots & \cdots & \cdots & \cdots & \cdots & a_{nn} \end{bmatrix}$$

(order $n \times n$)

$$X = \begin{bmatrix} x_1 \\ x_2 \\ x_3 \\ \cdots \\ \cdots \\ x_n \end{bmatrix} \qquad \text{and} \qquad B = \begin{bmatrix} b_1 \\ b_2 \\ b_3 \\ \cdots \\ \cdots \\ b_n \end{bmatrix},$$

(order $n \times 1$)   (order $n \times 1$)

then it can readily be established that

$$\begin{bmatrix} a_{11} & a_{12} & a_{13} & \cdots & \cdots & \cdots & \cdots & \cdots & a_{1n} \\ a_{21} & a_{22} & a_{23} & \cdots & \cdots & \cdots & \cdots & \cdots & a_{2n} \\ a_{31} & a_{32} & a_{33} & \cdots & \cdots & \cdots & \cdots & \cdots & a_{3n} \\ \cdots & \cdots & \cdots & \cdots & \cdots & \cdots & \cdots & \cdots & \cdots \\ \cdots & \cdots & \cdots & \cdots & \cdots & \cdots & \cdots & \cdots & \cdots \\ a_{n1} & a_{n2} & a_{n3} & \cdots & \cdots & \cdots & \cdots & \cdots & a_{nn} \end{bmatrix} \begin{bmatrix} x_1 \\ x_2 \\ x_3 \\ \cdots \\ \cdots \\ x_n \end{bmatrix} = \begin{bmatrix} b_1 \\ b_2 \\ b_3 \\ \cdots \\ \cdots \\ b_n \end{bmatrix}$$

or $AX = B$ $-(1)$

(The engineer should satisfy himself by matrix multiplication that this is so.)

Equation (1) illustrates perhaps best of all the purpose of using matrix algebra. A set of n linear simultaneous equations can be denoted by the shorthand form $AX = B$. Moreover by using the various processes for manipulating matrices, these equations can now be solved. Thus

$$AX = B \text{ and therefore } X = A^{-1}B.$$

The values of the elements in matrix X can therefore be obtained by premultiplying matrix B by the inverse of matrix A.

Consider the following numerical example.

$$\begin{aligned}
3x + 2y + z &= -4 \\
x + 3y + 2z &= 2 \\
4x + y + 3z &= -1
\end{aligned}$$

It is required to solve these equations for x, y, z.

In matrix form:

$$\begin{bmatrix} 3 & 2 & 1 \\ 1 & 3 & 2 \\ 4 & 1 & 3 \end{bmatrix} \begin{bmatrix} x \\ y \\ z \end{bmatrix} = \begin{bmatrix} -4 \\ 2 \\ -1 \end{bmatrix}$$

or $AX = B$.

Thus, $X = A^{-1}B$, or

$$\overset{(X)}{\begin{bmatrix} x \\ y \\ z \end{bmatrix}} = \overset{(A^{-1})}{\begin{bmatrix} 3 & 2 & 1 \\ 1 & 3 & 2 \\ 4 & 1 & 3 \end{bmatrix}^{-1}} \overset{(B)}{\begin{bmatrix} -4 \\ 2 \\ -1 \end{bmatrix}}$$

However, it has been shown above that

$$\begin{bmatrix} 3 & 2 & 1 \\ 1 & 3 & 2 \\ 4 & 1 & 3 \end{bmatrix}^{-1} = \frac{1}{20} \begin{bmatrix} 7 & -5 & 1 \\ 5 & 5 & -5 \\ -11 & 5 & 7 \end{bmatrix}$$

and therefore

$$\begin{bmatrix} x \\ y \\ z \end{bmatrix} = \frac{1}{20} \begin{bmatrix} 7 & -5 & 1 \\ 5 & 5 & -5 \\ -11 & 5 & 7 \end{bmatrix} \begin{bmatrix} -4 \\ 2 \\ -1 \end{bmatrix} = \frac{1}{20} \begin{bmatrix} -39 \\ -5 \\ 47 \end{bmatrix}$$

Hence

$$x = \frac{-39}{20} = -1.95$$

$$y = \frac{-5}{20} = -0.25, \text{ and } z = \frac{47}{20} = 2.35$$

since the corresponding elements in identical matrices must be equal. Matrix inversion thus provides a useful method of solving linear simultaneous equations.

If, therefore, a solution to an engineering problem can be achieved by using a method which results in the solving of linear simultaneous equations, matrix algebra can be used with great effect. Indeed the engineer should now be encouraged to devise such methods of solving his problems.

Further notes on matrix inversion

Since matrix inversion can be used to solve linear simultaneous equations as shown, it is clear that solving certain linear simultaneous equations might give an alternative method of obtaining the inverse of a matrix.

$AX = B$ represents a set of linear simultaneous equations and it has already been established that

$X = A^{-1}B$.

Now if A is a square matrix and B is made equal to the unit matrix I of the same order as A, then $X = A^{-1}I = A^{-1}$ and X is also a square matrix of the same order as A and I.

Thus the solution of X in the equation $AX = I$ gives the required inverse of matrix A. For example,

$$\text{if} \qquad A = \begin{bmatrix} a_{11} & a_{12} & a_{13} \\ a_{21} & a_{22} & a_{23} \\ a_{31} & a_{32} & a_{33} \end{bmatrix} \qquad \text{and } X = \begin{bmatrix} x_{11} & x_{12} & x_{13} \\ x_{21} & x_{22} & x_{23} \\ x_{31} & x_{32} & x_{33} \end{bmatrix}$$

$$\text{then} \qquad \begin{bmatrix} a_{11} & a_{12} & a_{13} \\ a_{21} & a_{22} & a_{23} \\ a_{31} & a_{32} & a_{33} \end{bmatrix} \begin{bmatrix} x_{11} & x_{12} & x_{13} \\ x_{21} & x_{22} & x_{23} \\ x_{31} & x_{32} & x_{33} \end{bmatrix} = \begin{bmatrix} 1 & 0 & 0 \\ 0 & 1 & 0 \\ 0 & 0 & 1 \end{bmatrix},$$

In ordinary algebra this represents three different sets of linear simultaneous equations, namely

$$\left. \begin{aligned} a_{11}x_{11} + a_{12}x_{21} + a_{13}x_{31} &= 1 \\ a_{21}x_{11} + a_{22}x_{21} + a_{23}x_{31} &= 0 \\ a_{31}x_{11} + a_{32}x_{21} + a_{33}x_{31} &= 0 \end{aligned} \right\} \begin{array}{l} \text{3 linear equations with} \\ \text{three unknowns} \\ x_{11}, x_{21}, \text{and } x_{31} \end{array}$$

$$\left. \begin{aligned} a_{11}x_{12} + a_{12}x_{22} + a_{13}x_{32} &= 0 \\ a_{21}x_{12} + a_{22}x_{22} + a_{23}x_{32} &= 1 \\ a_{31}x_{12} + a_{32}x_{22} + a_{33}x_{32} &= 0 \end{aligned} \right\} \begin{array}{l} \text{3 linear equations with} \\ \text{three unknowns} \\ x_{12}, x_{22}, \text{and } x_{32} \end{array}$$

and

$$\left. \begin{aligned} a_{11}x_{13} + a_{12}x_{23} + a_{13}x_{33} &= 0 \\ a_{21}x_{13} + a_{22}x_{23} + a_{23}x_{33} &= 0 \\ a_{31}x_{13} + a_{32}x_{23} + a_{33}x_{33} &= 1 \end{aligned} \right\} \begin{array}{l} \text{3 linear equations with} \\ \text{three unknowns} \\ x_{13}, x_{23}, \text{and } x_{33} \end{array}$$

These three sets of linear simultaneous equations can each be solved using one of the well-established methods, such as elimination, relaxation, etc.

Thus $X = A^{-1}$

$$\text{or} \quad A^{-1} = \begin{bmatrix} x_{11} & x_{12} & x_{13} \\ x_{21} & x_{22} & x_{23} \\ x_{31} & x_{32} & x_{33} \end{bmatrix}$$

In this way, powerful techniques such as the Gauss-Jordan Method have been devised (this particular method being based essentially on the solution of linear equations by elimination), and it is sophisticated methods of this type which are actually used in computer programs.

Matrix inversion has now virtually become a subject in its own right and it is clearly beyond the scope of this book to pursue this topic any further. However a bibliography is given at the end of the chapter for those readers who wish to extend their studies in this direction.

Use of the Computer

Library programs which will perform most of the standard operations of matrix arithmetic are generally readily available for any computer. For example, for the Elliott 4100 Computer System there is the 4100 Matrix Package[1], written in Algol. By following the instructions given for this library program, the various processes of matrix manipulation contained in the 'package' can be arranged to be carried out when required as basic steps in any program to be written.

Elaborate library programs such as A27 contain their own subroutines to perform the various operations of matrix arithmetic. Thus, in general, the engineer need not concern himself with the actual arithmetic which results from his using matrices in his engineering problems, as this can be carried out within his own special program on a computer merely by his specifying the arithmetical processes required according to some straightforward instructions.

Ill-Conditioned Problems

The engineer who uses the computer to solve his problems will come to expect— and rightly so—a high degree of accuracy in the solutions so obtained. Consequently it is important to realise that, in certain circumstances, this happy state of affairs can be very difficult to achieve. He must be aware of this possibility and be capable of recognising it and taking the appropriate action when it arises. For those engineering problems which can be reduced to the manipulation of matrices—in particular involving the solution of linear simultaneous equations or matrix inversion—it is possible for small errors, either in the values given for the elements in a matrix or in the actual process of solution which is used, to have a large effect on the accuracy of the solution. Such problems are said to be ill-conditioned.

Conversely, when small errors in the given data or in the process used have little effect on the solution, the problem is said to be well-conditioned, and it is clearly desirable for the engineer to strive at all times to ensure that this latter state is achieved if at all possible.

Examples of ill-conditioned problems

(i)   Consider the simultaneous equations

$$x + y = 10$$
and
$$999x + 1,001y = 20$$

If this problem were solved using analytical geometry, it reduces to finding the intersection of two straight lines which are almost parallel. The civil engineer is already well aware of this problem in surveying, where it is described as a 'poor intersection', and dealt with according to the dictum: 'prevention is better than cure.' These simultaneous equations are therefore ill-conditioned.

(ii)   In the equation $AX = B$ where A, X, and B are matrices, if there is an error of ΔB in B so that the equation becomes $AX = B + \Delta B$, then $X = A^{-1}(B + \Delta B) = A^{-1}B + A^{-1}\Delta B$, instead of $A^{-1}B$. The 'error' in X is therefore $A^{-1}\Delta B$, and could be large if A was nearly singular (i.e. det. $A = O$), even though ΔB is small. This is in fact the case in (i) above.

(iii)   The manipulation of matrices having zero or near zero elements on the leading diagonals always presents problems and should be avoided if at all possible, for example, by changing the sequence in which the rows are arrayed or by some other means.

These examples illustrate the nature of this trouble and draw the engineer's attention to the need for various check calculations or other measures in appropriate cases.

Conclusions

The reader has now some understanding of the role and function of matrices, and is therefore equipped to study the use of matrix methods in analysing engineering problems. Computers can both store the data given in matrices in an efficient way and perform extensive calculations in matrix arithmetic with speed and precision. In short matrix algebra is *the* mathematics for the computer.

Reference

1. 4100 Matrix Package. Elliott 4100 Computer Application Program. (All enquiries now to International Computers, Ltd., London).

Bibliography

A more intensive and detailed study of determinants, matrices, condition and accuracy can be carried out by reference to some of the following textbooks.

Pipes, L. A. Applied Mathematics for Engineers and Physicists. 2nd ed. McGraw. 1958

Aitken, A. C. Determinants and Matrices. 9th ed. Oliver and Boyd. 1958

Bowman, F. An Introduction to Determinants and Matrices. English Universities Press. 1962

Buckingham, R. A. Numerical Methods. Pitman. 1957

Noble, B. Numerical Methods: Iteration, programming and algebraic equations. Oliver and Boyd. 1964

McMinn, S. J. Matrices for Structural Analysis. 2nd ed. Spon. 1966

Fox, L. An Introduction to Numerical Linear Algebra. Oxford U.P. 1964

Fuller, L. E. and Bechtel, R. D. Introduction to Matrix Algebra. Wadsworth Pub. Co. 1967

Frazer, R. A. and others. Elementary Matrices. Cambridge U.P.

Gere, J. M. and Weaver, W. (Jr.). Matrix Algebra for Engineers. Van Nostrand. 1966

Thompson, E. H. Introduction to the Algebra of Matrices with some Applications. Hilger. 1969

Bickley, W. G. and Thompson, R. S. H. G. Matrices: Their Meaning and Manipulation. English Universities Press. 1968

Turnbull, H. W. The Theory of Determinants, Matrices and Invariants. Dover: Constable. 1945

Franklin, J. N. Matrix Theory. Prentice Hall. 1968

Hall, H. S. and Knight, S. R. Higher Algebra. 4th ed. Macmillan. 1891

McArthur, N. and Keith, A. Intermediate Algebra. 4th ed. Methuen. 1951

6
The Computer

Outline

The purpose of this chapter is to provide an introductory background to computer appreciation, discussing types of computers, ancillary equipment, computer programs and programming, etc., in extension to the mention already made in Chapter 1 (see Fig. 1.1, item 3) to computer technology. No details are given of any electronic engineering or solid-state physics here, the computer and its ancillary equipment being regarded as 'black boxes' used as tools to help the engineer solve his problems. In this context computer technology is therefore interpreted as a field of study intended to provide a basic understanding of what a computer is and how it is used. (Naturally the reader may study the electronic circuitry, units, computer logic, and working of a computer later if he so wishes, though this is not essential.)

Computer appreciation

This chapter could in fact be regarded as one on computer appreciation. Some items, however, have been singled out for special treatment, e.g. the digital computer, programming, etc., and these are dealt with under separate headings.

The Digital Computer

A diagrammatic representation of the basic parts of any digital computer is shown in Fig. 6.1. Data, programs, etc., are fed into the input, generally in a coded form acceptable to the computer, e.g. as punched tape or cards, on magnetic tape, as lines or symbols on the screen of a cathode ray tube (CRT), on a console, or perhaps even by depressing the keys of an on-line keyboard, thereby feeding the information in direct. A computer may have more than one input device for any of the different forms of input at the same or different locations. The use of teleprinter or Post Office lines can enable input (and output) stations to be at a considerable distance from the computer proper.

The results of the computer calculations are obtained at the output either as punched tape or cards, magnetic tape, or even as additional lines or symbols on the screen of the same CRT as used for input. However, in addition, it is also possible to obtain a direct print-out if a line printer has been connected on line with the computer system. Graphical and pictorial output can also be obtained using a digital or graphical plotter.

The working of the computer can be controlled by an operator at the control

170

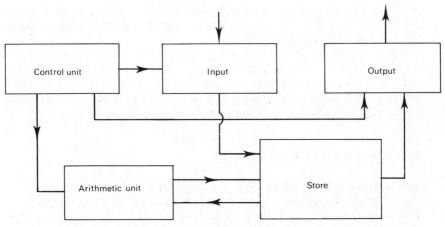

DIAGRAMMATIC REPRESENTATION OF
ANY COMPUTER

FIG. 6.1

unit. This generally consists of a control panel and a teleprinter, the operator giving his instructions to the computer or controlling its activity by using either piece of equipment. In addition, as a calculation proceeds, progress reports, news of errors, etc. can be indicated by lights, buzzers, loud-speakers or printed messages in the control unit (or sometimes at the output). However, much of the work of the control unit is carried out automatically within the system, though there is also the element of manual control outlined above.

All information, data, programs, etc., which are inserted are held in readiness in the store until required. As the calculation proceeds, information is extracted from the store, and the results are also held there as those become available, ready to be yielded at the output when called for. The size of the store in any given computer is an important factor which determines in no small measure how that particular computer can be used. It is also possible to apply data to, or remove from, the store without using the input or output if such data are held in an appropriate form, e.g. on magnetic tape. (Generally a store consists of electronic hardware which may or may not be supplemented by spools of magnetic tape.)

The actual calculations are carried out in the arithmetic unit, with a two-way transfer of data, information and results between it and the store. The control unit and the arithmetic unit are also connected, so that a calculation can be interrupted or held in abeyance as a result of an instruction given, for example, by the operator at the control unit.

The computer is not capable of undertaking any calculations unless detailed instructions are fed in at the input. Such instructions, of course, are termed a computer program, several examples of which have already been given in Chapters 3 and 4. Programs must be capable of dealing with all possibilities of input data, etc., since otherwise in certain cases the calculation just will not proceed; it is clearly quite unrealistic to imagine that at such times the computer itself can take over to put matters right. Programming (i.e. the art of writing programs) is therefore an important subject and merits a detailed study of its own beyond the scope of this book, although some introductory notes are given below and suitable references included.

Digital computers are used most widely to deal with problems in which both the input and output data are numerical. However, ancillary equipment is now available to convert input data from analogue to digital form and vice-versa for output, so that digital computers can now also be used to tackle problems which had hitherto been exclusively dealt with by analogue computers (q.v.).

Miscellaneous notes on digital computers

Many different types of digital computer are available throughout the world. While these are all generally constructed according to the descriptive outline given in diagrammatic terms above, the basic units need not be separate physical entities in all cases, and the actual appearance of the various computers can therefore be quite different. New computers are being designed and marketed all the time with considerable ingenuity and skill, and these developments, together with the re-organisation and re-grouping of the manufacturers within the computer industry, must present a perplexing picture to the civil engineer wishing to become a computer user. Because of such innovations, no attempt is made here to describe the individual features of any particular digital computer. In any case, this can best be done at this stage by the engineer inspecting any particular installation in person, while accompanied by the computer manager or skilled operator, and subsequently referring to the various manuals and drawings prepared for that computer. (See also Chapter 2.)

It has already been shown how well the computer is equipped to carry out a continuous series of calculations involving the evaluation of mathematical formulae or the manipulation of matrices, etc., in accordance with the instructions given in the computer program. However, the resulting numerical calculations are actually carried out within the computer by means of binary arithmetic. The input data in conventional numerical form is converted into binary number notation using teleprinters, etc., to produce punched tapes or cards which are then fed into the computer. All numbers within the computer are therefore in binary form and the calculations are carried out by means of binary arithmetic. The results are also given in binary form on punched tape or cards, etc., and these are then de-coded by means of teleprinters, etc., giving the output in conventional numerical form. Thus the engineer at this stage does not require to have any knowledge of binary numbers and binary arithmetic since this is handled by the computer and ancillary equipment. (He can, of course, extend his studies in this direction at a later date if he so wishes, and a bibliography is therefore given.)

The very nature of punched tapes and cards, and computer hardware, has made the use of binary arithmetic an obvious choice as the most attractive mathematical process to be used within the computer system. By using the binary notation, tapes and cards need only be punched to produce a set of circular or slotted holes. Thus for each channel on a tape or card there may or may not be a punched hole. These two possibilities can be applied to all channels on the tape or card to form a complete set of permutations, and each configuration can be coded to have a particular meaning or value.

In the case of five- and eight-channel tapes, there are $2^5 = 32$ and $2^8 = 256$ possible configurations, respectively. Examples of teleprinter codes for two computers manufactured by Elliotts are given in Fig. 6.2. These codes have been derived from the basic binary number notation, namely (for 4 binary 'bits'):

Basic Binary Number Notation

	2^3	2^2	2^1	2^0
$0 =$	0	0	0	0
$1 = 2^0$ which is denoted by:	0	0	0	1
$2 = 2^1$,, ,, ,, ,,	0	0	1	0
$3 = 2^1 + 2^0$,, ,, ,,	0	0	1	1
$4 = 2^2$,, ,, ,,	0	1	0	0
$5 = 2^2 + 2^0$,, ,, ,,	0	1	0	1
$6 = 2^2 + 2^1$,, ,, ,,	0	1	1	0
$7 = 2^2 + 2^1 + 2^0$,, ,,	0	1	1	1
$8 = 2^3$,, ,,	1	0	0	0
$9 = 2^3 + 2^0$,, ,,	1	0	0	1

etc.

Each of the four binary bits is either 0 or 1; these refer to the indices of the powers to which 2 is raised. Thus in the above table for any given number a particular index appears if 1 is given in the appropriate channel, and 0 is given if it does not. (In the case of punched tape, a punched hole in any channel denotes the appearance of that index.)

The result is that binary numbers can be manipulated in a similar way to logarithms, with multiplication being carried out by the addition of binary bits, etc. This advantage greatly simplifies calculations in binary arithmetic—this being another good reason for its use within the computer.

The design of the hardware for most computers generally consists of the setting up of a labyrinth of cells or 'switches'. Each switch may be either 'on' or 'off', thus providing the two possibilities, and a group of switches can therefore be used to store the various permutations. This arrangement is repeated very many times according to the store size of any given computer. The creation of such a system then gives rise to problems in logic of the following type.

Consider one set of five switches in a five-channel arrangement. Then if Switch 5 is closed and Switches 1 to 4 are open, go to Station A; in all other cases go to Station B.

Station A: Print the number 'one'.

Station B: If Switch 5 is open and Switches 1 to 4 are also open, go to station C; in all other cases go to Station D.

Station C: This produces blank tape (used at the beginning and end of data on punched tape, etc.)

Station D: If Switch 5 is . . . etc. . . . .

Figures and symbols (FIGURE SHIFT set):

1	2	3	4	5	Meaning
●	●		●	●	CARRIAGE RETURN
●	●	●	●		LINE FEED
●	●	●	●	●	LETTER SHIFT
●	●		●	●	FIGURE SHIFT
			●		1
		●			2
●			●	●	3
		●			4
●		●			5
●		●	●		6
		●	●	●	7
	●				8
●	●			●	9
●					0
		●	●		=
●				●	(
●			●		)
●	●			●	£
			●	●	*
●	●		●		@
		●	●	●	%
		●			$
		●	●		+
●		●	●		?
		●	●	●	/
●				●	▼
		●		●	,
		●	●		.
●		●			-
		●			:
●	●	●			SPACE
1	**2**	**3**	**4**	**5**	

All sequences of figures and symbols must be preceded by FIGURE SHIFT

Letters (LETTER SHIFT set):

1	2	3	4	5	Meaning
●	●	●	●	●	LETTER SHIFT
				●	A
			●		B
			●	●	C
		●			D
		●		●	E
		●	●		F
		●	●	●	G
	●				H
	●			●	I
	●		●		J
	●		●	●	K
	●	●			L
	●	●		●	M
	●	●	●		N
	●	●	●	●	O
●					P
●				●	Q
●			●		R
●			●	●	S
●		●			T
●		●		●	U
●		●	●		V
●		●	●	●	W
●	●				X
●	●			●	Y
●	●		●		Z
1	**2**	**3**	**4**	**5**	

All sequences of letters must be preceded by LETTER SHIFT

FIG. 6.2A

1 2 3 4 5	6 7 8	Meaning
......●..	●..	LINE FEED
●..●..	●●	CARRIAGE RETURN
●●..	..●	SPACE
.●..	..●	"
●●..	●●	½
.●..	●..	$
●●..	●..●	%
●●..	●●..	&
.●..	●●●	' (acute)
...●..	..●	(
●●●..	..●	)
●●●..	●..	*
.●●..	●●..	+
●●●●..	●..	,
●●●..	●..●	-
.●●..	●●..	.
●●●●..	●●●	/
.●●..		0
●.●●..	..●	1
●.●●..	●..	2
.●●..	●●..	3
●.●●..	●..	4
.●●..	●.●	5
.●●..	●●..	6
●.●●..	●●●	7
●.●●●		8
.●●●..	..●	9
.●●●..	●..	:
●.●●●..	●●..	;
.●●●..	●..	<
●.●●●●..	●..●	=
●.●●●●..	●●..	>
●●●●●..	●●●	DELETE

1 2 3 4 5	6 7 8	Meaning
●●●	●●●	10
●●..		` (grave)
.●..	..●	A
.●..	..●	B
●●..	●●	C
.●..●	●..	D
●●..●	●..●	E
●●..●	●●..	F
.●..●	●●●	G
.●..●.●		H
●●..●.●	..●	I
●●..●.●	●..	J
.●..●.●	●●..	K
.●..●.●	●..	L
.●..●.●	●..●	M
.●..●.●	●●..	N
●●..●.●	●●●	O
.●..●..●		P
●●..●..●	..●	Q
●●..●..●	●..	R
.●..●..●	●●..	S
●●..●..●	●..	T
.●..●..●	●.●	U
.●..●..●	●●..	V
●●..●..●	●●●	W
●●..●●..		X
.●..●●..	..●	Y
.●..●●..	●..	Z
●●..●●..	●●..	[
.●..●●..	●..	£
●●..●●..	●..●	]
●●..●●..	●●..	↑
.●..●●..	●●●	←

FIG. 6.2B

In this way the various permutations can be obtained and identified within the computer. The statements are quite explicit—one of two actions is always taken according to a stipulated condition.

This is known as *computer logic*. This phrase has now come to refer to either the process of philosophical reasoning along the above lines or to the actual hardware (i.e. printed circuits, relays, etc.) which carries out this process within the computer. The 'process of philosophical reasoning' can be interpreted mathematically and it is then known as *Boolean algebra* (see Bibliography). 'Logic' is also used (in both of its forms) in designing data logging systems (see Chapter 8), and the principles of Boolean algebra are implied in certain computer programs containing conditional statements (see below).

Instructions for operating any computer are best given by means of practical sessions with the computer in which the various operations are demonstrated by a skilled technician. This can be followed by a detailed study of the appropriate instruction manuals.

The Analogue Computer

The analogue computer makes use of the analogy which might exist between the actual problem to be solved and the particular problem whose properties and characteristics can be measured and studied directly by the equipment (i.e. the analogue computer) which has been designed specifically for this purpose.

The nature of the two problems can be quite different, possibly being from entirely different branches of science or engineering, yet each reducing to the same piece of mathematical analysis. Problems which are best tackled by analogue computers are essentially non-numerical in form and might, for example, deal with the variation or changing state of a physical system. These variations can all be studied on such a computer until the sought-after state is obtained, leading to the solution of the original problem. This process is frequently carried out by observing a trace on a CRT screen, etc., without the use of the punched tapes and cards required with a digital computer, and a permanent record need only be made of the final result.

Some analogue computers can be very simple and inexpensive. The accuracy of the results depends upon the design of the computer and should therefore be decided at that stage. On the other hand digital computers do tend to be more accurate but are also more expensive.

Examples

Seepage problems in soil mechanics generally reduce to the solution of Laplace's equation with different boundary conditions. An electrical circuit can be designed and constructed in which the variation of the electrical quantities is also dependent on Laplace's equation. The electrical quantities can be readily measured by appropriate instruments inserted in the circuit and the results obtained can lead, by analogy, to the solution of the particular seepage problem.

Vibration problems involve the solution of differential equations and once more the problems can be simulated by a suitable electrical circuit.

The analysis of shear stress in a section can be studied by forming a soap film over a cut-out having the same profile as the section. By measuring various parameters on this apparatus e.g. co-ordinates, direction cosines at various points, volumes between sections under the soap film, etc., the required shear stress distributions can be obtained by analogy. (The author was intrigued to learn recently that another author has suggested that shell-roof problems might be solved by using the analogy between the shell and shear stress distributions. The reader is left to draw his own conclusions!)

These brief notes on the analogue computer are intended to do no more than indicate the nature and features of such equipment and the type of problems which can be solved in this way.

The Hybrid Computer

This is a combination of both digital and analogue computers, and uses the hardware components and abstract techniques from each system. A digital computer coupled to an on-line graph plotter to give continuous traces as output represents one approach to the concept of a hybrid computer. So also would an analogue computer that had been modified to contain some element of storage or was coupled to an analogue-to-digital converter to give, for example, co-ordinates as output (i.e. in digital form) rather than the more usual graphical output (i.e. in analogue form).

A digital computer could be connected directly to an analogue computer to produce yet another type of hybrid computer or finally, a hybrid computer could be devised and designed as such from the outset.

Choosing a Computer System

When choosing a computer system the following considerations should be taken into account and then decisions made according to the particular circumstances prevailing within the potential user's organisation:

A.   *Physical properties*
  1. Size of store.
  2. Rate of operating and processing data.
  3. Choice of tapes or cards.
  4. Overall size of system and therefore accommodation required.
  5. Availability of software.
  6. Ability to extend system i.e. availability of ancillary equipment or additional hardware.

B.   *Managerial and costing considerations*
  1. Cost (including capital cost of system, running costs, depreciation, offset by financial benefits accruing to owners by its installation. (See items 3, 4, 5, and 6.)
  2. Stability of manufacturer (in respect to 'take-overs', mergers, etc.)
  3. Staffing and training of staff. (See Chapters 2 and 11.)

4. Day-to-day performance, breakdown, maintenance, availability of spare parts, obsolescence.
5. Useful life of computer.
6. Accommodation required, together with services (e.g. wiring, air conditioning, etc.) and supply of ancillary equipment. (See A above.)

C. *Owner's use of system*
1. How computer is used by owners.
2. Setting up of software library.
3. Distribution and location of input and output devices and consoles (if time-sharing).
4. Links with laboratory and site logging equipment, etc. (See Chapter 8.)
5. Organisation of staff. (See Chapter 2 and B above.)
6. Realisation by user that fresh fundamental thinking on his problems can now be undertaken.
7. Appreciation of need for precision in all operations involving computer hardware and software.

Brief notes only are now given on some of these points, since the listing of these items should itself provide sufficient information and food for thought. Various implications are considered later (see Chapter 11).

A. Physical properties

The size of the store dictates the magnitude of the problems which can be tackled by that computer. The engineer should therefore ensure that the store in the computer of his choice is both sufficiently large in its basic form for his immediate needs and is capable of being increased later to deal with possible future requirements. On the other hand the rate at which calculations are carried out on a given computer is generally immutable and determines both how long any calculation will take to be completed and how much work can be tackled in any given period of time. The engineer must therefore balance these considerations against cost with reference to his own particular circumstances.

Nevertheless, any computer can be used more efficiently by careful organisation and planning. For example, software and programs can be written or modified to provide only the actual results which are required. Again, the available machine time in any working day can be used more effectively by time-tabling and creating a scale of priorities, etc. (See Chapter 11 and C.)

Much time is also lost in producing print-outs of output from tapes or cards. While this may cause no hardship in certain circumstances, time lost in this way can be substantially reduced by introducing line-printers or other more appropriate output devices to the system. The availability of a plentiful supply of existing software and programs relevant to the needs of the engineer and which can be used directly with a particular computer must make that machine a most attractive proposition. Such material is possibly the result of years of effort by a team of experienced systems analysts and engineers, and it is therefore very tempting for any 'latecomer' to be provided with this type of opportunity when establishing himself in this field.

B.   Managerial and costing considerations

The list given in B above might be regarded as a memorandum to the engineer outlining the various items which have financial implications in setting up a computer system. Ultimately, the whole level of the enterprise will be decided by what finance can be made available. The engineer might also consider the possible consequences to his organisation if it does not set up such a system.

The stability of any computer manufacturer in this young, expanding, and vigorous industry presents the would-be purchaser of a system with a problem which is less concerned with the manufacturer's financial standing or ability to fulfil an order than with whether the manufacturer's products (including his supply of library programs) and method of doing things will endure after any take-over or merger involving that company.

Unfortunately, only a few civil engineering organisations (mainly consulting engineers) have purchased their own digital computers in the last two decades. The poor response by these potential customers has resulted in computer manufacturers reducing their commitment to provide and maintain an adequate supply of library programs and software in this field. This in turn must discourage future sales, and it is to be hoped that this vicious circle will be broken to the mutual benefit of manufacturers and users.

C.   Owners' use of the system

The development of computational techniques in an organisation can depend greatly on how its computer is used. A closed-shop system, in which only a few specialist staff have direct access to the computer and ancillary equipment, might well buy efficiency at the expense of discouraging potential participation from the rest of the staff. This method of operating creates an air of mystique and remoteness, with the output almost being regarded as messages from outer space! Some machine time should always be made available so that any member of staff can, under supervision, directly operate the computer to process his own data. Attempts should be also made to keep staff regularly informed of the latest developments by means of newsletters, demonstrations, lectures, discussion groups, etc.

A software library should be set up to store all tapes, cards, program print-outs and instructional leaflets. Booklets should be prepared for all computer programs, after these have been written and tested, so that any member of staff may use this existing material without any duplication of effort. All programs should be numbered or lettered and a register kept which also gives the title, author, source, date, and language used.

Flow Diagrams

Before any computer program can be written, the precise logical procedure for solving the particular engineering problem must be specified. This is independent of any particular computer or programming language. The final activity is the preparation of a flow diagram, now often referred to as 'systems analysis'. An example of what might loosely be described as a flow diagram is shown in Fig. 1.2. This gives a diagrammatic outline of the purpose of the book and includes the various alternatives open to the engineer at each stage. This particular illustration, however, is not complete in itself, each label merely denoting a process or a sequence of events, which is not uniquely or precisely described.

A flow diagram for Example 1 of Chapter 3 is shown in Fig. 6.3. This refers to Program DCTCES 1 and is virtually self-explanatory. Each step is straight-forward and precise, and from this preparatory work the computer program can be written.

The flow diagram for Example 2 of Chapter 3 is given in Fig. 6.4. This is the procedure used in preparing Program DCTCES 2. At various stages of this diagram there are two alternative steps according to whether $A < \frac{1}{2}L$ or $A \geqslant \frac{1}{2}L$. There is no ambiguity in the inequalities used; these are in accordance with the principles of Boolean algebra to which brief reference has already been made.

Flow diagrams should cater for all possible values of data and the various foreseeable difficulties which might be encountered by specifying the different procedures to be adopted in all such cases. A more-complicated flow diagram for Example 3 in Chapter 3 is shown in Fig. 6.5. Loops or jumps have been created to deal with the various alternatives which might arise, and procedures have been included to indicate the more obvious errors which might be made in input data.

Although a completed flow diagram might be very detailed and complicated, it must be realised that it is built up step by step, each step being clearly defined by a mathematical operation. This type of exercise gives an ideal opportunity of translating into mathematical terms, qualities which have been described as engineering intuition, judgment, and experience. Engineers are not really motivated by instinct—rather do they tend to make decisions rationally by considering all the possibilities open and rejecting each in turn until the final solution is obtained. This, of course, slightly over-simplifies the actual position, but the author would remind the reader that he, too, is an engineer!

This is where older, experienced engineers can make major contributions. If they can only be encouraged to record by a flow diagram how they set about solving the various problems in which they have so much expertise, then it would be possible to write computer programs in accordance with these priceless techniques. These methods could then endure, rather than being lost with the death of the engineer. The author therefore sees all engineers in due course as systems analysts in their own field, and is confident that considerable developments will take place in this direction within the next few years.

Finally, flow diagrams help to teach the engineer how to think analytically and to express all his activities in precise mathematical terms, though not at the expense of stifling the creative spirit. (See the reference to Orwell's '1984' in Chapter 1!) After a period of preparing flow diagrams to write computer programs, it is possible to acquire the ability to dispense with these altogether and to use instead mathematical statements of the form indicated in Chapter 3 (see also Appendices 1 and 2). This approach can save time but requires greater skill and knowledge, and there is a greater risk of making errors.

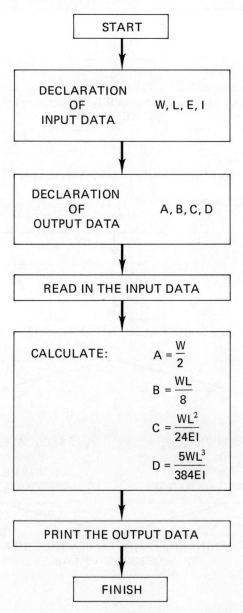

FLOW DIAGRAM FOR PROGRAM NO. DCTCES 1.
FIG. 6.3.

FLOW DIAGRAM (SIMPLIFIED) FOR PROGRAM NO. DCTCES 2.

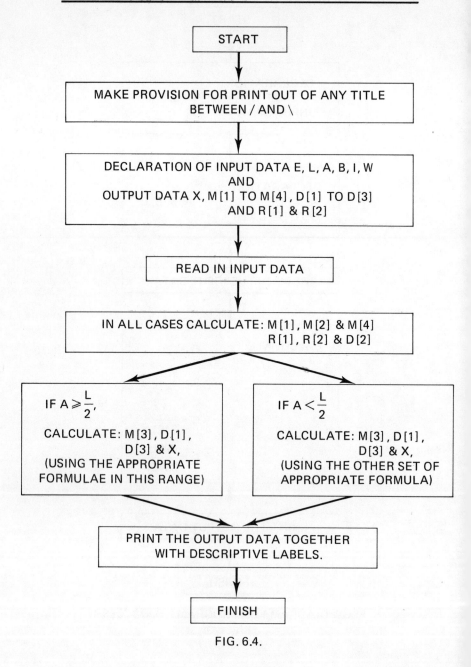

FIG. 6.4.

FLOW DIAGRAM (SIMPLIFIED) FOR PROGRAM NO. DCTCES 3.

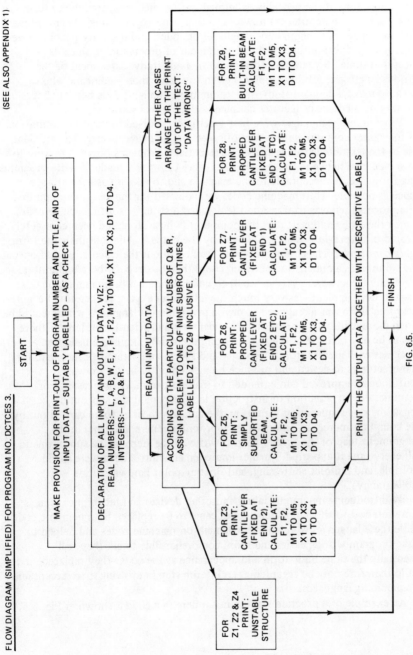

START

MAKE PROVISION FOR PRINT-OUT OF PROGRAM NUMBER AND TITLE, AND OF INPUT DATA – SUITABLY LABELLED – AS A CHECK

DECLARATION OF ALL INPUT AND OUTPUT DATA, VIZ:
REAL NUMBERS:- L, A, B, W, E, I, F1, F2, M1 TO M5, X1 TO X3, D1 TO D4.
INTEGERS:- P, Q & R.

READ IN INPUT DATA

ACCORDING TO THE PARTICULAR VALUES OF Q & R,
ASSIGN PROBLEM TO ONE OF NINE SUBROUTINES
LABELLED Z1 TO Z9 INCLUSIVE.

FOR
Z1, Z2 & Z4
PRINT:
UNSTABLE
STRUCTURE

FOR Z3,
PRINT:
CANTILEVER
(FIXED AT
END 2),
CALCULATE:
F1, F2,
M1 TO M5,
X1 TO X3,
D1 TO D4

FOR Z5,
PRINT:
SIMPLY
SUPPORTED
BEAM,
CALCULATE:
F1, F2,
M1 TO M5,
X1 TO X3,
D1 TO D4.

FOR Z6,
PRINT:
PROPPED
CANTILEVER
(FIXED AT
END 2 ETC),
CALCULATE:
F1, F2,
M1 TO M5,
X1 TO X3,
D1 TO D4.

FOR Z7,
PRINT:
CANTILEVER
(FIXED AT
END 1)
CALCULATE:
F1, F2,
M1 TO M5,
X1 TO X3,
D1 TO D4.

FOR Z8,
PRINT:
PROPPED
CANTILEVER
(FIXED AT
END 1, ETC),
CALCULATE:
F1, F2,
M1 TO M5,
X1 TO X3,
D1 TO D4.

FOR Z9,
PRINT:
BUILT-IN BEAM
CALCULATE:
F1, F2,
M1 TO M5,
X1 TO X3,
D1 TO D4.

IN ALL OTHER CASES
ARRANGE FOR THE PRINT
OUT OF THE TEXT:
"DATA WRONG"

PRINT THE OUTPUT DATA TOGETHER WITH DESCRIPTIVE LABELS

FINISH

FIG. 6.5.

183

Programming

The translation of the procedures defined in flow diagrams, etc., into a form acceptable to a computer is known as 'coding' or 'programming'. There are many different types of programming languages or codes, and it is only possible here to give a historical outline of the development of programming since its inception some 25 years ago, together with a qualitative assessment of the various languages. A bibliography is given to enable those engineers, who so wish, to make a detailed study of programming, though this is possibly best done anyway by a course of lectures coupled with practical sessions.

Programming languages were first specifically evolved to meet the limitations of the earlier computers in respect of speed and storage capacity. The efficient use of the restricted storage space was essential if problems of any size were to be solved, and hence in those early days every effort was made to write programs in a form directly acceptable to the machine and to take up as little storage space as possible. Thus machine codes were devised for the various computers. The process then was to translate the detailed instructions into machine code, punch the tape and insert it in the computer (whose store had been completely cleared of all previous work), followed by the numerical data in the same form *and nothing else.* The computer would then carry out the calculation and output the solution as a punched tape which could then be decoded on a teleprinter.

Machine codes were very difficult to understand, and programming at this stage of development was very much the job for an expert. It therefore became clear that there was a real need to devise programming languages which would be much more readily comprehensible to engineers and scientists at large if there was to be any increase in their use of computers. This implied a need for machines with greater storage capacity and speed, and created an incentive for manufacturers to design and produce them. Computer hardware was continuously improved and extended to make program writing easier. For example, the introduction of instruction labelling enabled a subroutine consisting of a sequence of various mathematical processes to be carried out by the computer merely by inserting an appropriate mnemonic label in the program, instead of the former need to write all the steps of the subroutine in full every time it appeared. Further improvements introduced were single, multiple, and indirect addressing, and floating-point hardware. (See Bibliography).

By introducing mnemonic labels and other devices and techniques such as those referred to above, machine codes came to be replaced by mnemonic codes. While these languages were an improvement on machine codes and helped to make program writing easier and more comprehensible to all, they had essentially the same basic form and appearance as the codes they replaced, and could therefore only be regarded as an interim step in evolving more acceptable programming languages.

An example of a program written in a mnemonic code is shown in Fig. 6.6.*

---

* This version of DCTCES 1 in NEAT is given solely to illustrate the nature and form of a program written in this mnemonic code and is not intended to be actually used by the reader on his own computer. The ALGOL version, however, can be so used.

FIG. 6.6

PRINT-OUT OF PROGRAM NO. DCTCES 1 WRITTEN IN NEAT

```
        % PROGRAM $ CES 1
        $ F DATA
$ W
$ A
$ L
$ B
$ X
$ Y
$ Z
        %  LDR:L $ 3
        %  JILX  $ P: OPEN * DEV * TRANSFER
        %  JILX  $ P:  STANDARD * TRANSFER
        %  LDR: L $ 3
        %  JILX $ P:  REAL * INPUT * TRANSFER
        %  WF $ W
        %  LDR: L $ 3
        %  JILX $ P: REAL * INPUT * TRANSFER
        %  WF $ L
        %  LDR: L $ 3
        %  JILX $ P: REAL * INPUT * TRANSFER
        %  WF $ E
        %  LDR: L $ 3
        %  JILX $ P: REAL * INPUT * TRANSFER
        %  WF $I
        %  LDR: L $ 3
        %  JILX $ P: CLOSE  * DEV * TRANSFER
        %  FL $ W
        %  FD: L $ 2
        %  WF $ A
        %  FL $ W
        %  FM $ L
        %  FD: L $ 8
        %  WF $ B
```

FIG. 6.6 (*continued*)

```
%  FL $ E
%  FM $I
%  WF $ Z
%  FM: L $ 24
%  WF $ Y
%  FL $ L
%  FM $ L
%  WF $ X
%  FM $ L
%  WF $ L
%  FL $ X
%  FM $ W
%  FD $ Y
%  WF $ C
%  FL $ Z
%  FM: L $ 384
%  WF $ Y
%  FL $ L
%  FM $ W
%  FM: L $ 5
%  FD $ Y
%  WF $ D
%  LDR: L $ 4
%  JIL X  $ P: OPEN * DEV * TRANSFER
%  JILX $ P: * STANDARD * TRANSFER
%  LDR: L $ 3
%  FL $ A
%  JILX $ P: REAL * OUTPUT * TRANSFER
%  LDR: L $ 3
%  FL $ B
%  JILX $ P: REAL * OUTPUT * TRANSFER
%  LDR: L $ 3
%  FL $ C
%  JILX $ P: REAL * OUTPUT * TRANSFER
```

```
%  LDR: L $ 3
%  FL $ D
%  JILX $ P: REAL * OUTPUT * TRANSFER
%  LDR: L $ 4
%  JILX $ P: CLOSE * DEV * TRANSFER
%  END;
```

This gives the print-out of Program DCTCES 1 of Chapter 3 written in NEAT—the assembly language or basic mnemonic code of the Elliott 4100 Computer System. It is interesting to compare the version of DCTCES 1 in ALGOL (in Chapter 3) with the one now given in NEAT (Fig. 6.6), and to note immediately that the version in ALGOL is by far the more comprehensible, since it appears to be written in 'normal' English and mathematical symbols.

A program in mnemonic code was normally used by punching the tape or set of cards for the user's program, clearing the store of the computer of all previous work, and then first inserting the systems programs relating to the particular mnemonic code, followed by the user's program and the data in the same form. The computer then carried out the calculations and gave the output in the usual way.

The systems programs for the particular mnemonic code used are generally supplied by the computer manufacturer, and the combination of these, together with the user's program, constitutes a set of instructions acceptable to that computer. The use of systems programs required greater storage capacity and speed, and machines with such features therefore had to be devised and marketed.

A machine code program is roughly similar in appearance to the corresponding program in mnemonic code, though the number of steps and sets of instructions are considerably greater in the machine code program as these must be written out in full in the form directly acceptable to the computer. Because of this similarity and because of the technical difficulty and futility of this task in these more enlightened times no print-out of DCTCES 1 in machine code is provided here.

The continuing development of computers enabled serious practical consideration to be given in the 1950's to the concept of writing a program in 'normal' English and 'straightforward' mathematical symbols, and then of arranging for the computer itself to first translate this into machine code before proceeding with the calculation proper. This required that the computer should have a store sufficiently large to hold the instructions to effect the translation—these instructions themselves constituting a program known as the compiler—as well as having enough capacity left for the calculation itself. Moreover, it was also recognised that this process would use up more machine time (and thus involve delay and expense) unless the operating speed of the computer could be increased. Thus as bigger, faster, computers were devised, programming languages of this type became a reality. These are termed symbolic programming languages, and their creation heralded the 'big break-through', bringing the writing and understanding of programming within the capability of the entire scientific and commercial world. In this way, the symbolic programming language known as FORTRAN became available in the USA around 1956, and at approximately the same time other symbolic programming languages known as AUTOCODES were devised for specific British computers such as the Pegasus, Mercury, and the 803.

The print-out of Program DCTCES 1 is given in FORTRAN in Fig. 6.7 and in 803 AUTOCODE in Fig. 6.8. These illustrations are primarily intended to give the engineer an indication of the nature and appearance of programs written in these symbolic programming languages to solve a problem already well known to him, and are not intended to be used by the reader on an actual computer. It can be seen that both programs are written in what might be regarded as quasi-scientific language, and the engineer should now really begin to appreciate the nature of programming. The reason why the Fortran program is shorter than that in 803 Autocode is mainly because, in the former, several mathematical operations can be specified on one line as a single instruction, whereas in the latter every individual mathematical operation has to be treated as a separate process.

The advantage which Fortran clearly exhibited over the different versions of Autocode had to be offset by the increase in the size and complexity of the Fortran Compiler Program over that required for any Autocode program, necessitating, in respect of size, either a bigger, faster computer, or the need to limit the problem to be solved to the space left in the store.

The ever increasing complexity of software such as the Fortran Compiler Program was resolutely faced in the USA by an intensive concentrated effort that was dominated and co-ordinated by one large computer manufacturer together with academic establishments such as the MIT and the University of Illinois. On the other hand, the British effort in this field tended to fragment, with several manufacturers competing in parallel with one another and consequently setting up an array of incompatible systems.

Many library programs were written, particularly in Fortran but also in various Autocodes. Those in Fortran have endured and most computer manufacturers nowadays are obliged to supply an appropriate Fortran Compiler Program with their machines, so virtually establishing Fortran as an international or universal symbolic programming language. Those in the various Autocodes could only be used on a particular machine from a particular manufacturer, and consequently all such library programs tended to have a severely restricted use and a lifespan decided by the useful life of the appropriate computer.

In Britain this process of evolution led to the eventual creation in 1960 of ALGOL. It was now gradually being realised that symbolic programming languages could be used, not only for writing programs, but also as a tool for devising and recording scientific and mathematical procedures in their entirety. Ordinary mathematics cannot be used alone to denote a sequence of operations—some words, phrases and sentences must generally be employed as well. For example, mathematical equations relate the various terms therein, but do not in themselves indicate what process is to be carried out with them, such instructions being normally given in words inserted between the different mathematical expressions. It was from this background that Algol was developed.

FIG. 6.7

PRINT-OUT OF PROGRAM NO. DCTCES 1 WRITTEN IN FORTRAN

```
CES1
        READ 1,W,P,E,R
        1 FORMAT(F3.0,F3.0,F8.0,F6.3)
        A=W/2
        B=W*P/8.0
        C=W*P**2/(24.0*E*R)
        D=5.0*W*P**3/(384.0*E*R)
        PUNCH 2,A,B,C,D
        2 FORMAT(F10.4)
END
```

FIG. 6.8

PRINT-OUT OF PROGRAM NO. DCTCES 1 WRITTEN IN 803 AUTOCODE

```
        SETV WLEIABCDF
        SETR 1
1)      READ W
        READ L
        READ E
        READ I
        A=0
        B=0
        C=0
        D=0
        F=0
        A=W/2
        B=W*L
        B=B/8
        F=W*L
        F=F*L
        F=F/E
        F=F/I
        C=F/24
        D=F*L
        D=5*D
        D=D/384
        PRINT A
          LINE
        PRINT B
          LINE
        PRINT C
          LINE
        PRINT D
          STOP
        START 1
```

FIG. 6.9

TRANSLATION OF 803 AUTOCODE VERSION OF PROGRAM NO. DCTCES 1
INTO AN ALGOL PROGRAM USING AUTALG.

```
AUTOCODE PROGRAM;
"BEGIN" "REAL" A,B,C,D,E,F,I,L,W;
       "INTEGER" SL,RDR,START,VAR;
       "BOOLEAN" "ARRAY" UNUSED[1:3];
       "REAL" PI;

"PROCEDURE" PROGRAM(REF); "VALUE" REF; "INTEGER" REF;
"BEGIN" "INTEGER" CT1,CT2,CT3,CT4,CT5,CT6,CT7,CT8,CT9,CT10;
       "SWITCH" LL:=L 1,TO,EXIT,TRIG;
       "PROCEDURE" READ(I,X,BOO,TRIGGER);"INTEGER" I;
       "REAL" X;"BOOLEAN" BOO;"LABEL" TRIGGER;
       "BEGIN" "REAL" NOR;"INTEGER" NOI;
       "IF" UNUSED[RDR] "THEN" "BEGIN" UNUSED[RDR]:="FALSE";
       ADVANCE(RDR) "END";
       RED: "IF" BUFFER(RDR,'.') "OR" BUFFER(RDR,'.') "OR"
       BUFFER(RDR,'.') "THEN" "BEGIN" ADVANCE(RDR);
       "GOTO" RED "END";
       "IF" BUFFER(RDR,')') "THEN" "BEGIN" WAIT;ADVANCE(RDR);
       "GOTO" RED "END";
       "IF" BUFFER(RDR,'=') "THEN" "BEGIN" "INTEGER" "ARRAY"
       TTL[1:100]; "INTEGER" CS;
       CS:=1; INSTRING(TTL,CS);
       CS:=1; OUTSTRING(TTL,CS);
       "END";
       "IF" BUFFER(RDR,'$') "THEN" "BEGIN" "IF" BOO "THEN"
       I:=1,100
       "ELSE" X:=1,100;
       "GOTO" GREEN;
       "END"$;
       "IF" BOO "THEN" "READ" NOI "ELSE" "READ" NOR;
       "IF" BUFFER(RDR,'(') "THEN" "BEGIN" REF:="IF" BOO
       "THEN" NOI "ELSE" NOR;
       ADVANCE(RDR); "GOTO" TRIGGER "END";
       "IF" BOO "THEN" I:=NOI "ELSE" X:=NOR;
```

```
GREEN:"END" READ;
TRIG:"GOTO" LL[REF];
L1:
READ(SL,W,"FALSE",TRIG);
READ(SL,L,"FALSE",TRIG);
READ(SL,E,"FALSE",TRIG);
READ(SL,I,"FALSE",TRIG);
A::=0;
B::=0;
C::=0;
D::=0;
F::=0;
A::=W/2;
B::=W*L;
B::=B/8;
F::=W*L;
F::=F*L;
F::=F/E;
F::=F/I;
C::=F/24;
D::=F*L;
D::=5*D;
D::=D/384;
"PRINT" A,'   ';
"PRINT" ''L'';
"PRINT" B,'   ';
"PRINT" ''L'';
"PRINT" C,'   ';
"PRINT" ''L'';
"PRINT" D,'   ';
STOP;


T 0:EXIT:"END" OF PROGRAM;
RDR::=1;VAR::=0;PI::=3.14159265;
UNUSED[1]:=UNUSED[2]:=UNUSED[3]:="TRUE";
DIGITS(4); SCALED(9); SAMELINE;
"PRINT" PUNCH(3),''L'IF ENTRY POINT IS STANDARD,
TYPE 0;','L'TO ENTER AT REFERENCE NUMBER N,TYPE N;
'S3'';
"READ" READER(3),START;
"IF" START=0 "THEN" PROGRAM(1) "ELSE" PROGRAM(START);

"END" OF TRANSLATED AUTOCODE PROGRAM;
```

191

Sequence No.								
0 0 0 0 1 0	IDENTIFICATION DIVISION.
0 0 0 0 2 0	PROGRAM-ID. BEAM01.
0 0 0 0 3 0	ENVIRONMENT DIVISION.
0 0 0 0 4 0	CONFIGURATION SECTION.
0 0 0 0 5 0	SOURCE COMPUTER. ABACUS-MARK1, MEMORY SIZE 20000 WORDS.
0 0 0 0 6 0	OBJECT COMPUTER. ABACUS-MARK1, MEMORY SIZE 20000 WORDS.
0 0 0 0 7 0	INPUT-OUTPUT SECTION.
0 0 0 0 8 0	FILE-CONTROL.
0 0 0 0 9 0	SELECT PAPERFILE1, ASSIGN TO PAPER-READER.
0 0 0 1 0 0	SELECT PAPERFILE2, ASSIGN TO PAPER-PUNCH.
0 0 0 1 1 0	DATA DIVISION.
0 0 0 1 2 0	FD PAPERFILE1; DATA RECORD IS BEAMDATA.
0 0 0 1 3 0	01 BEAMDATA.
0 0 0 1 4 0	02 W PICTURE IS 9(6).
0 0 0 1 5 0	02 L PICTURE IS 9(2).
0 0 0 1 6 0	02 E PICTURE IS 9(11).
0 0 0 1 7 0	02 I PICTURE IS 9(3).

FIG. 6.10

FORM C14/41(7.66)   © International Computers Limited 1967 Printed in Great Britain

Sequence No.						
000200	F.D. PAPERFILE2; DATA RECORD IS BEAMRESULT.					
000210	01. BEAMRESULT.					
000220	02 A PICTURE IS 9(10).					
000230	02 B PICTURE IS 9(10).					
000240	02 C PICTURE IS 9(10).					
000250	02 D PICTURE IS 9(10).					
000260	WORKING-STORAGE SECTION.					
000270	77 WORK1 PICTURE IS 9(10).					
000280	77 WORK2 PICTURE IS 9(10).					
000290	77 WORK3 PICTURE IS 9(10).					
000300	77 WORK4 PICTURE IS 9(4).					
000310	77 WORK5 PICTURE IS 9(10).					
000320	PROCEDURE DIVISION.					
000330	OPEN-FILES.					
000340	OPEN INPUT PAPERFILE1, OUTPUT PAPERFILE2.					
000350	GET-INPUT.					
000360	READ PAPERFILE1 RECORD.					
000370	DIVIDE 2 INTO W GIVING A.					
000380	DIVIDE 8 INTO L GIVING WORK1.					

FIG. 6.10 (continued)

ICL

Data processing

COBOL program sheet

title SIMPLY SUPPORTED BEAM WITH U.D.L. sheet number 03 OF 03

programmer

date

Identification

Sequence No.		
000400		MULTIPLY W BY WORK1 GIVING B.
000410		MULTIPLY E BY I GIVING WORK2.
000420		MULTIPLY 24 BY WORK2 GIVING WORK3.
000430		MULTIPLY L BY L GIVING WORK4.
000440		MULTIPLY W BY WORK3 GIVING WORK5.
000460		DIVIDE WORK3 INTO WORK5 GIVING C.
000460		MULTIPLY 384 BY WORK2.
000470		MULTIPLY L BY WORK4.
000480		MULTIPLY W BY WORK4.
000490		MULTIPLY 5 BY WORK4.
000500		DIVIDE WORK2 INTO WORK4 GIVING D.
000510	OUTPUT-RESULTS.	
000520		WRITE BEAMRESULT.
000530	CLOSE-FILES.	
000540		CLOSE PAPERFILE1, PAPERFILE2.

Algol is now used both as the main symbolic programming language for many computers from different manufacturers (especially in Britain), and also as a means of providing a complete and explicit description of the operations involved in any new computational process. Many examples of programs written in Algol are given in Chapters 3 and 4, and this language is referred to extensively throughout this book.

A program 'X' in any symbolic programming language such as Fortran, the Autocodes, Algol, etc., is normally used by punching the tape or cards for the program, clearing the computer store of all previous work, then first inserting in the input the appropriate compiler and systems programs (these sometimes being on several tapes or on several different sets of cards) followed by the program 'X' and the data. The computer then carries out the calculation and gives the output in the usual way.

With the introduction of Algol there immediately arose the problem of how to salvage the software already written at considerable expense and effort in the appropriate Autocodes and which would otherwise now be lost. In Britain some manufacturers had already made strenuous efforts to entice civil engineering firms and organisations to buy computers by writing several appropriate library programs in Autocode for this potential market, but as already stated this proved to be singularly unsuccessful. As a result of this experience, no manufacturer was prepared, at his own expense, to re-write, translate, or even provide to any extent, new library programs on civil engineering topics written in Algol.

The solution adopted in the 1960's by Elliott's, for example, to the problem of changing from the older 803 Computer and Autocode to the then more recent 4100 Computer and Algol, was to write both simulator and translator programs in Algol. The simulator program[1] was intended to make the 4100 Computer capable of undertaking all the functions of the 803 Computer which it replaced, so that existing library programs written in any language for the 803 Computer could be used on the 4100 Computer. This aim was clearly praiseworthy but, unfortunately, due to many difficulties proved on the whole to be unsatisfactory, and the user could well find himself spending hours trying in vain to obtain the results which would have been given so readily by the older 803 machine. Moreover this method required the user to maintain teleprinters (with print-out facilities) for the punching of tapes in both 803 and 4100 tape codes, so that at best this could only be regarded as a short-term expedient.

The translator program AUTALG[2], on the other hand, offered what appeared to be a much more attractive solution. Its purpose was to use the 4100 Computer to translate any existing 803 Autocode Program into Algol in a form which could then be used directly and repeatedly with the 4100 Computer. To illustrate this process the 803 Autocode version of Program DCTCES 1 (Fig. 6.8) has been translated by the Autalg program using a 4100 Computer, giving as output the Algol program shown in Fig. 6.9; this program does work and can in fact be used. Nevertheless, when this machine-translated Algol program is compared with the version specifically written in Algol (Chapter 3), it can be seen that the machine-produced version is more extensive and cumbersome, though this is to be expected since one is a translation and the other is not. While this machine-produced translation of DCTCES 1 does work, the program print-out is difficult to follow, and in translating large Autocode programs into

Algol by this method, the resulting program may be virtually incomprehensible (except by the more perceptive expert). Moreover, if the resulting program proves unsatisfactory and does not provide results, the correction of any errors is well-nigh impossible. If such an impasse were to occur, the only solution would be either to translate the existing Autocode program by hand or to rewrite the entire program. This is a difficult, arduous task and very much the job for an expert. It was, in fact, the method adopted to save Programs LC2 and LC3 for posterity. (See Chapter 3.)

Many lessons can be learned from this historical account of the evolution of programming, and perhaps it will now be appreciated why an abundance of library programs is not freely available for use on any computer.

In recent years the civil engineering profession itself has rightly begun to take over from the computer manufacturer some of the responsibility for co-ordinating the preparation of software in its own field. The Institutions of Civil Engineers and of Structural Engineers are both aware of this problem and have organised discussions and issued communications on this matter. Various consulting engineers and civil engineering departments in universities, colleges, and the like, have set up their own software libraries, making some of these facilities available to others. This activity will doubtlessly be extended.

All new programs should now be written in a symbolic programming language such as Algol or Fortran, though, for financial and commercial calculations, COBOL is possibly more appropriate. (A print-out of Program DCTCES 1 in Cobol is given (Fig. 6.10), only to illustrate the nature and appearance of this language in dealing once more with this familiar problem, and must not be used on any computer. Cobol is quite unsuitable for problems of this type, though particularly useful for programs on commercial data processing and financial calculations for which it is specifically intended.)

Algol and Fortran

There are several 'dialects' of Algol and Fortran, depending on the computer and the manufacturer, though it is generally simple to modify these to suit any other machine.

Programs in Algol and Fortran should provide a lasting record of the work carried out in their preparation in a form that can be readily adapted for use on any computers which might be designed during the reader's lifetime (at least!).

An Algol program consists essentially of a title, followed by a statement, and is terminated by a semi-colon. The statement can be either simple or compound, the latter being contained between the words "BEGIN" and "END". A compound statement in a simple program generally consists of at least the following information:

(a)   A specification (or declaration) of the quantities used and their nature, e.g. integers, real numbers, arrays, etc.;

(b)   Instructions for the reading-in of data;

(c)   Formulae and words indicating what calculations are to be carried out; and

(d)   Instructions regarding the printing of the output.

For example, consider the print-out of Program DCTCES 1 in Algol (see Chapter 3).

Algol program	Meaning
CES 1;	Title to identify the program.
"BEGIN"	Start of a compound statement
"REAL" W,L,E,I,A,B,C,D;	Specification of the quantities to be used in the program, namely 8 real numbers which, in due course, are assigned to stores designated as W,L,E,I,A,B,C,D.
"READ" W,L,E,I;	Instructions to read in the first 4 numbers on the data tape, and to assign these to their appropriate location, namely the first number to be inserted in the store known as W, the second in L, the third in E, and the fourth in I.
A: = W/2;	Assign to store A the numerical content of store W divided by 2.
B: = W*L/8;	Assign to store B the product of the numerical contents of stores W and L divided by 8.
C: = W*L↑2/(24*E*I);	Assign to store C the numerical value of $WL^2/24EI$ where $W,L,E,$ and I each represent the numerical contents of these stores.
D: = 5*W*L↑3/(384*E*I);	Assign to store D the numerical value of $5WL^3/384EI$.
"PRINT" A,B,C,D;	Provide as output the numerical contents of the stores A,B,C, and D in that order.
"END"	End of the compound statement.
,	End of program.

The individual quantities are separated by commas (though two spaces or carriage return followed by line feed (CRLF) can also be used). The various individual statements are terminated by a semi-colon, and

    * denotes multiply,
    / denotes divide, and
    ↑ denotes "raised to the power" (e.g. L↑2 denotes L^2).

This simple program in Algol can only give a glimpse of this language, and therefore the reader might examine more critically and appreciatively the print-outs of the programs in Algol in Chapter 3. He can thus acquire a closer insight into the process of writing programs in Algol and as a result may decide to study this language in detail.

    Fortran and Algol both fulfil the same role nowadays as international or universal symbolic programming languages (known as *high-level* languages). Moreover, Fortran is in many ways similar in appearance to Algol, so that a person experienced in the one should readily acquire competence in the other. On the other hand, a beginner is not advised to study both simultaneously, as this would lead to confusion.

One final point is worthy of note before this outline is completed. Brief details have already been given regarding the actual process of using the computer to solve a problem involving a program written in any language. All the processes so far described have utilised the technique known as the 'load and go' method.

However, an alternative technique described as the 'two pass' method exists, in which a single output tape or set of cards is first produced by the computer from the appropriate input tapes or cards (being essentially a translation of the program into machine code, or more precisely pseudo-machine code). The store of the computer is then completely cleared and this single tape or set of cards is inserted in the input followed by the numerical data in the same form. The computer then carries out the calculation and gives the output as usual. This method takes up less storage space and can therefore sometimes provide solutions when the 'load and go' method cannot, though it does mean that the store must be completely cleared every time this technique is used. (With the 'load and go' method it is only necessary to insert the appropriate compiler and systems programs once, different programs in the same language then being used in succession without the need to clear and reset on each occasion.)

Ancillary Equipment

No attempt is made here to describe the various proprietary products available as such knowledge can best be obtained through demonstrations and illustrative catalogues. Moreover, it would clearly be unwise to limit the value of these remarks to the uncertain lifetime of any particular piece of equipment.

The selection of peripheral equipment for any computer installation presents a problem on its own, and should generally reflect how that system is used. For instance, some items may be repeated many times when the emphasis of the main work load is channelled in one direction, or a wider diversification may be required when the precise nature of work is less well known. Some establishments may require either punched cards or tapes (5 or 8 channels), or both, either as input to or output from the computer. There are clearly many permutations of the possible arrangements, especially when multiple entry or exit is required. All these possibilities can be combined with other input and output devices as follows.

Device	Input or output	Remarks
Cathode ray tube	Both	—
Line printer	Output	Either on-line or off-line
Graph plotter	Output	Either on-line or off-line

Device	Input or output	Remarks
Analogue to digital converter	Input	Either on-line or off-line
Magnetic tapes	Both	—

Thus input and output devices should be chosen in such a way as to enable the bulk of the work of the particular computer system to be carried out with as little wastage as possible regarding time or cost. Within this concept the arrangement chosen can still be fairly flexible, with many options being left open for possible future needs, though in such cases the engineer should then endeavour to ascertain whether any manufacturer, or his products, will endure long enough for this proposal to be implemented.

Emanating from and associated with the choice of input and output devices are such essential items of equipment as teleprinters, card punching machines, and various facilities for copying, duplicating, editing, modifying, splicing tapes, etc. It is also possible to use output tapes or cards as a set of instructions or program with other pieces of ancillary equipment, such as automatic drafting machines and graph plotters in the design and drawing office, or with equipment such as data loggers, programmed loading devices, etc., in the laboratory or in the field. Ancillary equipment can also be used to carry out the reverse of the above process. For example, a digimeter (or graphical to digital converter) can be used in the design office to transfer the co-ordinates of significant points from a drawing of a survey to punched tape or cards, which can then be used as input data to be processed by the computer with an appropriate program. A further example is the use of data logging equipment in the laboratory or in the field to record scientific data on punched tapes or cards which again can be used as input to the computer. When ancillary equipment is not directly linked electrically to the computer, it is said to be off-line, and the examples just given are illustrations of this technique. On the other hand, when the ancillary equipment is linked electrically to the computer to enable a process to take place as an integrated action without the need for punched cards or tapes between the various units or devices, it is said to be on-line. Line printers and graph plotters, for example, can be used on-line with a computer when the output is obtained directly on these devices without the need to first produce punched tapes or cards.

On-line devices generally give results quicker and more directly, though the computer is then tied exclusively to this particular task and no tape or card records can normally be obtained at the same time. Off-line devices generally take longer to give results, and often much computer time is first required to produce the vast quantity of punched tape or cards normally necessary as input (to this off-line device) for even the simplest task. However, once such tapes or cards have been obtained, the off-line equipment can then be used independently of the computer, so releasing it for other duties.

Computer systems now tend to be modular in design and construction, with the various optional modules being connected by simple plug-and-socket type interfaces. Thus it is well-nigh impossible to decide where the computer ends and the ancillary equipment begins!

Hardware and Software

Hardware consists of all the actual physical equipment (mechanical, electrical, ancillary, etc.) appertaining to a computer system.

Software consists of the flow diagrams, programs, compilers, tapes, cards, papers, instruction manuals, brochures, etc., which can be used in conjunction with the hardware.

Any computer system or installation therefore consists of both hardware and software, both being important factors to be considered in any appraisal of such a system. It must be realised from the outset that the existence of software is due to the efforts of very many people either working alone or in teams comprising systems analysts, mathematicians, engineers, scientists and technologists from industrial companies, universities and similar organisations, and this can possibly represent as significant an investment in terms of wages, salaries, grants and overheads, as that required for the factories, materials, equipment and manpower needed to construct the hardware. Hardware and software have been developed and devised from or because of one another. For example, because the earlier computers had only a small storage capacity, software had to be developed to cope with this restriction, whereas when it was realised that, as a result of advances in software regarding symbolic programming languages, bigger and faster computers would be required to take full advantage of these benefits, there was a clear incentive to devise the necessary hardware.

Software is generally in the form of punched tapes or cards. Tapes give a smooth uninterrupted supply of information in a neat, compact form. They can be wound into rolls for storage and fed into tape reading heads by means of revolving wheels. Some tapes can be cut and spliced, and the material used is paper; they are relatively inexpensive, though perhaps a little fragile and likely to tear. Punched cards are more robust, bulkier and possibly more expensive, as is the equipment for their processing. However, cards can be particularly suitable and convenient to deal with certain problems. For example, in traffic engineering one card can be used to record all the information and data required for a given vehicle, or in commercial calculations one card can contain full details of the one customer or item of goods. Moreover, cards can readily be extracted from or replaced in a set, so that modifications can be made without the need to splice the material.

Many items of hardware, in keeping with current trends, are now being made smaller, and even computers themselves are being so treated. An example of such a digital computer is the Olivetti Programma Desk Top Computer (Fig. 6.11) which has a small store and uses special cards. Several library programs on civil engineering applications are already available for it, and the complete unit is reasonably portable, requiring no more than a conventional 13-amp plug, so that it can readily be 'borrowed' by the engineer for use in his own office or home.

Automatic Installations

Reference was briefly made in Chapter 1 to the concept of linking together the various processes shown in Fig. 1.2, though perhaps this can now be better appreciated. This concept can be implemented both for software and hardware.

FIG. 6.11

For example, it should be possible either to combine a series of different calculations to give a complete design process in structural engineering including the production of general-arrangement and working drawings, or to link together various items of equipment so that the output obtained from the system might be re-inserted as input from time to time until satisfactory results are achieved, with the processes in both of these examples proceeding automatically. Thus experimental laboratory measurements from electrical transducers and recorders could be inserted automatically into a computer by teleprinter lines so that calculations are performed and the results relayed back to the laboratory either as a print-out or as electrical pulses to actuate further equipment such as loading devices, and so on. All the electronic equipment in such an installation must be compatible and should preferably consist of a series of different modules linked together with appropriate interfaces.

Regarding software, computer programs which are complete in themselves can now be regarded as a series of subroutines to be joined together to form one large program to carry out, say, a complete design process. Substantial progress has already been made in the USA to develop such software to deal with structural engineering problems using various IBM computers. The better known of these software systems are STRESS[3], FRAN[4], and ICES STRUDL[5]. These use their own language to specify the various instructions and there is no doubt that as bigger and faster computers become more widely available in Britain, here, too, equivalent software will be compiled in due course. One example is the GENESYS system now being developed in Britain by Dr R. J. Allwood at the University of Technology, Loughborough.

References

1. 803 Simulator. Elliott 4100 Computer Application Program. (All enquiries now to International Computers, Ltd., London).
2. AUTALG. Elliott 4100 Computer Application Program. (See 1 above).
3. Fenves, S. J. STRESS—A Computer Programming System for Structural Engineering Problems. (MIT Department of Civil Engineering Report T63-2, June 1963.
4. "IBM 7090/7094 FRAN—Framed Structure Analysis Program". Application Directory, Publication H20-0082, IBM Corporation, August 1964.
5. Logcher, R. D., Flachsbart, B. B., Hall, E. J., Power, C. M., and Wells, R. A. (Jr.) ICES STRUDL-1. The Structural Design Language. Engineering User's Manual. Department of Civil Engineering, MIT (September 1967)

Bibliography

Further information can be obtained from the following textbooks.

Gould, I. H. and Ellis, F. S. Digital Computer Technology. Chapman & Hall, 1963
Jacobowitz, H. Electronic Computers Made Simple. Allen. 1967
Marchant, J. P. and Pegg, D. Digital Computers—a practical approach. Blackie. 1967
Maley, G. A. and Heilweil, M. V. Introduction to Digital Computers. Prentice Hall. 1968
Chandor, A. (with Graham, J. and Williamson, R.) A Dictionary of Computers. Penguin. 1970
Matthews, R. B. and Abrahams, J. R. Logic Tutors and their Applications. Harrap. 1968
Goodstein, R. L. Boolean Algebra. Pergamon. 1963
Halmos, P. R. Lectures on Boolean Algebras. Van Nostrand. 1963
MacNeal, R. H. Electric Circuit Analogies for Elastic Structures. Wiley. 1962
Korn. G. A. and Korn, T. M. Electronic, Analog, and Hybrid Computers. McGraw Hill. 1964
Gilbert, C. P. The Design and Use of Electronic Analogue Computers. Chapman & Hall. 1964
Stewart, C. A. and Atkinson, R. Basic Analogue Computer Techniques. McGraw Hill. 1967
Nenadál, Z. and Mirtes, B. Analogue and Hybrid Computers. Rev. ed. Iliffe. 1968
Bekey, G. A. and Karplus, W. J. Hybrid Computation. Wiley. 1969
Moon, B. A. M. Computer Programming for Science and Engineering. Butterworth. 1966
Sammet, J. E. Programming Languages: History and Fundamentals. Prentice Hall. 1969
Plumb, S. C. Introduction to Fortran. McGraw Hill. 1961.
Wooldridge, R. and Ractliffe, J. F. Introduction to Algol Programming. 3rd ed. English Universities Press. 1968
Nicol, K. Elementary Programming in Algol. McGraw Hill. 1965

7
Structural Analysis—An Outline of Methods Based on Modern Computational Processes

It is only now both appropriate and possible to study the theory related to modern methods of structural analysis. The recent development of this subject has been rapid and dramatic. Much has been done in a comparatively short time and the whole concept of structural analysis transformed in the process. This chapter can only provide a basic understanding of the elastic theory most commonly used, though the bibliography will enable any reader who wishes, to study various other aspects or specialist branches in greater depth. Accordingly, only brief details are given of the matrix methods specially devised for the elastic analysis of structures by computer.

The Stiffness Method

This method has been devised by Livesley[1,2] and is the basic theory used in library programs such as A27, A28, EAPF, LC7, LC8 and LC29 referred to in Chapter 4. The process is based essentially on the Slope Deflection method and includes an allowance for rib shortening of the members. This method provides a unique approach for the elastic analysis of any irregular structure constructed from members of uniform section. The structure may be a plane or space frame or a grid, the joints may be pinned or fixed, and the supports can be of any type, viz: fixed, pinned, on rollers, rigid or yielding.

Whilst the method can be developed in general terms from the outset, it is nevertheless convenient to be more restrictive at first, so that the new ideas can be absorbed more gradually.

Rigid-jointed plane frame

A single member of uniform sectional area A and second moment of area I is shown in Fig. 7.1. The left- and right-hand ends are numbered 1 and 2 respectively. Co-ordinate axes are chosen in which the origin is at end 1, with $0x$ being directed along the axis of the member from end 1 to end 2, and with $0y$ being located at an anticlockwise angle of $90°$ to $0x$. Loading is only applied to the member at its ends, but in its most general form for a plane frame, i.e. consisting of an axial force, a shearing force and a moment. The member is in equilibrium and both ends are free to deflect and rotate so that it can take up the deflected position as shown.

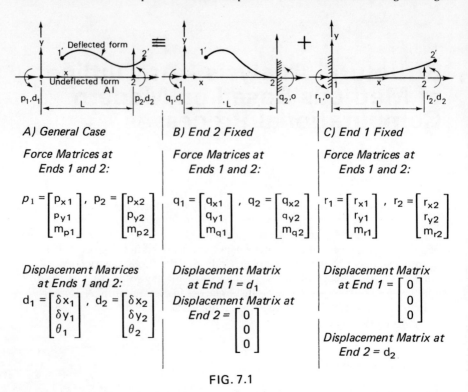

FIG. 7.1

By the principle of superposition the general case (A) can be replaced by the sum of cases (B) and (C), where end 2 is fully fixed in (B), and end 1 is fully fixed in (C).

It is now desired to obtain equations relating the various displacements and rotations to the forces and moments for both ends of the member in the general case (A). This is accomplished by obtaining such relationships for the simpler cases given in (B) and (C), and then combining these to obtain the results for the general case. Referring therefore to (B) in Fig. 7.1:

Axial force at end 1 = q_{x1}	Axial force at end 2 = q_{x2}
Shearing force at end 1 = q_{y1}	Shearing force at end 2 = q_{y2}
Moment at end 1 = m_{q1}	Moment at end 2 = m_{q2}

and

Displacement in the x direction at end 1 = δx_1	Displacement in the x direction at end 2 = 0
Displacement in the y direction at end 1 = δy_1	Displacement in the y direction at end 2 = 0
Rotation of joint at end 1 = θ_1	Rotation of joint at end 2 = 0

Using Mohr's Theorems, Slope Deflection, or otherwise, it can be shown that

$$q_{x1} = \frac{EA}{L} \cdot \delta x_1$$

$$q_{y1} = \frac{12EI}{L^3} \cdot \delta y_1 + \frac{6EI}{L^2} \cdot \theta_1$$

$$m_{q1} = \frac{6EI}{L^2} \cdot \delta y_1 + \frac{4EI}{L} \cdot \theta_1$$

These three equations can now be represented in matrix form thus:

$$
\begin{bmatrix} q_{x1} \\ q_{y1} \\ m_{q1} \end{bmatrix}
=
\begin{bmatrix} \dfrac{EA}{L} & 0 & 0 \\ 0 & \dfrac{12EI}{L^3} & \dfrac{6EI}{L^2} \\ 0 & \dfrac{6EI}{L^2} & \dfrac{4EI}{L} \end{bmatrix}
\begin{bmatrix} \delta x_1 \\ \delta y_1 \\ \theta_1 \end{bmatrix}
$$

$$3 \times 1 \qquad\qquad 3 \times 3 \qquad 3 \times 1$$

or as

$$q_1 = K_{11} d_1 \qquad\qquad\qquad \text{------(1)}$$

where

$$q_1 = \begin{bmatrix} q_{x1} \\ q_{y1} \\ m_{q1} \end{bmatrix} = \text{force matrix at end 1,}$$

$$d_1 = \begin{bmatrix} \delta_{x1} \\ \delta_{y1} \\ \theta_1 \end{bmatrix} = \text{displacement matrix at end 1,}$$

and

$$K_{11} = \begin{bmatrix} \dfrac{EA}{L} & 0 & 0 \\ 0 & \dfrac{12EI}{L^3} & \dfrac{6EI}{L^2} \\ 0 & \dfrac{6EI}{L^2} & \dfrac{4EI}{L} \end{bmatrix} = \text{stiffness matrix.}$$

Thus matrix equation (1) denotes three linear simultaneous equations in a compact form, i.e.

$$\begin{pmatrix} \text{force} \\ \text{matrix} \end{pmatrix} = \begin{pmatrix} \text{stiffness} \\ \text{matrix} \end{pmatrix} \times \begin{pmatrix} \text{displacement} \\ \text{matrix} \end{pmatrix}$$

This equation is already well known in its linear form, viz:

$$\text{force} = \text{stiffness} \times \text{displacement}$$

where it applies to the extension of a spring, and is in fact Hooke's Law. However, this concept is now applied to matrix equations and it is seen that there are three elements in both the force and displacement matrices for a joint in a rigid-jointed plane frame. These matrices thus enable the most general case of applied loading and deformation to be completely specified in a neat orderly manner for any point in a rigid-jointed plane frame.

The stiffness matrix is of order 3 x 3 and is symmetrical. Its elements are constants which depend solely on the sectional properties, the length and the material (i.e. E) of the member.

Finally it is seen from equation (1) that the three matrices are of compatible order.

Now returning to (B): for equilibrium of the member,

$$q_{x2} = -q_{x1}$$
$$q_{y2} = -q_{y1}$$
$$m_{q2} = q_{y1}L - m_{q1},$$

or

$$\begin{bmatrix} q_{x2} \\ q_{y2} \\ m_{q2} \end{bmatrix} = \begin{bmatrix} -1 & 0 & 0 \\ 0 & -1 & 0 \\ 0 & L & -1 \end{bmatrix} \begin{bmatrix} q_{x1} \\ q_{y1} \\ m_{q1} \end{bmatrix}$$

Hence

$$\begin{bmatrix} q_{x2} \\ q_{y2} \\ m_{q2} \end{bmatrix} = \begin{bmatrix} -1 & 0 & 0 \\ 0 & -1 & 0 \\ 0 & L & -1 \end{bmatrix} \begin{bmatrix} \dfrac{EA}{L} & 0 & 0 \\ 0 & \dfrac{12EI}{L^3} & \dfrac{6EI}{L^2} \\ 0 & \dfrac{6EI}{L^2} & \dfrac{4EI}{L} \end{bmatrix} \begin{bmatrix} \delta_{x1} \\ \delta_{y1} \\ \theta_1 \end{bmatrix}$$

$$\quad 3 \times 1 \qquad\qquad 3 \times 3 \qquad\qquad 3 \times 3 \qquad\qquad 3 \times 1$$

$$= \begin{bmatrix} \dfrac{-EA}{L} & 0 & 0 \\ 0 & \dfrac{-12EI}{L^3} & \dfrac{-6EI}{L^2} \\ 0 & \dfrac{6EI}{L^2} & \dfrac{2EI}{L} \end{bmatrix} \begin{bmatrix} \delta_{x1} \\ \delta_{y1} \\ \theta_1 \end{bmatrix} ,$$

or

$$q_2 = K_{21}d_1 \qquad\qquad \text{——— (2)}$$

where

$$K_{21} = \begin{bmatrix} \dfrac{-EA}{L} & 0 & 0 \\[2mm] 0 & \dfrac{-12EI}{L^3} & \dfrac{-6EI}{L^2} \\[2mm] 0 & \dfrac{6EI}{L^2} & \dfrac{2EI}{L} \end{bmatrix}$$

K_{21} is another stiffness matrix relating q_2 to d_1, and is skew symmetrical.
Referring now to (C), it can also be shown that

$$r_{x2} = \frac{EA}{L}\,\delta x_2$$

$$r_{y2} = \frac{12EI}{L^3}\,\delta y_2 - \frac{6EI}{L^2}\,\theta_2$$

$$m_{r2} = \frac{-6EI}{L^2}\,\delta y_2 + \frac{4EI}{L}\,\theta_2 ,$$

or in matrix form

$$\begin{bmatrix} r_{x2} \\[2mm] r_{y2} \\[2mm] m_{r2} \end{bmatrix} = \begin{bmatrix} \dfrac{EA}{L} & 0 & 0 \\[2mm] 0 & \dfrac{12EI}{L^3} & \dfrac{-6EI}{L^2} \\[2mm] 0 & \dfrac{-6EI}{L^2} & \dfrac{4EI}{L} \end{bmatrix} \begin{bmatrix} \delta x_2 \\[2mm] \delta y_2 \\[2mm] \theta_2 \end{bmatrix}$$

Thus

$$r_2 = K_{22}\cdot d_2 \qquad\qquad \text{——— (3)}$$

where r_2 and d_2 are the force and displacement matrices at end 2, and

$$K_{22} = \begin{bmatrix} \dfrac{EA}{L} & 0 & 0 \\[2mm] 0 & \dfrac{12EI}{L^3} & \dfrac{-6EI}{L^2} \\[2mm] 0 & \dfrac{-6EI}{L^2} & \dfrac{4EI}{L} \end{bmatrix}$$

is another symmetrical stiffness matrix.

For equilibrium of the member in (C)

$$r_{x1} = -r_{x2}$$

$$r_{y1} = -r_{y2}$$

$$m_{r1} = -r_{y2} \cdot L - m_{r2} \, ,$$

or

$$\begin{bmatrix} r_{x1} \\ r_{y1} \\ m_{r1} \end{bmatrix} = \begin{bmatrix} -1 & 0 & 0 \\ 0 & -1 & 0 \\ 0 & -L & -1 \end{bmatrix} \begin{bmatrix} r_{x2} \\ r_{y2} \\ m_{r2} \end{bmatrix}$$

Hence

$$\begin{bmatrix} r_{x1} \\ r_{y1} \\ m_{r1} \end{bmatrix} = \begin{bmatrix} -1 & 0 & 0 \\ 0 & -1 & 0 \\ 0 & -L & -1 \end{bmatrix} \begin{bmatrix} \dfrac{EA}{L} & 0 & 0 \\ 0 & \dfrac{12EI}{L^3} & \dfrac{-6EI}{L^2} \\ 0 & \dfrac{-6EI}{L^2} & \dfrac{4EI}{L} \end{bmatrix} \begin{bmatrix} \delta x_2 \\ \delta y_2 \\ \theta_2 \end{bmatrix}$$

$$= \begin{bmatrix} \dfrac{-EA}{L} & 0 & 0 \\ 0 & \dfrac{-12EI}{L^3} & \dfrac{6EI}{L^2} \\ 0 & \dfrac{-6EI}{L^2} & \dfrac{2EI}{L} \end{bmatrix} \begin{bmatrix} \delta x_2 \\ \delta y_2 \\ \theta_2 \end{bmatrix}$$

or

$$r_1 = K_{12} \cdot d_2 \qquad\qquad \text{(4)}$$

where r_1 is the force matrix at end 1 and

$$K_{12} = \begin{bmatrix} \dfrac{-EA}{L} & 0 & 0 \\ 0 & \dfrac{-12EI}{L^3} & \dfrac{6EI}{L^2} \\ 0 & \dfrac{-6EI}{L^2} & \dfrac{2EI}{L} \end{bmatrix}$$

is yet another stiffness matrix which again is skew symmetrical. Now referring to (A), (B), and (C) it can be seen that

$$p_1 = q_1 + r_1$$

and

$$p_2 = q_2 + r_2$$

Substituting in these equations for q_1, q_2, r_1, and r_2 as given in equations (1) to (4), then

$$p_1 = K_{11} \cdot d_1 + K_{12} \cdot d_2 \qquad\text{———(5)}$$

$$p_2 = K_{21} \cdot d_1 + K_{22} \cdot d_2 \qquad\text{———(6)}$$

or expressed in full:

$$
\begin{bmatrix} p_{x1} \\ p_{y1} \\ m_{p1} \end{bmatrix} =
\begin{bmatrix}
\dfrac{EA}{L} & 0 & 0 \\
0 & \dfrac{12EI}{L^3} & \dfrac{6EI}{L^2} \\
0 & \dfrac{6EI}{L^2} & \dfrac{4EI}{L}
\end{bmatrix}
\begin{bmatrix} \delta_{x1} \\ \delta_{y1} \\ \theta_1 \end{bmatrix} +
\begin{bmatrix}
\dfrac{-EA}{L} & 0 & 0 \\
0 & \dfrac{-12EI}{L^3} & \dfrac{6EI}{L^2} \\
0 & \dfrac{-6EI}{L^2} & \dfrac{2EI}{L}
\end{bmatrix}
\begin{bmatrix} \delta_{x2} \\ \delta_{y2} \\ \theta_2 \end{bmatrix}
$$

and

$$
\begin{bmatrix} p_{x2} \\ p_{y2} \\ m_{p2} \end{bmatrix} =
\begin{bmatrix}
\dfrac{-EA}{L} & 0 & 0 \\
0 & \dfrac{-12EI}{L^3} & \dfrac{-6EI}{L^2} \\
0 & \dfrac{6EI}{I^2} & \dfrac{2EI}{L}
\end{bmatrix}
\begin{bmatrix} \delta_{x1} \\ \delta_{y1} \\ \theta_1 \end{bmatrix} +
\begin{bmatrix}
\dfrac{EA}{L} & 0 & 0 \\
0 & \dfrac{12EI}{L^3} & \dfrac{-6EI}{L^2} \\
0 & \dfrac{-6EI}{L^2} & \dfrac{4EI}{L}
\end{bmatrix}
\begin{bmatrix} \delta_{x2} \\ \delta_{y2} \\ \theta_2 \end{bmatrix}
$$

Equations (5) and (6) are termed the *stiffness equations* for the member. These two matrix equations each represent three linear simultaneous equations and the convenience of the notation used in the shorthand form given above is readily apparent. The stiffness equations so formed are related to local axes which have been specially orientated parallel and perpendicular to the axis of this individual member as described above. However in the case of a complete, rigid-jointed frame, the various members will be inclined at different angles to any reference direction, and it is therefore clearly necessary to choose an overall system of axes for any given plane frame.

Thus in Fig. 7.2 let $x'0'y'$ be the overall axes for a plane frame of which 1-2 is an individual member. (These overall axes can also be described as frame axes, reference axes, or global axes.) $x0y$ are the local axes (also described as member axes) and α is the angle which $0x$ makes with a line through 0 and parallel to $0'x'$.

Then if the force and displacement matrices at ends 1 and 2 of the member with respect to the overall axes system $x'0'y'$ are denoted by P_1, D_1, P_2 and D_2,

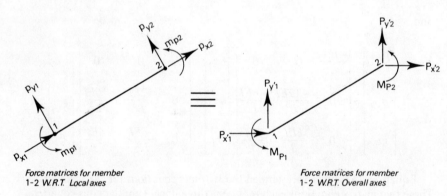

FIG. 7.2

where force matrices

$$P_1 = \begin{bmatrix} P_{x'1} \\ P_{y'1} \\ M_{F1} \end{bmatrix} \quad \text{and} \quad P_2 = \begin{bmatrix} P_{x'2} \\ P_{y'2} \\ M_{P2} \end{bmatrix}$$

and displacement matrices

$$D_1 = \begin{bmatrix} \Delta_{x'1} \\ \Delta_{y'1} \\ \theta_1 \end{bmatrix} \quad \text{and} \quad D_2 = \begin{bmatrix} \Delta_{x'2} \\ \Delta_{y'2} \\ \theta_2 \end{bmatrix}$$

then it can be shown that

$$P_{x'1} = p_{x1} \cdot \cos \alpha - p_{y1} \cdot \sin \alpha$$
$$P_{y'1} = p_{x1} \cdot \sin \alpha + p_{y1} \cdot \cos \alpha$$

and $\qquad M_{P1} = m_{p1}$, or in matrix form

$$\begin{bmatrix} P_{x'1} \\ P_{y'1} \\ M_{P1} \end{bmatrix} = \begin{bmatrix} \cos \alpha & -\sin \alpha & 0 \\ \sin \alpha & \cos \alpha & 0 \\ 0 & 0 & 1 \end{bmatrix} \begin{bmatrix} p_{x1} \\ p_{y1} \\ m_{p1} \end{bmatrix}$$

Therefore $P_1 = R \cdot p_1$ where

$$R = \begin{bmatrix} \cos \alpha & -\sin \alpha & 0 \\ \sin \alpha & \cos \alpha & 0 \\ 0 & 0 & 1 \end{bmatrix} = \text{rotation matrix}$$

It can also be similarly established that

$$P_2 = R \cdot p_2, \quad D_1 = R \cdot d_1 \quad \text{and} \quad D_2 = R \cdot d_2.$$

All these relationships are independent of any change of origin. The rotation matrix R thus enables a transformation of co-ordinate axes to be effected in a straightforward manner. In addition, R exhibits a property which can simplify its manipulation even further. This is given by the relationship $R^* = R^{-1}$, i.e. for the square matrix R, its transpose is equal to its inverse. Such a matrix is described as *orthogonal*.

Thus

$$R = \begin{bmatrix} \cos \alpha & -\sin \alpha & 0 \\ \sin \alpha & \cos \alpha & 0 \\ 0 & 0 & 1 \end{bmatrix}$$

Therefore

$$R^* = \begin{bmatrix} \cos \alpha & \sin \alpha & 0 \\ -\sin \alpha & \cos \alpha & 0 \\ 0 & 0 & 1 \end{bmatrix} = R^{-1}$$

Check:

$$RR^{-1} = \begin{bmatrix} \cos \alpha & -\sin \alpha & 0 \\ \sin \alpha & \cos \alpha & 0 \\ 0 & 0 & 1 \end{bmatrix} \begin{bmatrix} \cos \alpha & \sin \alpha & 0 \\ -\sin \alpha & \cos \alpha & 0 \\ 0 & 0 & 1 \end{bmatrix} = \begin{bmatrix} 1 & 0 & 0 \\ 0 & 1 & 0 \\ 0 & 0 & 1 \end{bmatrix} = I,$$

i.e. the unit matrix. This property therefore provides a neat method of obtaining the inverse of an orthogonal matrix.

It is now possible to obtain the stiffness equations with respect to the overall axes. From equations (5) and (6),

$$p_1 = K_{11} \cdot d_1 + K_{12} \cdot d_2 \qquad\text{——— (5)}$$

$$p_2 = K_{21} \cdot d_1 + K_{22} \cdot d_2 \qquad\text{——— (6)}$$

Also $P_1 = R \cdot p_1$ and $P_2 = R \cdot p_2$.

Therefore

$$P_1 = R \cdot p_1 = R \cdot K_{11} \cdot d_1 + R \cdot K_{12} \cdot d_2 \qquad\text{——— (7)}$$

and

$$P_2 = R \cdot p_2 = R \cdot K_{21} \cdot d_1 + R \cdot K_{22} \cdot d_2 \qquad\text{——— (8)}$$

Moreover $D_1 = R \cdot d_1$ and $D_2 = R \cdot d_2$.

Pre-multiply by R^{-1}. Therefore $R^{-1} \cdot D_1 = R^{-1} \cdot R \cdot d_1 = I \cdot d_1 = d_1$

and $R^{-1} \cdot D_2 = R^{-1} \cdot R \cdot d_2 = I \cdot d_2 = d_2$.

Hence, substituting in equations (7) and (8),

$$P_1 = R \cdot K_{11} \cdot R^{-1} \cdot D_1 + R \cdot K_{12} \cdot R^{-1} \cdot D_2$$

and

$$P_2 = R \cdot K_{21} \cdot R^{-1} \cdot D_1 + R \cdot K_{22} \cdot R^{-1} \cdot D_2$$

or

$$P_1 = Y_{11} \cdot D_1 + Y_{12} \cdot D_2 \qquad\text{——— (9)}$$

and

$$P_2 = Y_{21} \cdot D_1 + Y_{22} \cdot D_2 \qquad\text{——— (10)}$$

where

$$Y_{11} = R \cdot K_{11} \cdot R^{-1}, \qquad Y_{12} = R \cdot K_{12} \cdot R^{-1},$$
$$Y_{21} = R \cdot K_{21} \cdot R^{-1}, \quad\text{and}\quad Y_{22} = R \cdot K_{22} \cdot R^{-1}.$$

Thus equations (9) and (10) are the stiffness equations for the individual member 1-2 with respect to the overall axes; these altogether represent six linear equations.

The four Y matrices are all of order 3 x 3 and are the stiffness matrices for the member with respect to the overall axes (Note that the order in which the matrix multiplication is carried out is significant and consequently, for example, Y_{11} is not equal to K_{11}.)

It is now possible to consider the analysis of a complete frame. When following the method given below, it may be helpful to refer to the particular example shown in Fig. 7.3 (although the method applies to all rigid-jointed plane frames.)

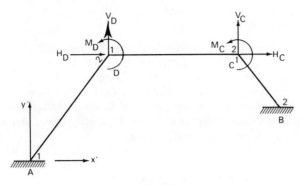

FIG. 7.3

Method

(In the following instructions the 'unit' is the matrix, so that 'equations' are matrix equations, 'forces' are force matrices, 'displacements' are displacement matrices, etc.)

1. Choose an overall axes system for the frame and specify ends 1 and 2 for all members.

2. Obtain the equilibrium equations for all joints in the frame by adding together all the forces acting at any joint, equating this summation to zero, and then repeating the process for all joints in the frame. (Care must be taken to distinguish between external actions and internal reactions when combining the appropriate force matrices in this way.)

3. Replace the internal forces in the equilibrium equations by the values given from the stiffness equations, and then insert all known values of external forces and displacements. In this way a set of equations is obtained in which the displacements of the joints are the unknowns.

4. Solve the appropriate equations in this set and so obtain the values of these displacements.

5. Insert these displacements in the various stiffness equations and so obtain the forces in the members and the reactions at the supports. (At this stage the forces in the members are, of course, all given with respect to the overall axes.)

6. Using the relationship involving the rotation matrices find the internal forces in the members with respect to the various individual member axes, so giving the axial and shearing forces, and the end moments for all members in the frame.

This completes the analysis of the frame. The following results have now been obtained.

(i) The axial and shearing forces and the end moments in all the members.

(ii) The reactions at the supports, i.e. two forces parallel to the overall axes and the fixing moment at each point of support.

(iii) The deflections of the joints, i.e. two linear components parallel to the overall axes and the rotation at each joint.

Example

Consider the elastic analysis of the stiff-jointed plane frame ABCD shown in Fig.
7.3. The supports at A and B are fully fixed. The applied loads at C and D
consist of horizontal and vertical forces and applied moments acting as shown. It
is required to obtain the results described in (i), (ii) and (iii) above.
Solution: Choose overall or frame axes $x'0'y'$ in which the origin $0'$ is at
joint A with the axes directed as shown, and then specify ends 1 and 2 for all
members. Then the applied force matrices are:

$$\text{At D, } F_D = \begin{bmatrix} H_D \\ V_D \\ M_D \end{bmatrix} \quad \text{and at C, } F_C = \begin{bmatrix} H_C \\ V_C \\ M_C \end{bmatrix}$$

F_C and F_D will now be referred to simply as 'forces'. (Note that the elements in
these matrices are positive when acting as shown.)
    Consider equilibrium at C and D. At D there are three forces, the external
applied force F_D, the internal force $(P_2)_{AD}$ at end 2 of member AD, and the
internal force $(P_1)_{CD}$ at end 1 of member CD. Hence the equilibrium equation
at D is:

$$F_D = (P_2)_{AD} + (P_1)_{CD} \qquad\qquad \text{------ (i)}$$

Similarly the equilibrium equation at C is:

$$F_C = (P_2)_{CD} + (P_1)_{BC} \qquad\qquad \text{------ (ii)}$$

    It is now necessary to use the appropriate stiffness equations in order to
substitute displacements for forces in equations (i) and (ii). Thus

$$(P_2)_{AD} = (Y_{21})_{AD} \cdot D_A + (Y_{22})_{AD} \cdot D_D$$
$$(P_1)_{CD} = (Y_{11})_{CD} \cdot D_D + (Y_{12})_{CD} \cdot D_C$$
$$(P_2)_{CD} = (Y_{21})_{CD} \cdot D_D + (Y_{22})_{CD} \cdot D_C$$
$$(P_1)_{BC} = (Y_{11})_{BC} \cdot D_C + (Y_{12})_{BC} \cdot D_B$$

Note that $(D_2)_{AD} = (D_1)_{CD} = D_D$ and $(D_2)_{CD} = (D_1)_{BC} = D_C$.
Also

$$(D_1)_{AD} = D_A = \begin{bmatrix} 0 \\ 0 \\ 0 \end{bmatrix} \text{ since joint A is fixed}$$

and

$$(D_2)_{BC} = D_B = \begin{bmatrix} 0 \\ 0 \\ 0 \end{bmatrix} \text{ since joint B is fixed.}$$

Now substitute all these values in equations (i) and (ii). Hence

$$F_D = (Y_{22})_{AD} \cdot D_D + (Y_{11})_{CD} \cdot D_D + (Y_{12})_{CD} \cdot D_C \qquad \text{------ (iii)}$$

and

$$F_C = (Y_{21})_{CD} \cdot D_D + (Y_{22})_{CD} \cdot D_C + (Y_{11})_{BC} \cdot D_C \quad\text{------ (iv)}$$

In equations (iii) and (iv) the only unknowns are D_C and D_D and their values can therefore be obtained. (F_C and F_D, the applied forces, are specified, and the Y matrices can all be evaluated using the values for E, L, A and I for the appropriate members.) Thus

$$\begin{bmatrix} F_D \\ F_C \end{bmatrix} = \begin{bmatrix} (Y_{22})_{AD} + (Y_{11})_{CD} & (Y_{12})_{CD} \\ (Y_{21})_{CD} & (Y_{22})_{CD} + (Y_{11})_{BC} \end{bmatrix} \begin{bmatrix} D_D \\ D_C \end{bmatrix}$$

and therefore

$$\begin{bmatrix} D_D \\ D_C \end{bmatrix} = \begin{bmatrix} (Y_{22})_{AD} + (Y_{11})_{CD} & (Y_{12})_{CD} \\ (Y_{21})_{CD} & (Y_{22})_{CD} + (Y_{11})_{BC} \end{bmatrix}^{-1} \begin{bmatrix} F_D \\ F_C \end{bmatrix}$$

$$6 \times 1 \qquad\qquad 6 \times 6 \qquad\qquad 6 \times 1$$

Note that the elements in this matrix equation are themselves matrices, and the values of D_D and D_C are obtained by first inverting a 6 × 6 matrix and then post-multiplying it by the 6 × 1 force matrix. (The inversion of a 6 × 6 matrix is reasonably straightforward if carried out by computer; if tackled manually the inversion could be avoided by alternative mathematical manipulation leading to the inversion of no more than two 3 × 3 matrices.) Then

$$\text{reaction at } A = (P_1)_{AD} = (Y_{12})_{AD} \cdot D_D$$

and

$$\text{reaction at } B = (P_2)_{BC} = (Y_{21})_{BC} \cdot D_C$$

As usual, these reactions are given with respect to the overall axes. To complete the analysis, the 'forces' in the three members will be obtained with respect to local axes for each member, so giving axial and shearing forces direct, as well as the various end moments. Thus:

Force at end 1 of member AD $= (p_1)_{AD} = R_{AD}^{-1} \cdot (P_1)_{AD} = R_{AD}^{-1} \cdot (Y_{12})_{AD} \cdot D_D$

Force at end 2 of member AD $= (p_2)_{AD} = R_{AD}^{-1} \cdot (Y_{22})_{AD} \cdot D_D$

Force at end 1 of member CD $= (p_1)_{CD}$
$$= R_{CD}^{-1} \cdot (Y_{11})_{CD} \cdot D_D + R_{CD}^{-1} \cdot (Y_{12})_{CD} \cdot D_C$$

Force at end 2 of member CD $= (p_2)_{CD}$
$$= R_{CD}^{-1} \cdot (Y_{21})_{CD} \cdot D_D + R_{CD}^{-1} \cdot (Y_{22})_{CD} \cdot D_C$$

Force at end 1 of member BC $= (p_1)_{BC} = R_{BC}^{-1} \cdot (Y_{11})_{BC} \cdot D_C$

Force at end 2 of member BC $= (p_2)_{BC} = R_{BC}^{-1} \cdot (Y_{21})_{BC} \cdot D_C$

A numerical illustration of this particular problem is shown in Fig. 7.4, and its solution by the Stiffness Method is given in Fig. 7.5. Simple numerical values have been chosen for all data in consistent but unspecified units to allow the manipulation of the various matrices to be followed with as little difficulty as possible.

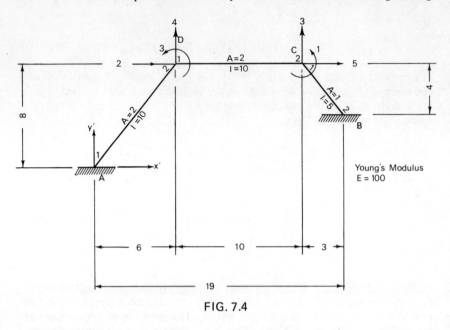

FIG. 7.4

As a check, another solution to this problem has been obtained using library Program A27 (see Chapter 4). This is also based on the Stiffness Method and the input to and output from the computer is shown in Fig. 7.6. The two different sets of results can be compared to check that they are in agreement.

FIG. 7.5.

Stiffness Method — Elastic Analysis of a Plane Frame —
Numerical Example

Referring to Fig. 7.4 and the algebraic solution already given above:—

Rotation Matrices R for the 3 Members AD, DC and CB:

$$R_{AD} = \frac{1}{5} \cdot \begin{bmatrix} 3 & -4 & 0 \\ 4 & 3 & 0 \\ 0 & 0 & 5 \end{bmatrix} \quad \text{and} \quad R_{AD}^{-1} = R_{AD}^{*} = \frac{1}{5} \cdot \begin{bmatrix} 3 & 4 & 0 \\ -4 & 3 & 0 \\ 0 & 0 & 5 \end{bmatrix}$$

$$R_{DC} = \begin{bmatrix} 1 & 0 & 0 \\ 0 & 1 & 0 \\ 0 & 0 & 1 \end{bmatrix} \quad \text{and} \quad R_{DC}^{-1} = \begin{bmatrix} 1 & 0 & 0 \\ 0 & 1 & 0 \\ 0 & 0 & 1 \end{bmatrix}$$

$$R_{BC} = \frac{1}{5} \cdot \begin{bmatrix} 3 & 4 & 0 \\ -4 & 3 & 0 \\ 0 & 0 & 5 \end{bmatrix} \quad \text{and} \quad R_{BC}^{-1} = R_{BC}^{*} = \frac{1}{5} \cdot \begin{bmatrix} 3 & -4 & 0 \\ 4 & 3 & 0 \\ 0 & 0 & 5 \end{bmatrix}$$

(the R matrices being orthogonal!)

FIG. 7.5. (*continued*)

Stiffness Matrices Y (i.e. W.R.T. Overall Axes):

$$(Y_{22})_{AD} = \tfrac{1}{25} \begin{bmatrix} 3 & -4 & 0 \\ 4 & 3 & 0 \\ 0 & 0 & 5 \end{bmatrix} \begin{bmatrix} 20 & 0 & 0 \\ 0 & 12 & -60 \\ 0 & -60 & 400 \end{bmatrix} \begin{bmatrix} 3 & 4 & 0 \\ -4 & 3 & 0 \\ 0 & 0 & 5 \end{bmatrix}$$

$$= \tfrac{1}{25} \begin{bmatrix} 60 & -48 & 240 \\ 80 & 36 & -180 \\ 0 & -300 & 2000 \end{bmatrix} \begin{bmatrix} 3 & 4 & 0 \\ -4 & 3 & 0 \\ 0 & 0 & 5 \end{bmatrix}$$

$$= \tfrac{1}{25} \begin{bmatrix} 372 & 96 & 1200 \\ 96 & 428 & -900 \\ 1200 & -900 & 10{,}000 \end{bmatrix} = \begin{bmatrix} 14.88 & 3.84 & 48 \\ 3.84 & 17.12 & -36 \\ 48 & -36 & 400 \end{bmatrix}$$

$$(Y_{11})_{CD} = (K_{11})_{CD} = \begin{bmatrix} 20 & 0 & 0 \\ 0 & 12 & 60 \\ 0 & 60 & 400 \end{bmatrix}$$

$$(Y_{12})_{CD} = (K_{12})_{CD} = \begin{bmatrix} -20 & 0 & 0 \\ 0 & -12 & 60 \\ 0 & -60 & 200 \end{bmatrix}$$

$$(Y_{21})_{CD} = (K_{21})_{CD} = \begin{bmatrix} -20 & 0 & 0 \\ 0 & -12 & -60 \\ 0 & 60 & 200 \end{bmatrix}$$

$$(Y_{22})_{CD} = (K_{22})_{CD} = \begin{bmatrix} 20 & 0 & 0 \\ 0 & 12 & -60 \\ 0 & -60 & 400 \end{bmatrix}$$

$$(Y_{11})_{BC} = R_{BC}(K_{11})_{BC}R_{BC}^{-1} = \tfrac{1}{25} \begin{bmatrix} 3 & 4 & 0 \\ -4 & 3 & 0 \\ 0 & 0 & 5 \end{bmatrix} \begin{bmatrix} 20 & 0 & 0 \\ 0 & 48 & 120 \\ 0 & 120 & 400 \end{bmatrix} \begin{bmatrix} 3 & -4 & 0 \\ 4 & 3 & 0 \\ 0 & 0 & 5 \end{bmatrix}$$

$$= \tfrac{1}{25} \cdot \begin{bmatrix} 60 & 192 & 480 \\ -80 & 144 & 360 \\ 0 & 600 & 2000 \end{bmatrix} \begin{bmatrix} 3 & -4 & 0 \\ 4 & 3 & 0 \\ 0 & 0 & 5 \end{bmatrix}$$

$$= \tfrac{1}{25} \begin{bmatrix} 948 & 336 & 2400 \\ 336 & 752 & 1800 \\ 2400 & 1800 & 10{,}000 \end{bmatrix}$$

$$= \begin{bmatrix} 37.92 & 13.44 & 96.00 \\ 13.44 & 30.08 & 72.00 \\ 96.00 & 72.00 & 400.00 \end{bmatrix}$$

FIG. 7.5. (*continued*)

$$\text{Hence } \begin{bmatrix} D_D \\ D_C \end{bmatrix} = \begin{bmatrix} 34.88 & 3.84 & 48 & -20 & 0 & 0 \\ 3.84 & 29.12 & 24 & 0 & -12 & 60 \\ 48 & 24 & 800 & 0 & -60 & 200 \\ -20 & 0 & 0 & 57.92 & 13.44 & 96 \\ 0 & -12 & -60 & 13.44 & 42.08 & 12 \\ 0 & 60 & 200 & 96 & 12 & 800 \end{bmatrix}^{-1} \begin{bmatrix} 2 \\ 4 \\ 3 \\ 5 \\ 3 \\ 1 \end{bmatrix}$$

The inverse of the above 6 x 6 matrix can now be obtained by computer using a library program such as the "matrix package" (see Chapter 5). Then:

$$\begin{bmatrix} D_D \\ D_C \end{bmatrix} = \begin{bmatrix} .04543383 & -.00666417 & -.00325060 & .02015564 & -.01271178 & -.00091554 \\ -.00666417 & .05052800 & .00130072 & .00120080 & .01716819 & -.00451640 \\ -.00325060 & .00130072 & .00177427 & -.00116106 & .00340075 & -.00045281 \\ .02015564 & .00120080 & -.00116106 & .03259543 & -.01071126 & -.00355058 \\ -.01271178 & .01716819 & .00340075 & -.01071126 & .03733298 & -.00141244 \\ -.00091554 & -.00451640 & -.00045281 & -.00355058 & -.00141244 & .00214919 \end{bmatrix} \begin{bmatrix} 2 \\ 4 \\ 3 \\ 5 \\ 3 \\ 1 \end{bmatrix}$$

$$= \begin{bmatrix} .11618649 \\ .24567802 \\ .00796862 \\ .16892411 \\ .11048163 \\ -.04109613 \end{bmatrix}$$

i.e.

$$D_D = \begin{bmatrix} .11618649 \\ .24567802 \\ .00796862 \end{bmatrix}$$

and

$$D_C = \begin{bmatrix} .16892411 \\ .11048163 \\ -.04109613 \end{bmatrix}$$

FIG. 7.5. (*continued*)

Thus:

Reaction at A = $\begin{bmatrix} -14.88 & -3.84 & -48 \\ -3.84 & -17.12 & 36 \\ 48 & -36 & 200 \end{bmatrix} \begin{bmatrix} .11618649 \\ .24567802 \\ .00796862 \end{bmatrix} = \begin{bmatrix} -3.0547523 \\ -4.3652935 \\ -1.6737332 \end{bmatrix}$

Reaction at B = $\begin{bmatrix} -37.92 & -13.44 & -96 \\ -13.44 & -30.08 & -72 \\ 96 & 72 & 200 \end{bmatrix} \begin{bmatrix} .16892411 \\ .11048163 \\ -.04109613 \end{bmatrix} = \begin{bmatrix} -3.9452469 \\ -2.6347061 \\ 15.952166 \end{bmatrix}$

Force at End = $\tfrac{1}{5}$
1 of AD $\begin{bmatrix} 3 & 4 & 0 \\ -4 & 3 & 0 \\ 0 & 0 & 5 \end{bmatrix} \begin{bmatrix} -14.88 & -3.84 & -48 \\ -3.84 & -17.12 & 36 \\ 48 & -36 & 200 \end{bmatrix} \begin{bmatrix} .11618649 \\ .24567802 \\ .00796862 \end{bmatrix}$

$= \begin{bmatrix} -5.3250862 \\ -.17537424 \\ -1.6737332 \end{bmatrix}$

Force at End = $\tfrac{1}{5}$
2 of AD $\begin{bmatrix} 3 & 4 & 0 \\ -4 & 3 & 0 \\ 0 & 0 & 5 \end{bmatrix} \begin{bmatrix} 14.88 & 3.84 & 48 \\ 3.84 & 17.12 & -36 \\ 48 & -36 & 400 \end{bmatrix} \begin{bmatrix} .11618649 \\ .24567802 \\ .00796862 \end{bmatrix}$

$= \begin{bmatrix} 5.3250862 \\ .17537424 \\ -.08000920 \end{bmatrix}$

Force at End =
1 of CD $\begin{bmatrix} 20 & 0 & 0 \\ 0 & 12 & 60 \\ 0 & 60 & 400 \end{bmatrix} D_D + \begin{bmatrix} -20 & 0 & 0 \\ 0 & -12 & 60 \\ 0 & -60 & 200 \end{bmatrix} D_C$

$= \begin{bmatrix} -1.0547524 \\ -.36529392 \\ 3.0800054 \end{bmatrix}$

Force at End =
2 of CD $\begin{bmatrix} -20 & 0 & 0 \\ 0 & -12 & -60 \\ 0 & 60 & 200 \end{bmatrix} D_D + \begin{bmatrix} 20 & 0 & 0 \\ 0 & 12 & -60 \\ 0 & -60 & 400 \end{bmatrix} D_C$

$= \begin{bmatrix} 1.0547524 \\ .36529392 \\ -6.7329399 \end{bmatrix}$

Force at End = $\tfrac{1}{5} \cdot$
1 of BC $\begin{bmatrix} 3 & -4 & 0 \\ 4 & 3 & 0 \\ 0 & 0 & 5 \end{bmatrix} \begin{bmatrix} 37.92 & 13.44 & 96 \\ 13.44 & 30.08 & 72 \\ 96 & 72 & 400 \end{bmatrix} D_C = \begin{bmatrix} .25938324 \\ 4.7370212 \\ 7.7329399 \end{bmatrix}$

FIG. 7.5. (*continued*)

and

Force at End $= \frac{1}{5} \cdot \begin{bmatrix} 3 & -4 & 0 \\ 4 & 3 & 0 \\ 0 & 0 & 5 \end{bmatrix} \begin{bmatrix} -37.92 & -13.44 & -96 \\ -13.44 & -30.08 & -72 \\ 96 & 72 & 200 \end{bmatrix} D_C$
2 of BC

$$= \begin{bmatrix} -.25938324 \\ -4.7370212 \\ 15.952166 \end{bmatrix}$$

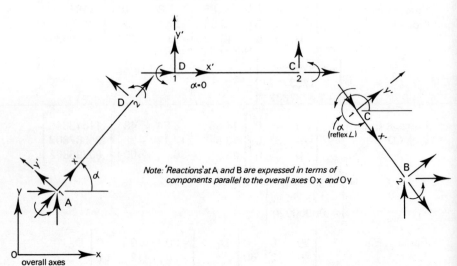

Note: 'Reactions' at A and B are expressed in terms of
components parallel to the overall axes O x and O y

FIG. 7.6

INPUT

/ EXAMPLE 1 CHAPTER 7 \

4	3	2	0	1	0	100	0	1
2								
0	0	6	8	16	8	19	4	
3								
1	2	0	2	10				
2	3	0	2	10				
3	4	0	1	5				
4								
1	1	1	1					
4	1	1	1					
6								

/ APPLIED FORCE MATRICES AT C AND D \

2			
2	2	-4	3
3	5	-3	1
7			
0			

FIG. 7.6. (*continued*)

OUTPUT

EXAMPLE 1   CHAPTER 7

YOUNG'S MODULUS =   100

MAX JOINT DIFFERENCE = 1

MEMBER PROPERTIES

NO.	END 1	END 2	LENGTH	SIN	COS	X-AREA	M OF $\mathbf{I}$
1	1	2	10.00	0.8000	0.6000	2.000_{10}^{+00}	1.000_{10}^{+01}
2	2	3	10.00	0.0000	1.0000	2.000_{10}^{+00}	1.000_{10}^{+01}
3	3	4	5.00	-0.8000	0.6000	1.000_{10}^{+00}	5.000_{10}^{+00}

JOINT LOADINGS   APPLIED FORCE MATRICES AT C AND D

JOINT NO.	X-FORCE	Y-FORCE	MOMENT
1	0.0000000_{10}^{+00}	0.0000000_{10}^{+00}	0.0000000_{10}^{+00}
2	2.0000000_{10}^{+00}	-4.0000000_{10}^{+00}	3.0000000_{10}^{+00}
3	5.0000000_{10}^{+00}	-3.0000000_{10}^{+00}	1.0000000_{10}^{+00}
4	0.0000000_{10}^{+00}	0.0000000_{10}^{+00}	0.0000000_{10}^{+00}

LOADCASE 1   APPLIED FORCE MATRICES AT C AND D

DEFORMATIONS

JOINT NO.	X-DIRECTION	Y-DIRECTION	ROTATION(DEGS.)
1	0.0000000	0.0000000	0.0000000
2	0.1161865	0.2456780	0.4565685
3	0.1689241	0.1104816	-2.3546346
4	0.0000000	0.0000000	0.0000000

FIG. 7.6. (*continued*)

LOADCASE 1   APPLIED FORCE MATRICES AT C AND D

FORCES AND MOMENTS ON MEMBERS

NO.	MEMBER END 1	MEMBER END 2	AXIAL FORCES END 1	AXIAL FORCES END 2	SHEAR FORCES END 1	SHEAR FORCES END 2	MOMENTS END 1	MOMENTS END 2
1	1	2	-5.33	-5.33	-0.18	0.18	-1.67	-0.08
2	2	3	-1.05	-1.05	-0.37	0.37	3.08	-6.73
3	3	4	0.26	0.26	4.74	-4.74	7.73	15.95

TOTAL HORIZONTAL FORCE = 7.00

TOTAL VERTICAL FORCE = -7.00

Miscellaneous notes on the Stiffness Method.

1.   Treatment of a pinned end (or ends):
If, in Fig. 7.1, end 2 is pinned, then in equation (6) $m_{p2} = 0$ and therefore

$$m_{p2} = 0 = \frac{6EI}{L^2} \cdot \delta_{y1} + \frac{2EI}{L} \cdot \theta_1 - \frac{6EI}{L^2} \cdot \delta_{y2} + \frac{4EI}{L} \cdot \theta_2$$

or

$$\theta_2 = \frac{-3}{2L} \cdot \delta_{y1} - \frac{1}{2} \cdot \theta_1 + \frac{3}{2L} \cdot \delta_{y2}.$$

The six linear equations in (5) and (6) can be written in full thus:

$$p_{x1} = \frac{EA}{L} \cdot \delta x_1 - \frac{EA}{L} \cdot \delta x_2$$

$$p_{y1} = \frac{12EI}{L^3} \cdot \delta y_1 + \frac{6EI}{L^2} \cdot \theta_1 - \frac{12EI}{L^3} \cdot \delta y_2 + \frac{6EI}{L^2} \cdot \theta_2$$

$$m_{p1} = \frac{6EI}{L^2} \cdot \delta y_1 + \frac{4EI}{L} \cdot \theta_1 - \frac{6EI}{L^2} \cdot \delta y_2 + \frac{2EI}{L} \cdot \theta_2$$

$$p_{x2} = \frac{-EA}{L} \cdot \delta x_1 + \frac{EA}{L} \cdot \delta x_2$$

$$p_{y2} = \frac{-12EI}{L^3} \cdot \delta y_1 - \frac{6EI}{L^2} \cdot \theta_1 + \frac{12EI}{L^3} \cdot \delta y_2 - \frac{6EI}{L^2} \cdot \theta_2$$

$$m_{p2} = \frac{6EI}{L^2} \cdot \delta y_1 + \frac{2EI}{L} \cdot \theta_1 - \frac{6EI}{L^2} \cdot \delta y_2 + \frac{4EI}{L} \cdot \theta_2$$

θ_2 can now be eliminated from these equations using the value given above. Then:

$$p_{x1} = \frac{EA}{L} \cdot \delta x_1 - \frac{EA}{L} \cdot \delta x_2$$

$$p_{y1} = \frac{3EI}{L^3} \cdot \delta y_1 + \frac{3EI}{L^2} \cdot \theta_1 - \frac{3EI}{L^3} \cdot \delta y_2$$

$$m_{p1} = \frac{3EI}{L^2} \cdot \delta y_1 + \frac{3EI}{L} \cdot \theta_1 - \frac{3EI}{L^2} \cdot \delta y_2$$

$$p_{x2} = \frac{-EA}{L} \cdot \delta x_1 + \frac{EA}{L} \cdot \delta x_2$$

$$p_{y2} = \frac{-3EI}{L^3} \cdot \delta y_1 - \frac{3EI}{L^2} \cdot \theta_1 + \frac{3EI}{L^3} \cdot \delta y_2$$

$$m_{p2} = 0$$

or in matrix form:

$$
\begin{bmatrix} p_{x1} \\ p_{y1} \\ m_{p1} \end{bmatrix}
=
\begin{bmatrix} \dfrac{EA}{L} & 0 & 0 \\ 0 & \dfrac{3EI}{L^3} & \dfrac{3EI}{L^2} \\ 0 & \dfrac{3EI}{L^2} & \dfrac{3EI}{L} \end{bmatrix}
\begin{bmatrix} \delta x_1 \\ \delta y_1 \\ \theta_1 \end{bmatrix}
+
\begin{bmatrix} \dfrac{-EA}{L} & 0 & 0 \\ 0 & \dfrac{-3EI}{L^3} & 0 \\ 0 & \dfrac{-3EI}{L^2} & 0 \end{bmatrix}
\begin{bmatrix} \delta x_2 \\ \delta y_2 \\ \theta_2 \end{bmatrix}
$$

and

$$
\begin{bmatrix} p_{x2} \\ p_{y2} \\ m_{p2} \end{bmatrix}
=
\begin{bmatrix} \dfrac{-EA}{L} & 0 & 0 \\ 0 & \dfrac{-3EI}{L^3} & \dfrac{-3EI}{L^2} \\ 0 & 0 & 0 \end{bmatrix}
\begin{bmatrix} \delta x_1 \\ \delta y_1 \\ \theta_1 \end{bmatrix}
+
\begin{bmatrix} \dfrac{EA}{L} & 0 & 0 \\ 0 & \dfrac{3EI}{L^3} & 0 \\ 0 & 0 & 0 \end{bmatrix}
\begin{bmatrix} \delta x_2 \\ \delta y_2 \\ \theta_2 \end{bmatrix}
$$

These two matrix equations can still be written in the shorthand forms

$$p_1 = K_{11} \cdot d_1 + K_{12} \cdot d_2$$

and

$$p_2 = K_{21} \cdot d_1 + K_{22} \cdot d_2$$

though the values of the four K stiffness matrices must now have the new values given immediately above. Thus in library programs such as A27, EAPF, etc., if any member is pinned at end 2, the stiffness matrices for that member should be evaluated in this way. Similar relationships could also be obtained for any member pinned at end 1 by inserting $m_{p1} = 0$ in equation (5). Finally there are various methods for dealing with a member which is pinned at both ends. The best solution is possibly to take I as zero for that member while maintaining the original values for the K matrices in equations (5) and (6).

2.   Stiffness matrices: Since the elements of the various K matrices for a given member are either zero or depend only on E, A, I and L, it is only necessary to first formulate EA/L and EI/L^3, say, readily to deduce the values of all the elements. This process is further simplified by the fact that the stiffness matrices are either symmetrical or can be obtained readily from one another.

Thus for any member, while there are four K matrices with 36 elements altogether, because of the forementioned properties, all this data can be readily obtained and stored in a concise, efficient way, requiring considerably less storage than that for the 36 real numbers which at first sight seem necessary.

3.   The method imposes no restrictions on the type of plane frame which can be analysed, provided that the individual members are of uniform section (though not necessarily the same section as each other). Thus problems such as side sway, sloping members, irregular features, occasional pinned joints, etc., can all be dealt with without undue difficulty.

4.   Sign convention: All forces, moments, displacements and rotations are positive when acting as in Fig. 7.1. Thus forces and displacements are positive when these are in the same sense as the axes, so that compression at end 1 and tension at end 2 of the member are positive. Moments and rotations are positive at both ends when anticlockwise. This is yet another sign convention for the engineer to learn, and it is important that he should be able to transfer the various values obtained from this method into tension or compression for forces, and into the 'beam' or 'moment-distribution' convention for moments as and when required.

Stiff-jointed space frames

The stiffness equations

$$p_1 = K_{11} \cdot d_1 + K_{12} \cdot d_2$$

and

$$p_2 = K_{21} \cdot d_1 + K_{22} \cdot d_2$$

and the method described above can be applied to stiff-jointed space frames, though obviously all the matrices will now be of a higher order. Thus the force matrices now have six elements, three forces parallel to three mutually perpendicular axes $0x, 0y$ and $0z$, and three moments acting in planes perpendicular to these axes, so that p_1 and p_2 now become:

$$p_1 = \begin{bmatrix} p_{x1} \\ p_{y1} \\ p_{z1} \\ m_{x1} \\ m_{y1} \\ m_{z1} \end{bmatrix} \quad \text{and} \quad p_2 = \begin{bmatrix} p_{x2} \\ p_{y2} \\ p_{z2} \\ m_{x2} \\ m_{y2} \\ m_{z2} \end{bmatrix}$$

Similarly, displacement matrices also now have six elements—three linear displacements parallel to the three axes and three rotations about the axes—so that d_1 and d_2 become

$$d_1 = \begin{bmatrix} \delta_{x1} \\ \delta_{y1} \\ \delta_{z1} \\ \theta_{x1} \\ \theta_{y1} \\ \theta_{z1} \end{bmatrix} \quad \text{and} \quad d_2 = \begin{bmatrix} \delta_{x2} \\ \delta_{y2} \\ \delta_{z2} \\ \theta_{x2} \\ \theta_{y2} \\ \theta_{z2} \end{bmatrix}$$

The stiffness matrices must now contain elements which cater for the bending of an individual member about two axes perpendicular to one another and parallel to the plane of the cross-section, and also for the torsion of the member about its own axis. In addition, these matrices must now all be of order 6 x 6 for compatibility in the matrix multiplications indicated in the stiffness equations. The stiffness equations therefore take the form shown on p. 227 (opposite) where I_y and I_z are the second moments of area of the section about the y and z axes, J is the torsional constant for the section about the x axis, and G is the Modulus of Rigidity for the material. The local axes x, y, z, constitute a right-hand orthogonal system with the origin 0 at end 1 and the x axis directed positively along the axis of the member from end 1 to end 2, as shown in Fig. 7.7. These two matrix stiffness equations now represent altogether twelve linear equations.

The transformation of the axes from local to overall axes $0x'$, $0y'$, $0z'$ is once more achieved by means of the rotation matrix R, which this time is of order 6 x 6.

If $0x$ has direction cosines $\quad \lambda_1, \quad \mu_1, \quad \nu_1 \quad$ with respect to $x'\, y'\, z'$,
and $0y$ has direction cosines $\quad \lambda_2, \quad \mu_2, \quad \nu_2 \quad$ with respect to $x'\, y'\, z'$,
and $0z$ has direction cosines $\quad \lambda_3, \quad \mu_3, \quad \nu_3 \quad$ with respect to $x'\, y'\, z'$,

respectively,

it can be shown that

$$R = \begin{bmatrix} \lambda_1 & \mu_1 & \nu_1 & 0 & 0 & 0 \\ \lambda_2 & \mu_2 & \nu_2 & 0 & 0 & 0 \\ \lambda_3 & \mu_3 & \nu_3 & 0 & 0 & 0 \\ 0 & 0 & 0 & \lambda_1 & \mu_1 & \nu_1 \\ 0 & 0 & 0 & \lambda_2 & \mu_2 & \nu_2 \\ 0 & 0 & 0 & \lambda_3 & \mu_3 & \nu_3 \end{bmatrix}$$

STIFFNESS EQUATIONS FOR A MEMBER IN THREE DIMENSIONS

$$
\begin{bmatrix}
-\dfrac{EA}{L} & 0 & 0 & 0 & 0 & 0 \\[4pt]
0 & -\dfrac{12EI_z}{L^3} & 0 & 0 & 0 & -\dfrac{6EI_z}{L^2} \\[4pt]
0 & 0 & -\dfrac{12EI_y}{L^3} & 0 & \dfrac{6EI_y}{L^2} & 0 \\[4pt]
0 & 0 & 0 & -\dfrac{GJ}{L} & 0 & 0 \\[4pt]
0 & 0 & -\dfrac{6EI_y}{L^2} & 0 & \dfrac{2EI_y}{L} & 0 \\[4pt]
0 & \dfrac{6EI_z}{L^2} & 0 & 0 & 0 & \dfrac{2EI_z}{L}
\end{bmatrix}
\begin{Bmatrix}\delta_{x2}\\\delta_{y2}\\\delta_{z2}\\\theta_{x2}\\\theta_{y2}\\\theta_{z2}\end{Bmatrix}
+
$$

$$
\begin{bmatrix}
\dfrac{EA}{L} & 0 & 0 & 0 & 0 & 0 \\[4pt]
0 & \dfrac{12EI_z}{L^3} & 0 & 0 & 0 & \dfrac{6EI_z}{L^2} \\[4pt]
0 & 0 & \dfrac{12EI_y}{L^3} & 0 & -\dfrac{6EI_y}{L^2} & 0 \\[4pt]
0 & 0 & 0 & \dfrac{GJ}{L} & 0 & 0 \\[4pt]
0 & 0 & -\dfrac{6EI_y}{L^2} & 0 & \dfrac{4EI_y}{L} & 0 \\[4pt]
0 & \dfrac{6EI_z}{L^2} & 0 & 0 & 0 & \dfrac{4EI_z}{L}
\end{bmatrix}
\begin{Bmatrix}\delta_{x1}\\\delta_{y1}\\\delta_{z1}\\\theta_{x1}\\\theta_{y1}\\\theta_{z1}\end{Bmatrix}
=
\begin{Bmatrix}p_{x1}\\p_{y1}\\p_{z1}\\m_{x1}\\m_{y1}\\m_{z1}\end{Bmatrix}
$$

and

$$
\begin{bmatrix}
\dfrac{EA}{L} & 0 & 0 & 0 & 0 & 0 \\[4pt]
0 & \dfrac{12EI_z}{L^3} & 0 & 0 & 0 & -\dfrac{6EI_z}{L^2} \\[4pt]
0 & 0 & \dfrac{12EI_y}{L^3} & 0 & \dfrac{6EI_y}{L^2} & 0 \\[4pt]
0 & 0 & 0 & \dfrac{GJ}{L} & 0 & 0 \\[4pt]
0 & 0 & \dfrac{6EI_y}{L^2} & 0 & \dfrac{4EI_y}{L} & 0 \\[4pt]
0 & -\dfrac{6EI_z}{L^2} & 0 & 0 & 0 & \dfrac{4EI_z}{L}
\end{bmatrix}
\begin{Bmatrix}\delta_{x2}\\\delta_{y2}\\\delta_{z2}\\\theta_{x2}\\\theta_{y2}\\\theta_{z2}\end{Bmatrix}
+
$$

$$
\begin{bmatrix}
-\dfrac{EA}{L} & 0 & 0 & 0 & 0 & 0 \\[4pt]
0 & -\dfrac{12EI_z}{L^3} & 0 & 0 & 0 & \dfrac{6EI_z}{L^2} \\[4pt]
0 & 0 & -\dfrac{12EI_y}{L^3} & 0 & -\dfrac{6EI_y}{L^2} & 0 \\[4pt]
0 & 0 & 0 & -\dfrac{GJ}{L} & 0 & 0 \\[4pt]
0 & 0 & \dfrac{6EI_y}{L^2} & 0 & \dfrac{2EI_y}{L} & 0 \\[4pt]
0 & -\dfrac{6EI_z}{L^2} & 0 & 0 & 0 & \dfrac{2EI_z}{L}
\end{bmatrix}
\begin{Bmatrix}\delta_{x1}\\\delta_{y1}\\\delta_{z1}\\\theta_{x1}\\\theta_{y1}\\\theta_{z1}\end{Bmatrix}
=
\begin{Bmatrix}p_{x2}\\p_{y2}\\p_{z2}\\m_{x2}\\m_{y2}\\m_{z2}\end{Bmatrix}
$$

SPACE FRAMES

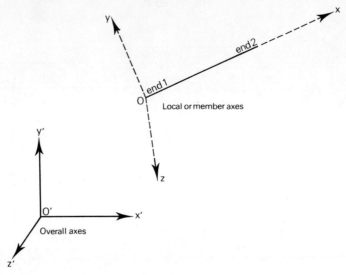

FIG. 7.7

Once more R is orthogonal, i.e. $R^* = R^{-1}$.
   Then, as before:

$$P_1 = R \cdot p_1, \ D_1 = R \cdot d_1, \ P_2 = R \cdot p_2, \ \text{and} \ D_2 = R \cdot d_2. \ \text{Also}$$

$$Y_{ij} = R \cdot K_{ij} \cdot R^{-1} \ \text{for} \ i = 1, 2, \ \text{and} \ j = 1, 2,$$

and the stiffness equations with respect to the overall axes x', y', z' are again

$$P_1 = Y_{11} \cdot D_1 + Y_{12} \cdot D_2$$
$$P_2 = Y_{21} \cdot D_1 + Y_{22} \cdot D_2,$$

although these now represent twelve linear equations. The matrix equilibrium equation for a given joint now contains six linear equations, compared with three in a rigid-jointed plane frame.

   The whole scale of the problem is now substantially increased. Space frames generally have more joints and members than plane frames, there are four sectional constants (I_y, I_z, J, and A) instead of two (I and A), two elastic moduli are required (E and G) instead of one (E), and six linear equilibrium equations now exist for each joint instead of three. In addition, the results required from the analysis are also more extensive—three forces and three moments must now be obtained at each end of a member instead of two forces and a single moment, and all joints now have six components of displacement instead of three. Consequently, bigger, faster, computers are required to tackle space-frame problems of even moderate size. At the time of writing, such computers are not widely available, and library programs for the elastic analysis of such frames even less so. In such circumstances it is therefore generally necessary to take any significant space-frame problem to a large, national, computer centre.

General case

The most general case to be considered in the elastic analysis of a frame—whether space or plane, rigid or pin-jointed, etc.,—is possibly a space frame with semi-rigid joints and non-uniform members, and it is apparent that the treatment of any frame then could be derived as a special case from this one general method of analysis. The engineer might therefore reason that, if a library computer program were to be written for this general category of frame, he could then analyse any particular frame using this single program. However, this is by no means a practical proposition for the average engineer at present, possibly because the more-popular existing computers are not big or fast enough to handle such a program. Nevertheless the situation could well soon change as improvements are achieved in computer hardware.

The analysis of the most general case by the Stiffness Method is not developed in this book—this is a topic for further study and accordingly a bibliography is given. However, what can still be considered are those special types of plane and space frames for which the analysis can be substantially simplified, so making it possible for library programs to be prepared for use with the more widely available, small-store, computers.

Special plane and space frames

1. Pin-jointed plane frames: The method given for rigid-jointed plane frames can be used by taking all second moments of area I as zero, and an example using this technique with **Program A27** is given in Problem 8 of Chapter 4. However, this process rather wastes computer storage space, and a saving can be made by writing a library program exclusively for pin-jointed frames if the engineer has a considerable number of such frames to analyse, each being too large to be dealt with by a rigid-jointed plane frame program such as A27.

In analysing such frames the various matrices can be simplified and their order reduced. The basic stiffness equations: $p_1 = K_{11} \cdot d_1 + K_{12} \cdot d_2$ and $p_2 = K_{21} \cdot d_1 + K_{22} \cdot d_2$, still apply, but here the force and displacement matrices need only be of order 2 x 1, so that

$$p_1 = \begin{bmatrix} p_{x1} \\ 0 \end{bmatrix}, p_2 = \begin{bmatrix} p_{x2} \\ 0 \end{bmatrix}, d_1 = \begin{bmatrix} \delta_{x1} \\ \delta y_1 \end{bmatrix}, \text{ and } d_2 = \begin{bmatrix} \delta x_2 \\ \delta y_2 \end{bmatrix}$$

with $p_{x1} = -p_{x2}$.

The stiffness matrices are now of order 2 x 2 and the stiffness equations, written in full, are

$$\begin{bmatrix} p_{x1} \\ 0 \end{bmatrix} = \begin{bmatrix} \dfrac{EA}{L} & 0 \\ 0 & 0 \end{bmatrix} \begin{bmatrix} \delta x_1 \\ \delta y_1 \end{bmatrix} + \begin{bmatrix} \dfrac{-EA}{L} & 0 \\ 0 & 0 \end{bmatrix} \begin{bmatrix} \delta x_2 \\ \delta y_2 \end{bmatrix}$$

and

$$\begin{bmatrix} p_{x2} \\ 0 \end{bmatrix} = \begin{bmatrix} \dfrac{-EA}{L} & 0 \\ 0 & 0 \end{bmatrix} \begin{bmatrix} \delta x_1 \\ \delta y_1 \end{bmatrix} + \begin{bmatrix} \dfrac{EA}{L} & 0 \\ 0 & 0 \end{bmatrix} \begin{bmatrix} \delta x_2 \\ \delta y_2 \end{bmatrix}$$

The rotation matrix R is now of order 2 x 2, where

$$R = \begin{bmatrix} \cos \alpha & -\sin \alpha \\ \sin \alpha & \cos \alpha \end{bmatrix} \text{ and is orthogonal as before.}$$

The stiffness equations with respect to the overall axes x', y' are the same as before, though the individual matrices are now of different order, with

$$Y_{11} = \frac{EA}{L} \cdot \begin{bmatrix} \cos^2 \alpha & \sin \alpha \cos \alpha \\ \sin \alpha \cos \alpha & \sin^2 \alpha \end{bmatrix} = Y_{22} = -Y_{12} = -Y_{21}.$$

The analysis can now continue as outlined for the rigid-jointed plane frame.
2.   *Grid frames*: This is a three-dimensional problem in which all members of the frame are co-planar and the force matrix for any joint consists of three elements—a force perpendicular to the plane of the frame, a moment and a torque. A member in a grid is shown in Fig. 7.8. The local axes, x, y, z for the member are a right-hand orthogonal set and are chosen so that the x axis lies along the axis of the member with the origin at end 1 and directed positively from end 1 to end 2, with the z axis in, and the y axis normal to, the plane of the grid. The force matrix at end 1 is then

$$p_1 = \begin{bmatrix} m_{x1} \\ p_{y1} \\ m_{z1} \end{bmatrix}$$

where m_{x1} is the torque,
    p_{y1} the shearing force, and
    m_{z1} the bending moment, at end 1, respectively.

The corresponding displacement matrix at end 1 is $d_1 = \begin{bmatrix} \theta_{x1} \\ \delta_{y1} \\ \theta_{z1} \end{bmatrix}$

where

θ_{x1} is the angle of twist of the section,

δ_{y1} the deflection of the joint perpendicular to the plane of the frame,

and θ_{z1} the rotation of the joint about the z axis, at end 1, respectively.

The three elements in both the force and displacement matrices are still given in

DETAILS OF GRID

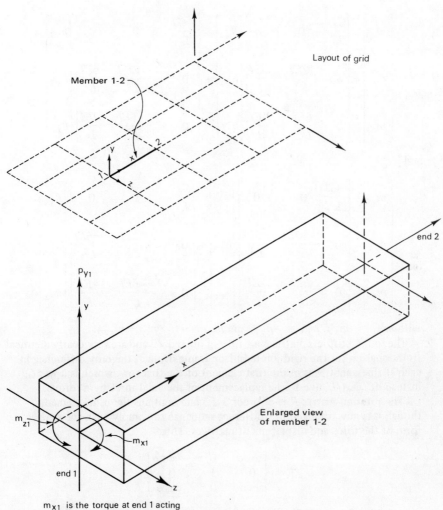

Layout of grid

Member 1-2

Enlarged view
of member 1-2

m_{x1} is the torque at end 1 acting
about the x axis, and m_{z1} is the moment
at end 1 acting about the z axis.

FIG. 7.8

the x, y, z sequence although these now denote different quantities from those given hitherto. Similar expressions also exist for the force and displacement matrices at end 2 (see below), and the sign conventions are in accordance with previous practice.

The basic stiffness equations: $p_1 = K_{11} \cdot d_1 + K_{12} \cdot d_2$ and $p_2 = K_{21} \cdot d_1 + K_{22} \cdot d_2$, still apply though now, when written in full, they become

$$
\begin{bmatrix} m_{x1} \\ p_{y1} \\ m_{z1} \end{bmatrix} = \begin{bmatrix} \dfrac{GJ}{L} & 0 & 0 \\ 0 & \dfrac{12EI}{L^3} & \dfrac{6EI}{L^2} \\ 0 & \dfrac{6EI}{L^2} & \dfrac{4EI}{L} \end{bmatrix} \begin{bmatrix} \theta_{x1} \\ \delta_{y1} \\ \theta_{z1} \end{bmatrix} + \begin{bmatrix} \dfrac{-GJ}{L} & 0 & 0 \\ 0 & \dfrac{-12EI}{L^3} & \dfrac{6EI}{L^2} \\ 0 & \dfrac{-6EI}{L^2} & \dfrac{2EI}{L} \end{bmatrix} \begin{bmatrix} \theta_{x2} \\ \delta_{y2} \\ \theta_{z2} \end{bmatrix}
$$

and

$$
\begin{bmatrix} m_{x2} \\ p_{y2} \\ m_{z2} \end{bmatrix} = \begin{bmatrix} \dfrac{-GJ}{L} & 0 & 0 \\ 0 & \dfrac{-12EI}{L^3} & \dfrac{-6EI}{L^2} \\ 0 & \dfrac{6EI}{L^2} & \dfrac{2EI}{L} \end{bmatrix} \begin{bmatrix} \theta_{x1} \\ \delta_{y1} \\ \theta_{z1} \end{bmatrix} + \begin{bmatrix} \dfrac{GJ}{L} & 0 & 0 \\ 0 & \dfrac{12EI}{L^3} & \dfrac{-6EI}{L^2} \\ 0 & \dfrac{-6EI}{L^2} & \dfrac{4EI}{L} \end{bmatrix} \begin{bmatrix} \theta_{x2} \\ \delta_{y2} \\ \theta_{z2} \end{bmatrix}
$$

with $m_{x1} = -m_{x2}$, $p_{y1} = -p_{y2}$, and $m_{z1} = -(L \cdot p_{y2} + m_{z2})$.

The four K stiffness matrices are now of order 3 x 3 and are very nearly identical to those given for the rigid-jointed plane frame. In fact, the only difference in each stiffness matrix is in the first element of the first row, which is now $\pm GJ/L$ instead of $\pm EA/L$, due to the replacement of the axial force by a torque.

The rotation matrix R is of order 3 x 3 and contains the usual elements, though it is now arranged in a different sequence to comply with the special form of the force and displacement matrices. Thus

$$
R = \begin{bmatrix} \cos \alpha & 0 & -\sin \alpha \\ 0 & 1 & 0 \\ \sin \alpha & 0 & \cos \alpha \end{bmatrix}
$$

and is still orthogonal. The relationship between the local and overall axes is given in Fig. 7.9, and it can be seen that $0'y'$ is parallel to $0y$, plane $z'0'x'$ is parallel to plane $z0x$, and the anticlockwise angle between $0'x'$ and a line in the plane $z'0'x'$ drawn through $0'$ and parallel to $0x$ is α.

As before, $P_1 = R \cdot p_1$, $D_1 = R \cdot d_1$,

$P_2 = R \cdot p_2$, and $D_2 = R \cdot d_2$.

CO-ORDINATE
AXES

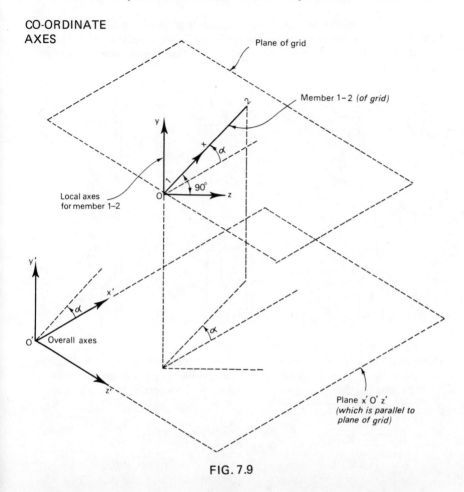

FIG. 7.9

Also $Y_{ij} = R \cdot K_{ij} \cdot R^{-1}$ for $i = 1, 2$ and $j = 1, 2$, the stiffness equations with respect to the overall axes x', y', z' are again: $P_1 = Y_{11} \cdot D_1 + Y_{12} \cdot D_2$ and $P_2 = Y_{21} \cdot D_1 + Y_{22} \cdot D_2$, and the analysis can proceed as before.

Library programs for analysing grid frames have been prepared for the more popular small-store computers, such as Programs LC8[3] (in Autocode) and A28[4] (in Algol) for the Elliott 803 and 4100 machines respectively. Some details of LC8 are given in Chapter 4.

3.   *Pin-jointed space frames*: Since there are no moments at the joints, the force matrix need only contain three elements (three forces perpendicular to one another). Moreover, the rotations of the joints have no significance in any 'normal' structural analysis, so that the displacement matrices can likewise be of order 3 x 1, each comprising only of three mutually perpendicular linear deformations. Finally, the members of such a frame are subjected solely to axial effects, and hence the various stiffness matrices need only contain elements involving A, E, and L. Bending and torsional constants (I, J, and G) are not therefore included.

It is now possible to proceed from the analysis given for rigid-jointed space frames using these simplifications. The stiffness equations are still: $p_1 = K_{11} \cdot d_1 + K_{12} \cdot d_2$ and $p_2 = K_{21} \cdot d_1 + K_{22} \cdot d_2$, although now, when written in full, take the form:

$$\begin{bmatrix} p_{x1} \\ p_{y1} \\ p_{z1} \end{bmatrix} = \begin{bmatrix} \dfrac{EA}{L} & 0 & 0 \\ 0 & 0 & 0 \\ 0 & 0 & 0 \end{bmatrix} \begin{bmatrix} \delta_{x1} \\ \delta_{y1} \\ \delta_{z1} \end{bmatrix} + \begin{bmatrix} \dfrac{-EA}{L} & 0 & 0 \\ 0 & 0 & 0 \\ 0 & 0 & 0 \end{bmatrix} \begin{bmatrix} \delta_{x2} \\ \delta_{y2} \\ \delta_{z2} \end{bmatrix}$$

and

$$\begin{bmatrix} p_{x2} \\ p_{y2} \\ p_{z2} \end{bmatrix} = \begin{bmatrix} \dfrac{-EA}{L} & 0 & 0 \\ 0 & 0 & 0 \\ 0 & 0 & 0 \end{bmatrix} \begin{bmatrix} \delta_{x1} \\ \delta_{y1} \\ \delta_{z1} \end{bmatrix} + \begin{bmatrix} \dfrac{EA}{L} & 0 & 0 \\ 0 & 0 & 0 \\ 0 & 0 & 0 \end{bmatrix} \begin{bmatrix} \delta_{x2} \\ \delta_{y2} \\ \delta_{z2} \end{bmatrix}$$

where $p_{y1} = p_{z1} = p_{y2} = p_{z2} = 0$.

(Note that the corresponding elements in the displacement matrices need not be zero since one end can be displaced and the member as a whole can be rotated.)

If $0x$ now has direction cosines λ_1, μ_1, ν_1, with respect to the overall axes x', y', z' respectively, then

$$P_{x'1} = \lambda_1 \cdot p_{x1}$$
$$P_{y'1} = \mu_1 \cdot p_{x1}$$

and

$$P_{z'1} = \nu_1 \cdot p_{x1}$$

or in matrix form

$$\begin{bmatrix} P_{x'1} \\ P_{y'1} \\ P_{z'1} \end{bmatrix} = \begin{bmatrix} \lambda_1 & 0 & 0 \\ \mu_1 & 0 & 0 \\ \nu_1 & 0 & 0 \end{bmatrix} \begin{bmatrix} p_{x1} \\ p_{y1} \\ p_{z1} \end{bmatrix}$$

i.e. $P_1 = R \cdot p_1$ where

$$R = \begin{bmatrix} \lambda_1 & 0 & 0 \\ \mu_1 & 0 & 0 \\ \nu_1 & 0 & 0 \end{bmatrix}$$

Similarly $P_2 = R \cdot p_2$.

However this time the rotation matrix R is singular (i.e. det. $R = 0$) and hence R^{-1} has no meaning and should therefore be avoided. This difficulty is overcome by making use of the relationship between force and displacement contained in the Principle of Contragredience, which states that, if $P = R \cdot p$ then $d = R^* \cdot D$, where R^* is the transpose of R. (Note that R need not be square by this definition.)

Hence the stiffness equations with respect to the overall axes can be obtained thus:

Pre-multiply both sides of equations (5) and (6) by R.

Therefore

$$R \cdot p_1 = R \cdot K_{11} \cdot d_1 + R \cdot K_{12} \cdot d_2$$

and

$$R \cdot p_2 = R \cdot K_{21} \cdot d_1 + R \cdot K_{22} \cdot d_2.$$

Now by the Principle of Contragedience, $d_1 = R^* \cdot D_1$ and $d_2 = R^* \cdot D_2$.

Hence

$$R \cdot p_1 = R \cdot K_{11} \cdot R^* \cdot D_1 + R \cdot K_{12} \cdot R^* \cdot D_2$$

and

$$R \cdot p_2 = R \cdot K_{21} \cdot R^* \cdot D_1 + R \cdot K_{22} \cdot R^* \cdot D_2$$

Thus

$$P_1 = Y_{11} \cdot D_1 + Y_{12} \cdot D_2$$

and

$$P_2 = Y_{21} \cdot D_1 + Y_{22} \cdot D_2,$$

as before, although on this occasion $Y_{ij} = R \cdot K_{ij} \cdot R^*$ for $i = 1, 2$ and $j = 1, 2$.

Now

$$R = \begin{bmatrix} \lambda_1 & 0 & 0 \\ \mu_1 & 0 & 0 \\ \nu_1 & 0 & 0 \end{bmatrix} \quad \text{and therefore} \quad R^* = \begin{bmatrix} \lambda_1 & \mu_1 & \nu_1 \\ 0 & 0 & 0 \\ 0 & 0 & 0 \end{bmatrix}$$

Thus

$$Y_{11} = \begin{bmatrix} \lambda_1 & 0 & 0 \\ \mu_1 & 0 & 0 \\ \nu_1 & 0 & 0 \end{bmatrix} \begin{bmatrix} \dfrac{EA}{L} & 0 & 0 \\ 0 & 0 & 0 \\ 0 & 0 & 0 \end{bmatrix} \begin{bmatrix} \lambda_1 & \mu_1 & \nu_1 \\ 0 & 0 & 0 \\ 0 & 0 & 0 \end{bmatrix}$$

$$= \frac{EA}{L} \begin{bmatrix} \lambda_1{}^2 & \lambda_1 \cdot \mu_1 & \lambda_1 \cdot \nu_1 \\ \lambda_1 \cdot \mu_1 & \mu_1{}^2 & \mu_1 \cdot \nu_1 \\ \lambda_1 \cdot \nu_1 & \mu_1 \cdot \nu_1 & \nu_1{}^2 \end{bmatrix}$$

and hence $Y_{11} = Y_{22} = -Y_{12} = -Y_{21}$.

The equilibrium equations can now be obtained and the analysis can proceed as before. Library programs could now be prepared for the more popular small-store computers (though the author knows of no such existing programs!).

The Flexibility Method

Outline

This is intrinsically a unit-load method of elastic analysis, and the basic theory is developed here with reference to plane structures. Most two-dimensional structural systems (i.e. structure + loading) which are indeterminate (i.e. have redundant supports or members, stiff joints, etc.), can, by the Principle of Superposition, be subdivided into a number of component systems, the sum of which is equal to the original system, and various examples of this procedure are shown in Fig. 7.10.

In each of these examples the given system is replaced by the sum of a statically-determinate one (including all the applied loading) and others involving the redundancies (either considered separately or together). The various structural quantities such as bending moment, shearing force, axial force, etc., at any point can then be obtained as required for each of the component systems in terms of the unknown redundancies (except for the statically-determinate system where the actual values can be quoted), and separate bending-moment, shearing-force and axial-force diagrams can then be drawn for the component systems when each redundancy (generally a force or a moment) is given a value of unity.

From these results, using Castigliano's First Theorem, a set of linear simultaneous equations can be obtained and solved by matrix methods to give the actual values of the various redundancies. These values can then be

THE FLEXIBILITY METHOD

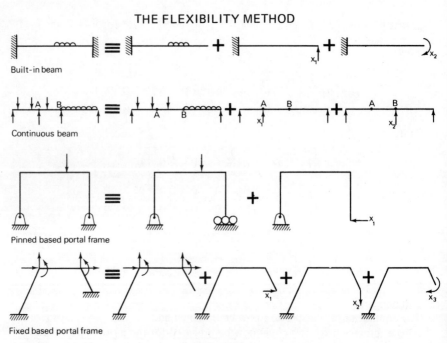

Built-in beam

Continuous beam

Pinned based portal frame

Fixed based portal frame

FIG. 7.10

substituted into the expressions already obtained for bending moments, shearing forces and axial forces to give the numerical values. The analysis is then completed by determining the deflections at specified locations using an extension of this basic theory.

Plane frames

Consider a two-dimensional system with n redundancies (i.e. forces or moments) from any of the illustrations given in Fig. 7.10. Let the n redundancies be $x_1, x_2, \ldots x_r, \ldots x_n$.

If, at any given section, M is the total bending moment in the given structural system, m_o the bending moment in the statically-determinate system, and m_r the bending moment in the component system involving redundancy x_r when $x_r = 1$, (i.e. m_r is the general term),

$$\text{then } M = m_o + m_1 x_1 + m_2 x_2 + \ldots + m_r x_r + \ldots + m_n x_n.$$

The various values of m_r viz., $m_1, m_2, m_3, \ldots m_n$, can readily be obtained from the component systems by equating the appropriate value of x_r to unity. There are thus n unknowns, $x_1, x_2, \ldots x_n$, and the total bending moment at any section can therefore be expressed in terms of these unknowns. Similarly, expressions for the shearing or axial force at any section can also be obtained in terms of these same unknowns (by drawing shearing-force and axial-force diagrams for values of x_r equal to unity, etc.)

Now, if U is the total strain energy in the given system, then

$$U = \int_s \frac{M^2 ds}{2EI} + k \int_s \frac{S^2 ds}{2GA} + \int_s \frac{P^2 ds}{2EA}$$

$$= U_{\text{bending}} + U_{\text{shear}} + U_{\text{axial}}$$

with the integrals being evaluated over the entire system. Also P is the axial force and S the shearing force at the same section as M, A is the cross sectional area, I the second moment of area of the section, E is Young's Modulus, G the Modulus of Rigidity, and k a factor depending on the shape of the cross section. If all the redundancies act as rigid supports, then, by Castigliano's First Theorem,

$$\frac{\partial U}{\partial x_1} = \frac{\partial U}{\partial x_2} = \cdots = \frac{\partial U}{\partial x_n} = 0,$$

and these n equations can now be solved for $x_1, x_2, \cdots x_r, \cdots x_n$.

Moreover, if the given system is such that only the strain energy due to bending is significant, then

$$U = \int_s \frac{M^2 ds}{2EI}, \quad \text{and} \quad U = \int_s \frac{1}{2EI} (m_0 + m_1 x_1 + m_2 x_2 + \ldots m_n x_n)^2 ds.$$

Therefore

$$\frac{\partial U}{\partial x_1} = \int_s \frac{m_1}{EI}(m_0 + m_1 x_1 + m_2 x_2 + \ldots + m_n x_n)ds = 0.$$

This equation can now be re-written as

$$x_1 \cdot \int_s \frac{m_1{}^2 ds}{EI} + x_2 \cdot \int_s \frac{m_1 m_2 ds}{EI} + \ldots + x_n \cdot \int_s \frac{m_1 m_n ds}{EI} + \int_s \frac{m_1 m_0 ds}{EI} = 0$$

or as $f_{11}x_1 + f_{12}x_2 + \ldots + f_{1n}x_n = -u_1$, where

$$f_{11} = \int_s \frac{m_1{}^2 ds}{EI}, f_{12} = \int_s \frac{m_1 m_2 ds}{EI}, \ldots \; f_{1n} = \int_s \frac{m_1 m_n ds}{EI},$$

and

$$u_1 = \int_s \frac{m_1 m_0 \cdot ds}{EI}.$$

This process can now be repeated for $\dfrac{\partial U}{\partial x_2} = \dfrac{\partial U}{\partial x_3} = \ldots = \dfrac{\partial U}{\partial x_n} = 0$, so obtaining

the complete set of n equations,

$$f_{11}x_1 + f_{12}x_2 + \ldots + f_{1r}x_r + \ldots + f_{1n}x_n = -u_1$$
$$f_{21}x_1 + f_{22}x_2 + \ldots + f_{2r}x_r + \ldots + f_{2n}x_n = -u_2$$
$$\cdots \quad \cdots \quad \cdots \quad \cdots \quad \cdots \quad \cdots \quad \cdots$$
$$f_{q1}x_1 + f_{q2}x_2 + \ldots + f_{qr}x_r + \ldots + f_{qn}x_n = -u_q$$
$$\cdots \quad \cdots \quad \cdots \quad \cdots \quad \cdots \quad \cdots \quad \cdots$$
$$f_{n1}x_1 + f_{n2}x_2 + \ldots + f_{nr}x_r + \ldots + f_{nn}x_n = -u_n$$

These linear simultaneous equations can be expressed in matrix form thus

$$\begin{bmatrix} f_{11} & f_{12} & f_{13} & \cdots & \cdots & \cdots & f_{1r} & \cdots & \cdots & \cdots & f_{1n} \\ f_{21} & f_{22} & f_{23} & \cdots & \cdots & \cdots & f_{2r} & \cdots & \cdots & \cdots & f_{2n} \\ \cdots & \cdots & \cdots & \cdots & \cdots & \cdots & \cdots & \cdots & \cdots & \cdots & \cdots \\ f_{q1} & f_{q2} & f_{q3} & \cdots & \cdots & \cdots & f_{qr} & \cdots & \cdots & \cdots & f_{qn} \\ \cdots & \cdots & \cdots & \cdots & \cdots & \cdots & \cdots & \cdots & \cdots & \cdots & \cdots \\ f_{n1} & f_{n2} & f_{n3} & \cdots & \cdots & \cdots & f_{nr} & \cdots & \cdots & \cdots & f_{nn} \end{bmatrix} \begin{bmatrix} x_1 \\ x_2 \\ \cdots \\ x_q \\ \cdots \\ x_n \end{bmatrix} = - \begin{bmatrix} u_1 \\ u_2 \\ \cdots \\ u_q \\ \cdots \\ u_n \end{bmatrix}$$

and hence

$$F \cdot x = -u, \qquad\qquad \text{------ (11)}$$

where

$$F = \begin{bmatrix} f_{11} & f_{12} & f_{13} & \cdots & \cdots & \cdots & f_{1r} & \cdots & \cdots & \cdots & f_{1n} \\ f_{21} & f_{22} & f_{23} & \cdots & \cdots & \cdots & f_{2r} & \cdots & \cdots & \cdots & f_{2n} \\ \cdots & \cdots & \cdots & \cdots & \cdots & \cdots & \cdots & \cdots & \cdots & \cdots & \cdots \\ f_{q1} & f_{q2} & f_{q3} & \cdots & \cdots & \cdots & f_{qr} & \cdots & \cdots & \cdots & f_{qn} \\ \cdots & \cdots & \cdots & \cdots & \cdots & \cdots & \cdots & \cdots & \cdots & \cdots & \cdots \\ f_{n1} & f_{n2} & f_{n3} & \cdots & \cdots & \cdots & f_{nr} & \cdots & \cdots & \cdots & f_{nn} \end{bmatrix}$$

$$x = \begin{bmatrix} x_1 \\ x_2 \\ \cdots \\ \cdots \\ \cdots \\ x_r \\ \cdots \\ \cdots \\ \cdots \\ x_n \end{bmatrix}, \quad \text{and} \quad u = \begin{bmatrix} u_1 \\ u_2 \\ \cdots \\ \cdots \\ \cdots \\ u_r \\ \cdots \\ \cdots \\ \cdots \\ u_n \end{bmatrix}$$

Matrix F is termed the *flexibility matrix.* The elements of this matrix, i.e. f_{11}, f_{12}, f_{13}, etc., are termed flexibility influence coefficients, and it is shown later that each coefficient has a physical interpretation as a deflection per unit force, or rotation per unit moment, etc.

The general element in the flexibility matrix is f_{qr}, where

$$f_{qr} = \int_s \frac{m_q m_r}{EI} \, ds, \qquad\qquad \text{------ (12)}$$

and that in the matrix u is u_r, where

$$u_r = \int_s \frac{m_r m_o}{EI} \, ds. \qquad\qquad \text{------ (13)}$$

From equation (11), $x = -F^{-1} u$. This gives the values of the n redundancies and so leads to the values of M, S, and P.

The engineer should now appreciate that the formation of the flexibility matrix F and the matrix u for the given structural system is an important preliminary process or subroutine in the Flexibility Method which merits due attention on its own account.

Evaluation of the integrals

The definite integrals in equations (12) and (13) should be evaluated for all values of q and r in the particular system, so providing the numerical elements for both matrices. This can be accomplished precisely by direct integration where possible, or approximately by interpreting each integral as an area under a curve (which can then be determined by Simpson's Rule, etc.). Each integral involves the product of two bending moments and, since the shapes of the more-common bending-moment diagrams are well known, it is possible to draw up a table giving some of the algebraic values of $\int_s m_q m_r ds$. Such a table is shown in Fig. 7.11. Certain values of the integral $\int_s m_r m_o ds$ can also be obtained from this source. In those cases where the strain energy due to shearing and/or to axial force (i.e. U_{shear} and U_{axial} as shown above) is significant and should be included, it is then also necessary to draw shearing-force and axial-force diagrams for the various members in the system and hence to obtain values for the

THE FLEXIBILITY METHOD

TABLE OF PRODUCT INTEGRALS $\int_s m_q m_r ds$ AND $\int_s p_q p_r ds$

m_q AND m_r, p_q AND p_r	⬜ L ⟍a	◺ ⟟—L—⟟ a	◿ ⟟—L—⟟ a	△ a, L	◹ a⟍b, L	PARABOLIC a, L
⬜ L, c	Lac	$\frac{1}{2}Lac$	$\frac{1}{2}Lac$	$\frac{1}{2}Lac$	$\frac{1}{2}L(a+b)c$	$\frac{2}{3}Lac$
◺ ⟟—L—⟟ c	$\frac{1}{2}Lac$	$\frac{1}{3}Lac$	$\frac{1}{6}Lac$	$\frac{1}{4}Lac$	$\frac{1}{6}L(2a+b)c$	$\frac{1}{3}Lac$
◿ ⟟—L—⟟ c	$\frac{1}{2}Lac$	$\frac{1}{6}Lac$	$\frac{1}{3}Lac$	$\frac{1}{4}Lac$	$\frac{1}{6}L(a+2b)c$	$\frac{1}{3}Lac$
△ c, L	$\frac{1}{2}Lac$	$\frac{1}{4}Lac$	$\frac{1}{4}Lac$	$\frac{1}{3}Lac$	$\frac{1}{4}L(a+b)c$	$\frac{5}{12}Lac$
◹ c⟍d, L	$\frac{1}{2}La(c+d)$	$\frac{1}{6}La(2c+d)$	$\frac{1}{6}La(c+2d)$	$\frac{1}{4}La(c+d)$	$\frac{1}{6}L[a(2c+d)+b(c+2d)]$	$\frac{1}{3}La(c+d)$
PARABOLIC c, L	$\frac{2}{3}Lac$	$\frac{1}{3}Lac$	$\frac{1}{3}Lac$	$\frac{5}{12}Lac$	$\frac{1}{3}L(a+b)c$	$\frac{8}{15}Lac$

FIG. 7.11

corresponding integrals

$$\int_s s_q s_r ds \text{ and } \int_s s_r s_o ds \text{ for shearing force, and}$$

$$\int_s p_q p_r ds \text{ and } \int_s p_r p_o ds \text{ for axial force,}$$

where s_q, s_r, s_o and p_q, p_r, p_o are the appropriate unit shearing-force and axial-force diagrams defined similarly to the bending-moment diagrams, so that

$$S = s_o + s_1 x_1 + s_2 x_2 + \ldots s_r x_r + \ldots s_n x_n,$$

and

$$P = p_o + p_1 x_1 + p_2 x_2 + \ldots p_r x_r + \ldots p_n x_n.$$

Again, it should be realised that the values given for the integrals can now also be used to determine appropriate values of shearing or axial force when the shapes of the respective diagrams are comparable.

The sign conventions when using the Flexibility Method are illustrated in Fig. 7.12. The conventions for bending moment and shearing force are essentially 'beam' conventions, although it is now important to indicate the positive direction of the length co-ordinate s for each member and to establish explicit arrangements for vertical or sloping members. (Note that a positive value of s for each straight member coincides with the member axis, and this is also normally taken to be the local x axis for the member.) For axial force, tension is positive.

Some other writers (see Bibliography) use different sign conventions when dealing with the Flexibility Method, though the engineer should be well prepared to cope with this difficulty! Such other conventions place the theory on a more strict mathematical basis so that it can be more readily used to compile a computer program to effect the elastic analysis of any frame. Since no programs are given here, such complications can be avoided at this stage, and the reader is therefore free to make a basic understanding of the method using a sign convention with which he is already familiar.

Physical interpretation of flexibility influence coefficients

If the redundancies do not now act at rigid supports, then by Castigliano's First Theorem, in the case of x_1

$$\frac{\partial U}{\partial x_1} = \delta_1,$$

where δ_1 is the total displacement of the point of application of x_1 in the direction of x_1. (Displacement δ_1 can be either a deflection or a rotation depending on whether redundancy x_1 is a force or a moment.) Thus

$$\delta_1 = \frac{\partial U}{\partial x_1} = x_1 \cdot \int_s \frac{m_1^2}{EI} ds + x_2 \cdot \int_s \frac{m_1 m_2 ds}{EI} + \ldots x_n \cdot \int_s \frac{m_1 m_n ds}{EI} + \int_s \frac{m_1 m_o ds}{EI},$$

FLEXIBILITY METHOD
SIGN CONVENTIONS

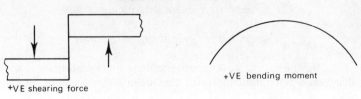

FIG. 7.12

or

$$\delta_1 = x_1 f_{11} + x_2 f_{12} + \ldots + x_n f_{1n} + u_1. \hspace{2cm} (14)$$

This equation is valid for all values of $x_1, x_2, \ldots x_n$, and u_1.

Now if all the redundancies are removed so that $x_1 = x_2 \ldots = x_n = 0$, then from equation (14)

$$\delta_1 = u_1,$$

that is, the displacement of the point of application of x_1 in the direction of x_1, due solely to the applied loading is u_1. Next, if the applied loading and all the redundancies except x_1 are removed, so that $x_2 = x_3 = \ldots = x_n = u_1 = 0$,

$$\text{then } \delta_1 = x_1 f_{11} \text{ or } f_{11} = \delta_1/x_1.$$

Thus if $x_1 = 1$, then $f_{11} = \delta_1$, so that the displacement of the point of application of x_1 in the direction of x_1 due solely to unit value of the redundancy x_1 is f_{11}. This process can be repeated with corresponding deductions for all influence coefficients in equation (14). For the general case it can similarly be shown that

$$\delta_q = f_{q1} x_1 + f_{q2} x_2 + \ldots + f_{qr} x_r + \ldots + f_{qn} x_n + u_q \hspace{1.5cm} (15)$$

Thus f_{qr} is the displacement of the point of application of x_q in the direction of x_q, due solely to unit value of the redundancy x_r, and u_q is the displacement of the point of application in the direction of x_q, due solely to the applied loading, with the three necessary reactions being provided in all cases at the same locations and in the same direction, no matter what the values of q and r are. Finally, f_{qr} and u_q are displacements, which are either deflections or rotations according to whether x_q is a force or a moment.

Determination of deflection at any point in the system

The deflection in a specified direction at any point X in the structural system can be obtained using Castigliano's first theorem. If a force P is applied at X, then the deflection δ in the direction P is given by $\delta = \partial U/\partial P$, where U is the total strain energy in the system.

If 'bending' energy only need be considered, then

$$U = U_{\text{bending}} = \int_s \frac{1}{2EI} (m_o + m_1 x_1 + \ldots + m_n x_n + m'P)^2 \, ds$$

where m' is the bending moment at the same section in the system as m_o and the various values of m_r due solely to unit value of P. Thus

$$\delta = \frac{\partial U}{\partial P} = \int_s \frac{m'}{EI} (m_o + m_1 x_1 + \ldots + m_n x_n + m'P) \, ds$$

If P is now made equal to zero, then

$$\delta = \int_s \frac{m'}{EI} (m_o + m_1 x_1 + \ldots + m_n x_n) \, ds.$$

But $M = m_o + m_1 x_1 + \ldots + m_n x_n$, and therefore

$$\delta = \int_s \frac{m'M}{EI} \, ds \qquad \qquad \text{———(16)}$$

Equation (16) can be used to obtain the deflection in a specified direction at any point in the system. The evaluation of the integral depends on information already obtained together with moment m' due to a unit load acting at the point and in the direction of the required deflection.

A similar method can also be evolved for rotation, and the process modified to allow for 'shear' and 'axial' strain energy if required. (Details can be obtained from the references in the Bibliography.)

Numerical example

The problem depicted in Fig. 7.4, already evaluated by the Stiffness Method, is now analysed using the Flexibility Method. The solution is shown in Fig. 7.13 and it can be immediately verified that, on taking account of the differing sign conventions, complete agreement is obtained with the solution already obtained in Fig. 7.5 by the Stiffness Method.

FIG. 7.13. NUMERICAL EXAMPLE

Note: Throughout this calculation, extensive use will be made of the formulae given in Fig. 7.11 (q.v.).

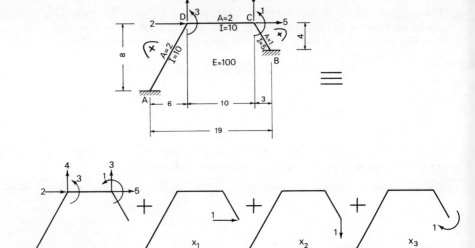

There are 3 released at B — x_1, x_2 and x_3. (See also Fig. 7.10.)

Bending Moments:

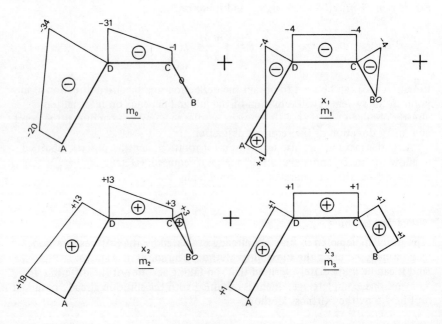

FIG. 7.13. (*continued*)

$EI_{AD} = EI_{DC} = 1000; EI_{BC} = 500.$

$$f_{11}x_1 + f_{12}x_2 + f_{13}x_3 = -u_1 \qquad \text{------ (1)}$$
$$f_{21}x_1 + f_{22}x_2 + f_{23}x_3 = -u_2 \qquad \text{------ (2)}$$
$$f_{31}x_1 + f_{32}x_2 + f_{33}x_3 = -u_3 \qquad \text{------ (3)}$$

Both bending moments *and* axial forces will be taken into account when computing the values of the f and u terms above.

Direct Forces:

AXIAL FORCES
IN MEMBERS →

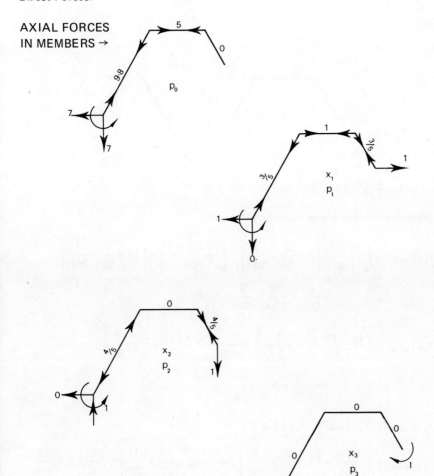

FIG. 7.13. (*continued*)

AXIAL FORCE
DIAGRAMS
(TENSION + VE)

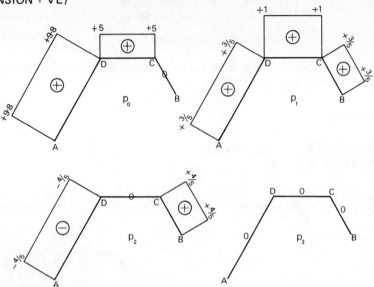

For AD and DC, EA = 200; for CB, EA = 100

f elements — components from Axial Forces:

$$[f_{11}]_{axial} = \int_s \frac{p_1^2\,ds}{EA} = \left[5 \times \frac{9}{25} \times \frac{1}{100}\right] + \left[\frac{10 \times 1 \times 1}{200}\right] + \left[10 \times \frac{9}{25} \times \frac{1}{200}\right]$$

$$= \frac{9}{500} + \frac{10}{200} + \frac{9}{500} = 0.086$$

$$[f_{12}]_{axial} = \int_s \frac{p_1 p_2\,ds}{EA} = \left[5 \times \frac{3}{5} \times \frac{4}{5} \times \frac{1}{100}\right] + 0 + \left[10 \times \frac{3}{5} \times \frac{-4}{5} \times \frac{1}{200}\right]$$

$$= \frac{12}{500} + 0 - \frac{12}{500} = 0$$

$$[f_{13}]_{axial} = [f_{21}]_{axial} = [f_{23}]_{axial} = 0$$

$$[f_{22}]_{axial} = \int_s \frac{p_2^2\,ds}{EA} = \left[5 \times \frac{16}{25} \times \frac{1}{100}\right] + 0 + \left[10 \times \frac{16}{25} \times \frac{1}{200}\right] = \frac{16}{500} + 0 + \frac{16}{500}$$

$$= 0.064$$

$$[f_{31}]_{axial} = [f_{32}]_{axial} = [f_{33}]_{axial} = 0$$

FIG. 7.13 (*continued*)

f elements — components from Bending Moments.

$$[f_{11}]_{B.M.} = \int_s \frac{m_1^2 ds}{EI} = \left[\frac{5}{500} \times \frac{1}{3} \times -4^2\right] + \left[\frac{10}{1000} \times -4^2\right]$$

$$+ \left[\frac{10}{3000} \{16 - 16 + 16\}\right] = +\frac{16}{300} + \frac{16}{100} + \frac{16}{300}$$

$$= 0.26667$$

$$[f_{12}]_{B.M.} = \int_s \frac{m_1 m_2 ds}{EI} = \left[\frac{5}{500} \times \frac{1}{3} \times -12\right] + \left[\frac{-40}{2000}(13 + 3)\right]$$

$$+ \left[\frac{10}{6000} \{4(38 + 13) - 4(19 + 26)\}\right]$$

$$= -\frac{4}{100} - \frac{32}{100} + \frac{4}{100} = -0.32$$

$$[f_{13}]_{B.M.} = \int_s \frac{m_1 m_3 ds}{EI} = \left[\frac{5 \times -4 \times 1}{1000}\right] + \left[\frac{10 \times 1 \times -4}{1000}\right] + \left[\frac{10}{2000}\{-4 + 4\}1\right]$$

$$= -\frac{2}{100} - \frac{4}{100} + 0 = -0.06$$

$$[f_{21}]_{B.M.} = \int_s \frac{m_2 m_1 ds}{EI} = [f_{12}]_{B.M.} = -0.32$$

$$[f_{22}]_{B.M.} = \int_s \frac{m_2^2 ds}{EI} = \left[\frac{5 \times 9}{3 \times 500}\right] + \left[\frac{10}{3000}(9 + 39 + 169)\right]$$

$$+ \left[\frac{10}{3000}(169 + 247 + 361)\right] = \frac{3}{100} + \frac{217}{300} + \frac{777}{300} = 3.34333$$

$$[f_{23}]_{B.M.} = \int_s \frac{m_2 m_3 ds}{EI} = \left[\frac{5 \times 3 \times 1}{1000}\right] + \left[\frac{10}{2000}(13 + 3)1\right] + \left[\frac{10}{2000}(19 + 13)\ 1\right]$$

$$= \frac{3}{200} + \frac{8}{100} + \frac{16}{100} = 0.255$$

$$[f_{31}]_{B.M.} = \int_s \frac{m_3 m_1 ds}{EI} = [f_{13}]_{B.M.} = -0.06$$

$$[f_{32}]_{B.M.} = \int_s \frac{m_3 m_2 ds}{EI} = [f_{23}]_{B.M.} = 0.255$$

$$[f_{33}]_{B.M.} = \int_s \frac{m_3^2 ds}{EI} = \left[\frac{5 \times 1 \times 1}{500}\right] + \left[\frac{10 \times 1 \times 1}{1000}\right] + \left[\frac{10 \times 1 \times 1}{1000}\right] = 0.03$$

FIG. 7.13. (*continued*)

u elements – components from Bending Moments

$$[u_1]_{B.M.} = \int_s \frac{m_1 m_0 \, ds}{EI} = 0 + \left[\frac{10 \times -4}{2000}(-32)\right]$$

$$+ \left[\frac{10}{6000}\{4(-40-34) - 4(-20-68)\}\right]$$

$$= 0 + \frac{64}{100} + \frac{56}{600} = 0.73333$$

$$[u_2]_{B.M.} = \int_s \frac{m_2 m_0 \, ds}{EI} = 0 + \left[\frac{10}{6000}\{13(-62-1) + 3(-31-2)\}\right]$$

$$= 0 - \frac{918}{600} - \frac{2550}{600} = -5.78$$

$$[u_3]_{B.M.} = \int_s \frac{m_3 m_0 \, ds}{EI} = 0 + \left[\frac{10 \times 1}{2000}(-32)\right] + \left[\frac{10 \times 1}{2000}(-54)\right] = 0 - \frac{16}{100} - \frac{27}{100}$$

$$= -0.43$$

u elements – components from Axial Forces:

$$[u_1]_{axial} = \int_s \frac{p_1 p_0 \, ds}{EA} = 0 + \left[\frac{10 \times 1 \times 5}{200}\right] + \left[10 \times 9.8 \times \frac{3}{5} \times \frac{1}{200}\right]$$

$$= 0 + \frac{25}{100} + \frac{29.4}{100} = 0.544$$

$$[u_2]_{axial} = \int_s \frac{p_2 p_0 \, ds}{EA} = 0 + 0 + \left[10 \times -\frac{4}{5} \times 9.8 \times \frac{1}{200}\right] = -0.392$$

$$[u_3]_{axial} = \int_s \frac{p_3 p_0 \, ds}{EA} = 0$$

The total values of the f and u elements can now be obtained by addition. Thus:

$$f_{11} = [f_{11}]_{B.M.} + [f_{11}]_{axial} = 0.26667 + 0.086 = +0.35267$$

$$f_{12} = \quad \text{etc.} \qquad = -0.32 + 0 \qquad = -0.32$$

$$f_{13} = \quad \text{etc.} \qquad = -0.06 + 0 \qquad = -0.06$$

FIG. 7.13. (*continued*)

$f_{21} =$ etc. $= -0.32 + 0$ $= -0.32$

$f_{22} =$ etc. $= 3.34333 + 0.064 = +3.40733$

$f_{23} =$ etc. $= 0.255 + 0$ $= +0.355$

$f_{31} =$ f_{13} $= -0.06$

$f_{32} =$ f_{23} $= +0.255$

$f_{33} = [f_{33}]_{B.M.} + [f_{33}]_{axial} = 0.03 + 0$ $= +0.03$

And:

$u_1 = [u_1]_{B.M.} + [u_1]_{axial} = 0.73333 + 0.544 = +1.27733$

$u_2 =$ etc. $= -5.78 - 0.392 = -6.172$

$u_3 =$ etc. $= -0.43 + 0$ $= -0.43$

Hence equations (1), (2) and (3) become:

$$35.267x_1 - 32x_2 - 6x_3 = -127.733 \qquad \text{------ (1)}$$
$$-32x_1 + 340.733x_2 + 25.5x_3 = +617.2 \qquad \text{------ (2)}$$
$$-6x_1 + 25.5x_2 + 3x_3 = +43 \qquad \text{------ (3)}$$

solving equations (1), (2) and (3) for x_1, x_2 and x_3 gives:

$$x_1 = -3.95; \quad x_2 = 2.635 \quad \text{and} \quad x_3 = -15.9520$$

(From the computer results given in Fig. 7.6 using Program No. A27 the following values of x_1, x_2 and x_3 can be obtained:

$$x_1 = -3.9451731; \quad x_2 = 2.6347023 \quad \text{and} \quad x_3 = -15.951982,$$

all of which are in good agreement with the values given above.)

Now:—

Bending Moment at C in member CB $= -4x_1 + 3x_2 + x_3$ $= 7.7328$

Bending Moment at C in member DC $= 6.7328$

Bending Moment at D in member DC $= -31 - 4x_1 + 13x_2 + x_3 = 3.0798$

Bending Moment at D in member AD $= 0.0798$

Fixing Moment at A $= -20 + 4x_1 + 19x_2 + x_3 = -1.6733$

Fixing Moment at B $= +x_3$ $= -15.9667$

FIG. 7.13. *(continued)*

Axial force in CB $= \frac{3}{5}x_1 + \frac{4}{5}x_2$ $= -0.25934$

Axial force in CD $= 5 + x_1$ $= 1.0548269$

Axial force in AD $= 9.8 + 0.6x_1 - 0.8x_2$ $= 5.32513$

Shearing force in CB $= 0.8x_1 - 0.6x_2$ $= -4.73696$

Shearing force in CD $= 3 - x_2$ $= 0.3652977$

Shearing force in AD $= -1.4 - \frac{4}{5}x_1 - \frac{3}{5}x_2 = 0.17531710$

These values can now be rounded off to realistic practical results and then compared with the computer results in Fig. 7.6, where, once more, it can be seen that they are in good agreement.

Deformations:

Vertical Deflection at D. δ_{D_v}

$$\delta_{D_v} = \int_s \frac{m'M ds}{EI} + \int_s \frac{p'P ds}{EA}$$

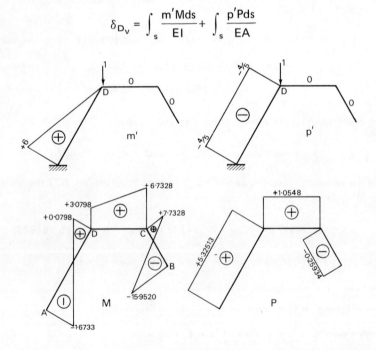

(Note: The diagrams for M and P apply to all calculations for any deflection. See below.)

$$\delta_{D_v} = \frac{10 \times 6}{6 \times 1000}\{-3.34666 + 0.0798\} + \left\{10 \times -\frac{4}{5} \times \frac{5.32513}{200}\right\} = -0.24567$$

$$= 0.24567 \text{ upwards}$$

FIG. 7.13. (*continued*)

Horizontal Deflection at D. δ_{DH}

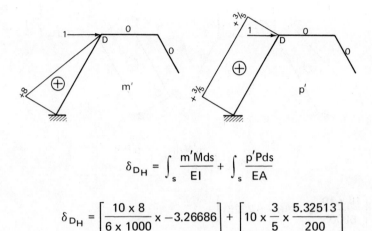

$$\delta_{DH} = \int_s \frac{m'M ds}{EI} + \int_s \frac{p'P ds}{EA}$$

$$\delta_{DH} = \left[\frac{10 \times 8}{6 \times 1000} \times -3.26686 \right] + \left[10 \times \frac{3}{5} \times \frac{5.32513}{200} \right]$$

$$= +0.11619 \text{ (i.e. from left to right.)}$$

Rotation at D. θ_D

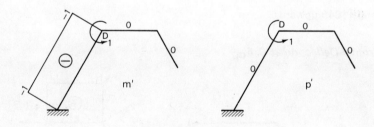

$$\theta_D = \int_s \frac{m'M ds}{EI} + \int_s \frac{p'P ds}{EA}$$

$$\therefore \theta_D = \frac{10 \times -1}{2000} [-1.6733 + 0.0798]$$

$$= +0.007968 \text{ rads (i.e. anticlockwise)}$$

FIG. 7.13. (*continued*)

Vertical Deflection at C. δ_{C_V}

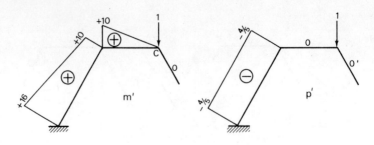

$$\delta_{C_V} = \int_s \frac{m'Mds}{EI} + \int_s \frac{p'Pds}{EA}$$

$$\delta_{C_V} = \left[\frac{10 \times 10}{6000}(6.1596 + 6.7328)\right] +$$

$$+ \left[\frac{10}{6000}\{(-3.34666 + 0.0798)\,16 + (.1596 - 1.6733)\,10\,\}\right]$$

$$+ \left[10 \times -\frac{4}{5} \times \frac{5.32513}{200}\right]$$

$$= -0.1104768$$

$$= 0.11048 \text{ upwards.}$$

Horizontal Deflection at C. δ_{C_H}

$$\delta_{C_H} = \int_s \frac{m'Mds}{EI} + \int_s \frac{p'Pds}{EA}$$

$$\delta_{C_H} = -\frac{4.355813}{100} + \frac{15.97539}{100} + \left[\frac{10 \times 1 \times 1.0548}{200}\right] = +0.11619 \text{ (i.e. from left to right.)}$$

FIG. 7.13. (*continued*)

Rotation at C. θ_C

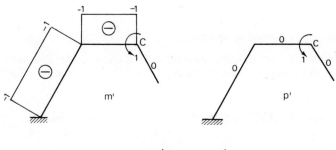

$$\theta_C = \int_s \frac{m'\,M\,ds}{EI} + \int_s \frac{p'\,P\,ds}{EA}$$

$$\theta_C = +0.007968 - \left[\frac{10 \times 1}{2000}\,(3.0798 + 6.7328)\right] = -0.041095 \text{ radians}$$

$$= 0.041095 \text{ rads. clockwise.}$$

Once more a comparison can be made between these results for the deformations at C and D with those already obtained by computer (Fig. 7.6), from which it can be seen that close agreement has been obtained.

In this problem, and indeed in all rigid-jointed frames analysed by the Flexibility Method, the strain energy due to axial forces is generally significant and should be included, so that $U = U_{\text{bending}} + U_{\text{axial}}$, with

$$f_{qr} = \int_s \frac{m_q m_r\,ds}{EI} + \int_s \frac{p_q p_r\,ds}{EA}$$

$$u_r = \int_s \frac{m_r m_o\,ds}{EI} + \int_s \frac{p_r p_o\,ds}{EA}.$$

However, the strain energy due to shearing forces, viz $U_{\text{shear}} = k \int_s \frac{S^2\,ds}{2GA}$, can normally be omitted. (If U_{shear} is to be included, it is necessary to specify values for k and G, and this particular numerical problem has not been defined clearly enough to allow the reader to insert appropriate values!) Nevertheless it is interesting to realise that such values were not required in obtaining the solution by the Stiffness Method, the inference being that the effect of shear lag is not included in that method. However, since the two solutions given are identical, it appears that the approximations made in each case are essentially equivalent.

On examining the calculation in Fig. 7.13, the reader will appreciate that it is much more difficult to program this process, which is perhaps why no library program based on the Flexibility Method is widely available at present. Consequently it is not possible to provide a solution by computer at this stage.

Treatment of pin-jointed plane frames

If all the external 'forces' acting on a pin-jointed frame are applied at the joints, the only strain energy induced in the system is that due to the axial forces in the

members, and thus, $U = U_{\text{axial}}$, with $f_{qr} = \int_s \dfrac{p_q p_r ds}{EA}$ and $u_r = \int_s \dfrac{p_r p_o ds}{EA}$. The

analysis can then proceed essentially as outlined above.

Space frames

Each member in a rigid-jointed space frame is subjected to an axial force, two shearing forces mutually perpendicular, two bending moments acting in perpendicular planes, and finally a torque acting in a plane parallel to the cross section. The total strain energy in the system has components from all these six sources, and the reader is already familiar with U_{bending}, U_{shear}, and U_{axial}. There is now in addition

$$U_{\text{torque}} = \int_s \frac{T^2 ds}{2GJ},$$

where T is the torque at a given section in a member, and J the torsional constant of the section. (This is the polar second moment of area only in the case of circular sections.) T is generally considered positive when clockwise when looking along the axis of the member in the direction of increasing s (i.e. the right-hand screw rule applies). The f and u elements now include terms of the form

$$\int_s \frac{t_q t_r ds}{GJ}   \text{and}   \int_s \frac{t_r t_o ds}{GJ},$$

respectively, where t_q, t_r, and t_o, are the appropriate unit torques at the same section as m_q, m_r, and m_o etc.

The nature and number of redundancies can now become quite complex and, although the analysis can proceed by a process similar to that outlined for plane frames, the entire undertaking generally proves to be fairly extensive and complicated.

Energy Methods

Some reference to energy methods is essential, although the treatment here must of necessity be very brief. The books by Argyris and Kelsey[5], and Neal[6], deal with this subject extensively in a context appropriate to the modern methods, and the engineer should profit from a detailed study of these if he so wishes.

The reader will have already observed how Castigliano's First Theorem was used in developing the Flexibility Method from energy considerations. McMinn[7] has shown how modern methods can be devised from basic energy theorems to deal with structural components such as box girders, changed sections and cut-outs, as well as the treatment of initial and thermal stresses. (See also Argyris and Kelsey[5]). Neal[6] and Livesley[2] also indicate how energy methods can be developed to analyse non-linear structures, particularly with reference to the plastic analysis of plane frames. The author considers that such an approach offers attractive possibilities for a systematic analysis on a trial-and-error basis, leading in due course to the provision of library programs for limit-state and plastic-theory analysis. These programs would then obviously be as useful to the civil engineer as A27, EAPF, LC29, etc., are in their respective fields.

Energy methods can also be used to evaluate the stiffness matrices of the individual elements in the Finite Element Method.

Finite Element Method

All the methods previously described are directed chiefly at analysing beams and frames. The Finite Element Method has been derived primarily for analysing shell, slab, and plate structures, arches, and structures of varying section and complexity (such as are used in reinforced concrete construction, etc.), and represents an attempt to effect a more realistic, theoretical, analysis of these types of structural form than those possible by any crude simulation from a so-called 'equivalent' system composed entirely of beams and columns.

The very nature of this approach ensures that the calculations must be complex and lengthy, and practical applications invariably require the use of the computer. Indeed it is the very existence of computers that has encouraged many techniques based on this method to be devised. A practical illustration of the elastic analysis of a plate with a hole carried out on a Univac computer using a program based on the Finite Element Method, is given in Chapter 4, and a study of this problem and its solution at this stage will give the engineer an immediate appreciation of this process.

The analysis of structures by the Finite Element Method can be either elastic or plastic, though in the introductory notes given here elastic theory is considered almost exclusively using a version of the Stiffness Method.

Idealisation of the structure

An important preliminary activity is the idealisation of the structure. This is accomplished by dividing it into a series of discrete, finite, elements which are interconnected in an appropriate way, and the entire accuracy of the method depends in no small measure on how faithfully the actual structure is portrayed by this simulation. Even in the case of a conventional plane frame, for example, where the 'members' are connected to the 'joints' by means of an axial force, a shearing force, and a moment, this approach has been unwittingly adopted. Moreover, if the 'members' are of uniform section, each individual 'member' can then be regarded as a discrete finite 'element' and the idealisation of the structure can be completely and satisfactorily defined by a standard line diagram.

An extension of this process has already been illustrated in Example 9 of Chapter 4 where the elastic analysis of a curved arch rib is accomplished using Program A27 by a process of simulation in which the actual structure is replaced by an idealised form consisting of a set of straight members of uniform section.

In the more general case of an irregular continuous structure or continuum (such as, for example, a bridge deck or a shell roof), its division into a series of elements cannot normally be accomplished in as obvious a manner as in the previous examples because of the infinite number of possibilities both in the shape and size of the elements and the nature of the interconnections between any arbitrarily chosen element of the continuum and the surrounding material. To obtain a satisfactory numerical solution in such a system, the infinite number of restraints between elements must be reduced to a finite number of nodal points or nodes. Thus Turner, et al.[8] suggest that, if the connections can be confined to a selected number of nodal points, located generally to coincide with the joints or 'corners' of the elements, the stiffness properties of the resulting elements can then be obtained.

The basis of the method therefore is to represent the actual continuum by an approximate structure consisting of a finite number of individual elements (of one or more convenient mathematical shapes or forms), interconnected only at their nodes in an appropriate manner (according to the nature of the structure), and to carry out the calculations on the resulting idealised structure. Equivalent nodal 'forces' (i.e. force matrices) are then made to replace the internal stress distribution in each element and, similarly, the external 'forces' are also assumed to act only at the nodes. The nodal 'displacements' (i.e. displacement matrices) of the elements now become the unknown parameters, and it is further assumed that the 'displacement' within any element can be expressed in terms of the nodal displacements of that element. The selection of this displacement function is an important basic step, and the accuracy of any resulting solution depends critically on this choice. This topic is discussed at length in several pub-lications. [9], [10], [11]

Outline of the method

The calculation now follows a similar pattern to that described for the analysis of conventional frames by the Stiffness Method. The basic steps in the Finite Element Method are as follows.

1.   Idealisation of the structure—subdivision of the continuum into elements with interconnecting nodes at all joints or 'corners'.

2. Assumption of displacement function and evaluation of stiffness properties for all elements.
3. Transformation of the stiffness matrices for all elements from local to overall axes.
4. Assembly of the one overall stiffness matrix for the entire continuum.
5. Introduction of boundary conditions.
6. Solution of the stiffness equation for the entire continuum to give the nodal displacements.
7. Use of the results to obtain stress, strain, etc., at desired locations in the continuum.

It is quite outside the scope of this book to give a detailed exposition of the Finite Element Method. The engineer can pursue his studies in this direction using the publications listed in the References and Bibliography. This outline is therefore completed merely by a note on the properties of an individual element, since this is a key factor in any qualitative appraisal of the method.

Evaluation of properties for each element

Unlike the simple, linear, 'one-dimensional' elements in conventional frame analysis, the finite elements in a continuum can have a variety of shapes or forms. For example, in a two-dimensional system such as a bridge deck or slab floor, square, rectangular, triangular, or segmental elements of uniform thickness might be chosen. In three-dimensional structures, appropriately shaped elements should again be chosen, though this time these might well be tetragonal, orthorhombic, or more complex, prisms, or perhaps tetrahedra or curved segments (in the case of a shell), etc. In all cases the entire continuum will be divided into a set of elements, and it is clearly desirable that this should be done using as few different mathematical shapes or forms as possible when dealing with the one continuum.

Determining the stiffness matrix for any such element, relating nodal forces and displacements, now becomes more complicated and various techniques have been documented by Przemieniecki.[10] The first step is the assumption of a displacement function. In addition, it is desirable that the stress-equilibrium equation for any element should be satisfied, and details of this aspect of the problem can be found in other published work.[12],[13]

Thus, in a similar manner to the process described for frames, the stiffness matrix k of a finite element in a continuum relates the generalised nodal forces P to the generalised nodal displacements Δ by the matrix equation $P = k\Delta$. The assumption is then made regarding the deformation patterns that can occur within the element. The resulting displacement function should satisfy the requirements discussed above, and it is normally convenient to assume that the element has as many deformation patterns as it has degrees of freedom (by virtue of the supports provided at the nodes). Now let the displacements at any point in the element be designated by the matrix equation $\delta = A\alpha$ where α is a column matrix of constants. Then, for the particular case of the nodal displacements Δ, $\Delta = B\alpha$ and therefore $\alpha = B^{-1}\Delta$. Hence $\delta = AB^{-1}\Delta$, i.e. the displacements at any point in the element may be expressed in terms of the nodal displacements for that element.

The generalised strain ϵ at any point in the element may also be expressed in terms of the nodal displacements as $\epsilon = C\Delta$, and if the stress-strain relationship is given by $\sigma = D\epsilon$, then $\sigma = DC\Delta$, where D is a matrix of material properties. (In the above equations A, B, C, and D are matrices which can be of any appropriate order.)

The stiffness matrix k for the element can now be determined using the Principle of Virtual Displacements. Let Δ be any arbitrary virtual displacement of the nodes. Then W_e, the external virtual work done by the nodal forces P is given by $W_e = \overline{\Delta}*P$, where $\overline{\Delta}*$ is the transpose of $\overline{\Delta}$. The corresponding virtual strain $\overline{\epsilon}$ is then given by $\overline{\epsilon} = C\Delta$, and the internal virtual work $W_{int.}$ sustained by the three-dimensional element is

$$W_{\text{int}} = \int_v \overline{\epsilon}*\sigma\, dv = \int_v C*\overline{\Delta}*D\epsilon\, dv = \int_v C*\overline{\Delta}*DC\Delta\, dv = \overline{\Delta}* \int_v C*DC\, dv.\Delta,$$

with the integration being taken over the entire volume of the element. By the Principle of Virtual Work, $W_e = W_{int}$, and hence

$$\overline{\Delta}*P = \overline{\Delta}* \int_v C*DC\, dv\, \Delta,$$

so that

$$k = \int_v C*DC\, dv,$$

thus giving the stiffness matrix of the element with respect to the local co-ordinate axes. The manner in which external forces are allocated to the nodes may also be obtained by a virtual displacement technique as described by Zienkiewicz.[9] The method is then completed by means of steps 3-7 as listed above (see References and Bibliography).

Appraisal of Various Methods

For a complete structural system a single matrix equation can be obtained relating the elastic 'displacements' of the joints to the applied 'forces' at the joints, and the analysis is effected in all cases by solving this equation. In the Stiffness Method (also known as the Equilibrium or Displacement Method) this equation is given as:

$$W = K\Delta$$

whereas in the Flexibility Method (also known as the Compatibility or Force Method) this same equation is given as

$$\Delta = FW$$

where
  W is the force matrix for the entire system, giving the applied 'forces' on all
    the joints,
  Δ is the displacement matrix for the entire system, giving the 'displacements'
    of all the joints, and
  K and F are the stiffness and flexibility matrices respectively for the entire
    system.
It can therefore be seen from the above equations that $K = F^{-1}$ and $F = K^{-1}$ so
that the stiffness matrix is the inverse of the flexibility matrix and vice versa.

The engineer will clearly be wondering which is the better approach, and
consequently wish to receive some guidance regarding the advantages and
disadvantages of the two methods. This is already a well-established bone of
contention and many experts have made useful pronouncements extolling the
merits of the method they personally favour. The reader is therefore advised to
ask himself once more the question posed in Chapter 2, i.e. 'With due regard to
my own special circumstances, what is the best approach to my own
computational problems?' In the absence of any other adverse consideration, the
availability of suitable library programs which can be used by the engineer on
the particular computer to which he has access will doubtlessly substantially
influence the answer.

However, beyond this sensible initial attitude it is useful to study what
various experts have said on this matter. Accordingly. Argyris[14] gives a detailed
analogy between the Stiffness and Flexibility Methods, (though this is possibly
best described by Pestel and Leckie[15] while Livesley[2], Laursen[16],
Rubinstein[17], and Gallagher[18] all give useful commentaries on the differences
between these methods.

The traditional methods are generally based on reactions being the
unknowns—this essentially being the Force or Flexibility Method—and consequently it is understandable for readers to favour this approach. However, when
this method is applied to redundant structures it is necessary to know how many
redundant 'forces' there are, and to specify these as unknowns. This process is
generally difficult to program and, whilst the arithmetic can be simpler than by the
Stiffness Method when there are only a few unknowns, the converse is true when
there are many redundancies. With the Stiffness Method, on the other hand,
there is no need to specify the number of unknowns and a unique approach
which is easier to program can be adopted, though the calculations can
sometimes be more extensive. However, since these are effected by computer in
any case this only becomes a real obstacle when there is a shortage of computer
storage capacity or when the machine time taken becomes excessive.

At present there are more library programs available based on the Stiffness
Method than on the Flexibility Method, so that the former has become more
attractive to the practical engineer.

Energy methods provide attractive possibilities for analysing non-linear
structures involving plastic theory, yield-line theory, limit state and dynamic
analysis, and there will doubtlessly be considerable developments in this field
during the next decade or so—especially if new mandatory clauses are introduced
into design codes. Finally the Finite Element Method is possibly the modern
method *par excellence*, and enables the elastic, plastic, or dynamic analysis of

structures having an irregular or unconventional form to be effected, so that the day may not be too far distant when the structural engineer will positively welcome a progressive architect's 'brainchild'.

Extension of the Process

The computer, together with transducers and data-logging devices that automatically record various physical quantities such as stress, strain, deflection, rotation, load, pressure, wind velocity, settlement, levels, lengths, angles, temperature, etc., with regard to the static or dynamic response of a structure, component, or model, under load either on site, in the factory, or in the laboratory, provides a unique opportunity to co-ordinate both theoretical analysis and practical load tests, and so to 'prove' any such structure or indicate what additional stiffening or bracing may be required. Load-time charts during service may also be obtained and the information evaluated by computer to assist in maintenance, etc.

These activities can be integrated or interchanged so that, for example, the various elements in a flexibility matrix may be obtained by experimental measurements on the full-scale structure or a model, in place of, or for comparison with, theoretical calculations. However, before such techniques can be properly developed, it is necessary for the engineer to have some background knowledge of automatic techniques in the laboratory and on site, and this topic is therefore considered in the next chapter.

References

1. Livesley, R. K. Analysis of rigid frames by an electronic digital computer. Engineering, London, 1953, 176 (Aug. 21), pp 230-233
2. Livesley, R. K. Matrix Methods of Structural Analysis. Pergamon. 1964
3. Program Number LC8—Structural Analysis of a Grid. Sept. 1962. Elliott 803 Computer Application Program. (All enquiries now to International Computers, Ltd., London.)
4. Program Number A28—Elastic Analysis of a Grid. G. Maunsell & Partners, London
5. Argyris, J. H. & Kelsey, S. Energy Theorems in Structural Analysis. Butterworth. 1960
6. Neal, B. G. Structural Theorems and their applications. Pergamon. 1964
7. McMinn, S. J. Matrices for Structural Analysis. 2nd ed. Spon. 1966
8. Turner, M. J., Clough, R. W. Martin, H. C. & Topp, L. J. Stiffness and Deflection Analysis of Complex Structures. Journal of Aero Sciences Vol. 23, No. 9, Sept. 1956. pp 805-823, 854
9. Zienkiewicz, O. C. The Finite Element Method in Structural and Continuum Mechanics. McGraw Hill
10. Przemieniecki, J. S. Theory of Matrix Structural Analysis. McGraw Hill. 1963
11. Zienkiewicz, O. C. and Holister, G. S. Stress Analysis. Wiley. 1965
12. Melosh, R. J. Basis for Derivation of Matrices for the Direct Stiffness Method. Journal of Am. Inst. Aero. & Astro. 1, 1631-1637 (1963)

13. Bazeley, G. P., Cheung, Y. K., Irons, B. M., and Zienkiewicz, O. C. Triangular Elements in Plate Bending: Conforming and Non Conforming Solutions. Proc. Conf. Matrix Methods in Struct. Mechs., Wright-Patteson Air Force Base, Ohio. Oct. 26-28, 1965, AFFDL TR 66-80, 1966
14. Argyris, J. H. Recent Advances in Matrix Methods of Structural Analysis. Pergamon. 1964
15. Pestel, E. C., & Leckie, F. Matrix Methods in Elastomechanics. McGraw-Hill. 1963
16. Laursen, H. I. Matrix Analysis of Structures. McGraw-Hill. 1966
17. Rubinstein, M. F. Matrix Computer Analysis of Structures. Prentice Hall. 1966
18. Gallagher, R. H. A Correlation Study of Methods of Matrix Structural Analysis. Pergamon. 1964

Bibliography

A more detailed study of various modern methods of analysis can be made by referring to the textbooks listed above, together with:

Morice, P. B. Linear Structural Analysis. Thames & Hudson. 1959
Hall, A. S., & Woodhead, R. V. Frame Analysis. 2nd ed. Wiley. 1967
Wang, C. K. Matrix Methods of Structural Analysis. International Textbook Company. 1966
Fenves, S. J. Computer Methods in Civil Engineering. Prentice Hall. 1967
Tuma, J. J. Theory and Problems of Structural Analysis. McGraw Hill. 1969
Jenkins, W. M. Matrix and Digital Computer Methods in Structural Analysis. McGraw Hill. 1969
Rogers, G. L. Dynamics of Framed Structures. Wiley. 1959
Gibson, J. E. The Design of Shell Roofs. Spon. 1968
Blaszkowiak, S., & Kaczkowski, Z. Iterative Methods in Structural Analysis. Pergamon. 1966
Ziegler, H. Principles of Structural Stability. Blaisdell. 1968
Marshall, W. T. and Nelson, H. M. Structures. Pitman. 1969

8
Automatic Techniques in the Laboratory and on Site

Parallel with the development of computer hardware and software, has been that of data logging devices, transducers and instruments to effect and record physical measurements in the laboratory and on site in keeping with the trend towards automation discussed in Chapter 1. The image of the site or research engineer, laboriously recording experimental measurements manually with pencil and paper from fluctuating pointers on dial gauges, etc., should disappear in the not-too-distant future, to be replaced by a 'space age' outlook involving the automatic recording and subsequent processing of such data. The author—in conjunction with J. P. Cole—has already philosophised on this theme elsewhere[1], outlining a policy and a method used in establishing automatic data processing techniques in civil engineering laboratories.

However, there is much more to this than merely applying the benefits of labour-saving techniques or recording abundant data without the prejudice of or disturbance from a human observer. Providing such devices introduces a unique opportunity to co-ordinate and extend various engineering methods such as those employed in structural analysis, to embrace both theoretical and experimental work in a single process as a standard procedure. The engineer should therefore remember this while studying these introductory notes on data logging equipment and its use.

Data Logging of Experimental Results

There are three main methods of recording data: manually, by observing instruments and writing down the results; graphically, using pen or ultra-violet galvanometer recorders, and digitally, using automatic data acquisition systems involving punched tapes or cards, direct print-outs, etc. All the methods have inherent advantages and disadvantages. For instance, the first is invaluable in providing students and young engineers with a background of knowledge about measurement, using mechanical as well as electronic or electrical instruments. Its disadvantages have already been briefly referred to[1] and extend from the tedium of taking many readings with the probable loss of accuracy, to the need for static physical conditions with a long-term stability while the measurements are taken. The second method is ideal for recording the continuous or intermittent variation of physical quantities with time. Most modern pen recorders have been designed for measurements at low frequencies, (though some can now also be used at higher values). On the other hand, the ultra-violet recorder is intended to measure vibrations, etc., up to the lower audio range of frequencies, and still higher frequencies may be measured and recorded by using a cathode ray oscilloscope (known as a C.R.O.) in conjunction with a high-speed camera. The third method is the more recent development in data logging and makes use of a digital voltmeter as the basic measuring instrument in a circuit. Since it plays an important role in data logging, some of the properties of this instrument are now considered in greater detail.

The digital voltmeter (D.V.M.)

The D.V.M. (Fig. 8.1) measures voltage and has many advantages over other electrical measuring instruments, among which are its ability to measure very low level static signals with great accuracy, to display the result in 'real language' (i.e. in digital form on a screen) and to output a replica of the reading in a re-usable coded form. The latter facility is particularly useful when the results are to be processed by an automatic data acquisition system.

A further advantage is the wide range of readings which can be shown as numerical digits. Some D.V.M.s read to six digits with a full-scale value of 999999 either positive or negative, so that the potential range of such an instrument is twice this value. At present, however, because of practical

DIGITAL VOLTMETER

FIG. 8.1

hardware limitations it is unlikely that the full range can be utilised, but this does indicate the order of magnitude of readings and the implied accuracy which might be achieved in due course (if required!). At present the more practical and less costly instruments are generally limited to four digits.

Accuracies of 0.01% of the indicated reading are quite common, with almost perfect stability over long periods. An ultimate resolution of reading of one microvolt is possible with the more expensive instruments. Thus, with a strain gauge in a Wheatstone Bridge circuit (Fig. 8.2), a four digit D.V.M. having a resolution of one microvolt would provide a full-scale reading of 9999 microstrain when the bridge supply voltage (across the active arm) is 2 volts and the gauge factor (for both active and dummy gauges) is 2.0.

WHEATSTONE BRIDGE CIRCUIT

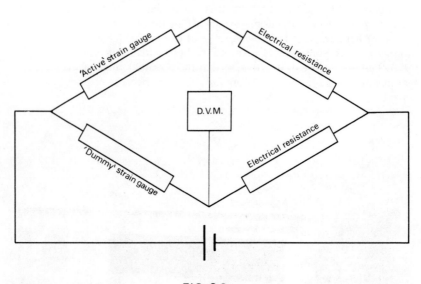

FIG. 8.2

Recording of data

Along with the visual display, a coded output from the D.V.M. is produced simultaneously as electrical 'pulses' in binary form by a fan-out unit generally located at the rear of the instrument. There is a four-wire output for each character displayed on the screen—these normally being polarity (+ or −) and the D.V.M. output in units of 10^3, 10^2, 10^1 and 10^0 (for a four-digit instrument).

When the D.V.M. forms part of a complete data-logging system (a typical layout described below being shown in Fig. 8.3) this coded output is generally transferred to punched paper tape (identical to that used with computers), and it is therefore customary to introduce other characters to represent system

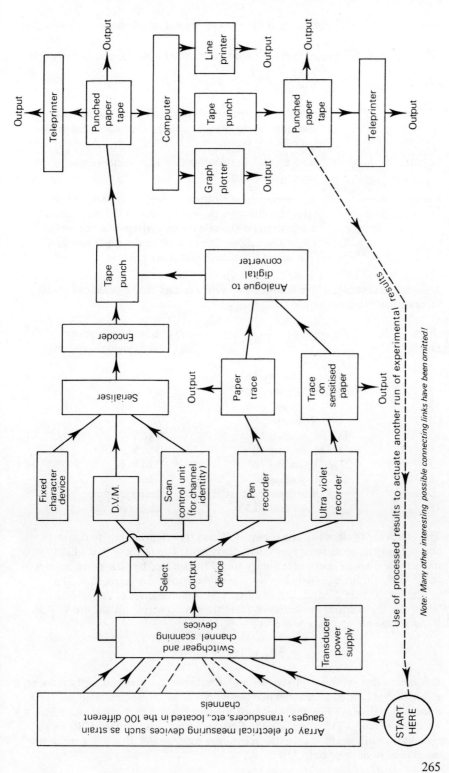

Note: Many other interesting possible connecting links have been omitted!

FIG. 8.3

information, data separation symbols, and teleprinter drive signals. For example:

Character	Use
10^1 and 10^0	For channel identification, specifying 100 different channels (0–99) which could be used for strain gauges, transducers, etc.
Comma	As a means of separating different numbers in a sequence of data.
Carriage return (and) line feed	As a means of separating different numbers, and to allow for the subsequent print-out of the data tape on a teleprinter. (The teleprinter carriage will not return to its original position or turn up the paper to a new line unless these instructions are given.)

A complete data 'word' containing all these characters can be produced on 10 separate four-wire outputs:

Output number	Description	Actual signal given (in binary coded form)
1.	Channel identity 10^1	0 to 9
2.	Channel identity 10^0	0 to 9
3.	Comma	,
4.	Polarity	+ or −
5.	D.V.M. reading 10^3	0 to 9
6.	D.V.M. reading 10^2	0 to 9
7.	D.V.M. reading 10^1	0 to 9
8.	D.V.M. reading 10^0	0 to 9
9.	Carriage return (CR)	Carriage return
10.	Line feed (LF)	Line feed

The two polarity signs (4), the comma (3), and the teleprinter drive signals (9 and 10) are known as *fixed characters,* there being five of these in all. Each has its own fixed coded output (in binary form) transmitted by four wires, as with the D.V.M. output. Similarly the two channel identity characters (1 & 2) are output by the scan control unit (see Fig. 8.3). All characters in a complete 'word' are then output in succession from the scan control unit and the D.V.M.

An example of a data 'word' is

$$82, - 2157 \text{ (CRLF)}$$

This means that the value of the reading taken from the strain gauge, transducer, etc., which is located at channel number 82 is −2157. The corresponding information displayed on the screens of the scan control unit and the D.V.M. would be 82 and −2157 respectively. Carriage return and line feed (CRLF) are coded instructions which drive the teleprinter carriage and do not appear in any visual display or print-out.

The immediate advantage of this arrangement is that when such a 'word' is transferred to paper tape, it can be immediately identified by a computer as two numbers, since numbers on a data tape used as input can be separated by a comma or CRLF. Moreover all subsequent 'words' in a series will be similarly treated, so that it is possible to use the resulting tape directly as a data tape which can be applied as input to a computer along with an appropriate program. It is also possible, of course, to print-out the paper tape on a teleprinter first to check the experimental data so recorded.

The evolution of a data-logging system

A diagrammatic layout of a system built from basic compatible instruments and devices is shown in Fig. 8.3. This 'theoretical' arrangement is intended to record a series of readings from up to 100 different electronic or electrical devices measuring various physical quantities e.g. strain, deflection, pressure, temperature, humidity, etc. The devices can be strain gauges, transducers, thermisters, counters, etc., all calibrated to give the relationship between the physical quantity measured and the electrical signal induced, the various devices being attached to specimens or sited in appropriate locations as required.

A scanning cycle consists of recording one measurement from each device in turn, and it is generally necessary for steady or static conditions to prevail in each experiment or test while the results are being recorded. Each device is therefore assigned to a channel identified by a number from 0 to 99. The data 'word' (where appropriate) from any particular device then consists of the channel identity number and the reading of the physical quantity, each device being switched into and out of the recording circuit in turn, either manually or automatically using the switchgear and channel scanning devices. The type of output is then selected from a D.V.M., a pen recorder or an ultra-violet recorder.

With a D.V.M., the electrical 'quantities' or 'messages' from the measuring devices are directed into the unit, and all the characters in a 'word' are output in succession as electrical pulses in coded binary form. The complete 'word' is then scanned in sequence by the serialiser and transmitted character by character to the encoder where the binary characters are converted into whatever punched tape code is specified by energising on demand any one of 16 control lines, each representing a different character. Thus, a 'word' written as given above might be 79, +1279 followed by carriage return line feed. If this is to be translated into, say, Elliott 4100 8-channel punched paper code by the encoder, the process can be followed by referring to the appropriate code converter matrix or diode matrix given in Fig. 8.4. This shows the input control lines from the encoder and the resulting output code in Elliott 4100 8-channel punched paper tape code. The 'word' would then be re-coded as 8, 10, 13, 12, 2, 3, 8, 10, 14, 15. This output can then be punched in sequence on to tape by the tape punch so giving a record of this and all other 'words'.

With pen or ultra-violet recorders, continuous traces of the results from any devices are produced directly on paper as graphs of experimental readings against time. These traces can be regarded as the final record of the experimental readings or be converted into digital form and recorded once more as punched tape.

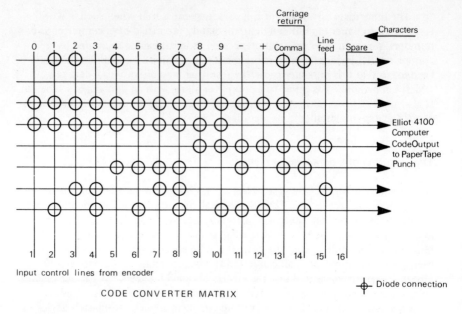

CODE CONVERTER MATRIX

FIG. 8.4

Thus it is possible for all three types of output to eventually be recorded on the same type of punched tape and in the same code. These records can then either be printed out on a teleprinter or inserted directly into a computer as a data tape—to be used with an appropriate library or special program—so that the results may be processed and evaluated in whatever way the engineer may decide.

Following this description of the synthesis of a basic data logging system to be used for the automatic measurement, recording and perhaps processing of experimental readings, the engineer should study the various possibilities indicated in the flow diagram in Fig. 8.3, and realise that many more interconnections could be inserted between the units to modify or improve this basic proposal according to his own particular requirements. The author and J. P. Cole have already given details elsewhere[1],[2] of a data-logging system they have devised and constructed, and the reader might now care to study that practical implementation of the above system. (Further details of the basic units and other useful background information are given by Cole elsewhere[3],[4].)

The reader should not have gained any false impression that it is normal practice for the various pieces of equipment to be chosen or acquired piecemeal; it is important that they are carefully matched and that the various electrical quantities and power supplies are compatible with one another.

Layout and Operation of a Large Data-Logging System

In reference (1), the author and J. P. Cole indicated their ideal layout of a logging system, involving a central control room and permanent wiring providing adequate channels to several different laboratories. The author is pleased to report that they have now been able to implement this proposal in the civil engineering laboratory block of the Department of Civil Engineering of the Dundee College of Technology, where the facilities will be used to instruct B.Sc. degree course students as well as for research, industrial and other purposes.

A general layout of the first phase of this 2,000 channel data-logging system is shown in Fig. 8.5, which deals with the arrangement for the main floor (having an approximate area of 1,000 m^2) of the civil engineering laboratory block. It can be seen that the instrumentation laboratory containing the data logger, assembled from basic units in the form of a control console, is located in the centre of the building, and that permanent wiring, housed in metal trunking, extends from the control console to the various laboratories, with several pick-up points for electrical devices such as transducers, strain gauges, etc. being provided by the station outlet boxes A, B, C, ———— I. By this arrangement the one set of instruments can be used to cover all the laboratories, so avoiding the expense incurred by duplicating any equipment. The data logger shown in Fig. 8.6, was designed by J. P. Cole to scan the various channels using a time-sharing process. Two station outlet boxes (F and E) are shown in Fig. 8.7.

It is intended to extend this logging system in due course to cover the various other laboratories in the top floor and basement of the block by providing additional trunking from the instrumentation laboratory.

Technical description and operation of the system

This is more fully dealt with in the manual[5], and details of the design and construction of the system will be given elsewhere later. A flow diagram for the system is given in Fig. 8.8. The custom-built system has been designed to serve up to ten discrete locations in the various laboratories. Each of the nine station outlet boxes at present in use (Fig. 8.5) has a logging capacity of 200 channels, the data logger being able to accept readings in sequence from up to 2,000 channels during any scanning cycle.

A plan of the control console containing the logger is shown in Fig. 8.9, and can be used in conjunction with Fig. 8.6 to identify the various units. Station and channel selection are both made on eight-channel punched tape (known as the program tape) which is prepared on the program tape punch (H) using the keyboard (G), and is then inserted in the program tape reader (F). F, G and H are all components of the scan control unit (E) which has been designed to effect automatically whatever scanning cycle has been programmed, the entire process being controlled by switches, etc., on E, although these have been kept to a minimum.

The sampling units (Fig. 8.8) each contain all the necessary switching circuits for up to 20 channels and are plugged into the station outlet boxes. Any sampling unit can readily be located on a test rig so enabling the connecting leads to the various measuring devices to be kept as short as possible. Power supplies are provided in each unit and two-pole switching for all types of transducers is standard throughout the system. (Arrangements can also be made for four-pole switching from any two outlet boxes.)

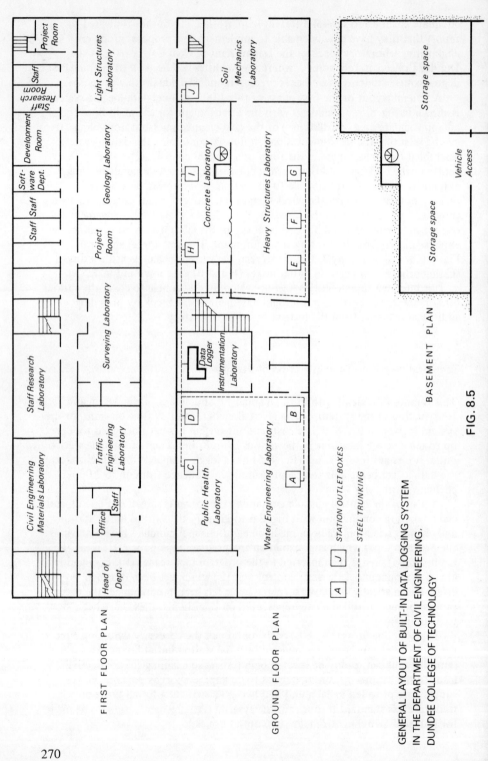

FIRST FLOOR PLAN

Project Room

Staff Research Room

Development Room

Soft-ware Dept.

Staff

Staff

Light Structures Laboratory

Geology Laboratory

Staff Research Laboratory

Civil Engineering Materials Laboratory

Surveying Laboratory

Project Room

Traffic Engineering Laboratory

Head of Dept.

Office

Staff

GROUND FLOOR PLAN

J

Soil Mechanics Laboratory

I

Concrete Laboratory

H

Heavy Structures Laboratory

E

F

G

Data Logger

Instrumentation Laboratory

D

C

Public Health Laboratory

B

A

Water Engineering Laboratory

A J STATION OUTLET BOXES

– – – – – STEEL TRUNKING

BASEMENT PLAN

Storage space

Storage space

Vehicle Access

FIG. 8.5

GENERAL LAYOUT OF BUILT-IN DATA LOGGING SYSTEM
IN THE DEPARTMENT OF CIVIL ENGINEERING,
DUNDEE COLLEGE OF TECHNOLOGY

DATA LOGGER

FIG. 8.6

The station outlet boxes (Fig. 8.7; see also Figs. 8.5 and 8.8) are the pick-up points for the sampling units in any particular laboratory. Each box also has manual entry facilities and an isolating switch, so that it may be taken out of the system and programmed locally using a specially devised manual entry unit. This arrangement enables the operator to actually be in attendance at a particular test rig in any laboratory to carry out zero checks on transducer readings, etc. He would accomplish this by using an auxiliary D.V.M. but would not interfere with the logging operations elsewhere in the system. The various measuring devices are then positioned as required and are connected to the sampling units via range and calibration devices (designed for specific transducers). Useful basic information on transducers is given in Fig. 8.10, and reference can also be made to other publications.[1],[3],[4]

STATION OUTLET BOXES

FIG. 8.7

The electrical 'signals' from the transducers are transmitted successively to the main D.V.M. (D) on the data logger. Time-sharing facilities embodied in the system regulate this process, so that one reading in turn is transmitted from each specified channel to the logger for recording during a scanning cycle. The readings are displayed visually on the screen of the D.V.M. and, as already described, the binary coded output is transferred via the serialiser/encoder (C) to the data output punch (A) where a record of the data 'words' is produced in Elliott 4100 eight-channel punched paper tape code. The operator can then edit this tape as required using the manual entry keyboard (B) on the console.

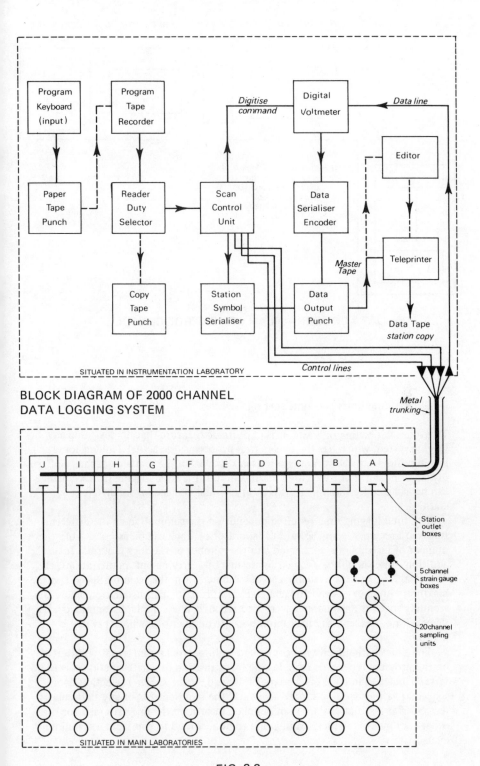

BLOCK DIAGRAM OF 2000 CHANNEL
DATA LOGGING SYSTEM

FIG. 8.8

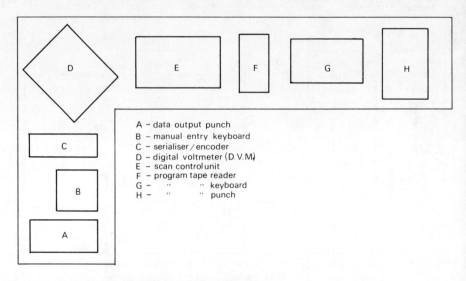

A – data output punch
B – manual entry keyboard
C – serialiser / encoder
D – digital voltmeter (D.V.M.)
E – scan control unit
F – program tape reader
G –   "   " keyboard
H –   "   " punch

DATA LOGGER – PLAN OF CONTROL CONSOLE

FIG. 8.9

Total operating times and time-sharing facilities

To avoid overloading or waste, it is important to ensure that, in designing any logging system which has the one central processor, the actual demand for logging time from the various stations is kept within the available capacity of the installation, and that a weighted distribution of the total logging time available can be effected by the operator to meet unequal demands from different stations.

The total logging time required depends on the summation of the different time-requirements from the various laboratories. Each can be assessed from the number of instruments connected and the number of readings required. In addition, time should be allowed for setting up, carrying out modifications, etc. The total logging time available can be obtained from the scanning speed and the maximum number of available channels. In this particular installation where the scanning speed is one channel every two seconds, one complete scanning cycle takes 67 minutes, and the maximum possible number of readings per day is 43,200.

The time-sharing facility is controlled and operated by information inserted on the program tape. This includes the identification of whatever station is to be served (punched in code on the tape) followed by the channel selection data spaced at various specified intervals to suit the requirements of any particular test. Similar instructions are then punched in turn on the program tape for the other stations from which readings are to be recorded to complete a scanning cycle.

PHYSICAL QUANTITY TO BE MEASURED	TRANSDUCER TYPES	DATA LOGGING SYSTEM		RESOLUTION	FULL SCALE OUTPUT	CIRCUIT POWER SUPPLY
		DYNAMIC (SPEED)	STATIC			
STRAIN	Electrical – Wire, Foil, or Semi-Conductor Gauges	Slow to Fast	Also	Infinite	Millivolts	AC or DC
LOAD	Load Cells with internal Potentiometric or Strain Gauge Bridge Circuits	Slow	Also	1% to Infinite	Millivolts to Volts	AC or DC
PRESSURE	As Load Cells	Slow	Also	As Above	As above	AC or DC
DISPLACEMENT	Potentiometric, Strain Gauge, * Inductive, * Capacitive	Depending on degree of Displacement	Also	As above	As above	AC or DC except * AC only
SPEED	Tachogenerators	Slow	Also	As above	Volts	Self Generating
	Photo Electric Counting				Integrated	DC
ACCELERATION	Piezo Electric Potentiometric	Fast		As above	Millivolts to volts	DC
FLOW	Tachogenerators	Slow	Also	As above	Volts	DC
	Propellor Flowmeter Counter				Integrated	
LEVEL	Potentiometric * Inductive * Capacitive	Depending on degree of Displacement	Also	As above	Millivolts to Volts	AC or DC * AC only
TEMPERATURE	Resistive (Thermistors etc.)	Depending on mass	Also	Infinite	Millivolts	AC or DC
	Thermocouples					Self Generating
HUMIDITY	As Temperature	As above	Also	Infinite	Millivolts	AC or DC

DETAILS OF ELECTRICAL MEASURING DEVICES (TRANSDUCERS, ETC.)

FIG. 8.10

When the same scanning cycle is to be repeated throughout the logging operation, the program tape for the cycle can be inserted in the tape reader (F), the ends spliced together, and the tape driven repeatedly through the reading head until all measurements have been recorded. On the other hand, when the successive scanning cycles are varied, a continuous program tape must be prepared for the entire duration of the logging operation (a 24-hour program being approximately 120 metres long.)

When the system is in operation, readings from up to 2,000 channels in the various laboratories will be recorded on punched tape in the one logging process according to the scanning program which is inserted (as punched tape) at the control console, the entire process being virtually fully automatic. It is envisaged that daily scanning programs will be arranged with staff and students booking time for the logging of readings from the various measuring devices with which they are concerned.

Finally, it is clear that an installation of this sort can best be used in conjunction with equipment which can automatically and successively vary the relevant steady-state physical conditions on a particular test from one value or level to another (e.g. automatic programmed loading or temperature control devices, etc.). This matter is now discussed.

Automatically operated testing equipment

Along with the development of data-logging systems has been that of automatically operated testing equipment. Much pioneer work in this field has been carried out within the aircraft industry in constructing various test rigs simulating on the ground conditions of loading, temperature, etc., encountered in flight by any particular aircraft or component. A typical diagram giving the change in loading or temperature sustained during an average flight might be as shown in Fig. 8.11. A mechanism would then be devised to apply such a loading or temperature cycle repeatedly and automatically. This is the basis used to conduct some fatigue tests, and the number of cycles completed before the level

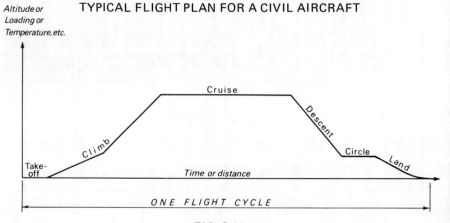

FIG. 8.11

of structural cracking or damage becomes unacceptable gives the number of flights in the unfactored fatigue life of the aircraft. The aircraft or the component would be fully instrumented with measuring devices such as strain gauges, transducers, etc., the readings being logged in an appropriate manner while the simulated flight loading cycles are automatically applied.

The construction of such ad hoc mechanisms has revealed the considerable ingenuity of aeronautical engineers, and the reader might care to study the report[6] on the 'tank' test on a Comet aircraft, carried out about twenty years ago, to appreciate this more fully for himself! The loading devices could consist of servomechanisms, solenoids, levers, cylinders, jacks, or even simple water pressure, and the multi-level cyclic loading produced was controlled automatically by 'black boxes', magnetic tape units, or by a system of specially cut, toothed cams.

Some manufacturers of testing machines (e.g. Losenhausen, Amsler, M.A.N., and Schenk), have been engaged in designing and supplying stepped and random loading equipment from these same early days, and their products are extremely useful when used in a laboratory with a modern data-logging system.

The ability to hold to close-tolerance steady-state conditions during the logging of experimental data is a prerequisite of many laboratory tests. Thus, for example, the loading applied by hydraulic jacks which derive their pressure from manually-operated pumps is generally insufficiently steady during a data-logging sequence because of fluid leakage and 'bleeding', and consequently those testing machines which can hold the load within fine limits (by automatic compensation) will become essential items of equipment in modern engineering laboratories.

Other examples of automatically operated equipment used in civil engineering are as follows.

  (i) Automatically controlled traffic lights at complex road junctions. The flow of vehicles and the build-up of congested conditions can be recorded automatically, this information being transmitted instantaneously to a computer which analyses the data and outputs an appropriate signal to actuate the lights.

  (ii) Heat treatment in a furnace. Any specified conditions of temperature/time at different values could be achieved by such equipment.

 (iii) Curing, testing and shrinkage of concrete. This is similar to (ii) above. Programmed variations in temperature, pressure and humidity, can be applied automatically to a test chamber containing the specimens.

 (iv) Hydrological models simulating full-scale conditions (see example in Chapter 9).

  (v) The opening and closing of sluice gates and valves according to a specified program.

 (vi) The operation of safety devices such as fire alarms, warnings of crane overloading, flooding, etc.

(vii) Control of earth-moving equipment, such as that used in cutting ground to a given profile.

These are just a few examples and there will no doubt be a great upsurge of activity in this field within the next few years.

Automatically operated testing equipment and data-logging systems complement one another and can jointly enable complicated, unattended, tests or processes to be carried out even at the most inconvenient times e.g. by night, at week-ends and during holiday periods. Moreover, with such facilities vast numbers of readings can be taken quickly and effectively with the same precision throughout, so that the population (of readings or samples, etc.) for any statistical analysis can be made large enough to ensure that subsequent calculations on curve fitting and probability, say, can have significance and credibility.

Data Logging on Site

Although the notes given above relate primarily to installations for engineering laboratories, workshops, factories and test-houses, these can also—with some qualification—refer to arrangements for site and field measurements. Thus in a data-logging system which produces a punched tape output, the main problem on site is to provide adequate electrical power in the absence of a mains supply. Brief notes are therefore given here on this and other problems that arise when using tape logging systems in this context.

Power supplies

Tape punches use a great deal of power and if readings are to be taken over a long period its supply can be both difficult and expensive. For example, the power requirement for a low-speed tape punch is of the order of 75 watts at 240 volts. A further 75 watts is required for the rest of the data logging system, with yet another 50 watts, say, for contingencies, so that the total power supply necessary would be 200 watts (at 240 volts). Two methods can be used to provide such power over a long period.

(i) A low-voltage D.C. petrol generating plant with automatic start and shut-down facilities, charging a 50-volt bank of either lead/acid or Ni/Fe batteries, which in turn drive a solid state convertor producing 240 volts A.C. With this arrangement the engine of the generator would be started and stopped according to the state of the charge existing in the batteries.

(ii) An on-line petrol alternator with automatic stop/start facilities providing an output of 240 volts A.C., together with a 32-volt D.C. charging circuit maintaining a 24-volt lead/acid or Ni/Fe battery (to be used exclusively for starting the engine). In this case the automatic start/stop device would be operated by the data logger 'start scan' clock.

The power supply and data-logging equipment should be housed in well-ventilated temporary buildings, making the provision of such a system rather expensive at present. The second type of power supply can also be used for short-term, manned projects, the automatic start/stop facilities not then being necessary.

Data records

The supply of paper tape—normally available in 1,000-ft. rolls—can also present
a problem when scanning transducers at frequent intervals. For example, if
readings from 100 channels are to be recorded once per hour, 24 hours per day,
this will consume approximately 200-ft of paper tape per day or one complete
roll in 5 days. Thus either the site must be visited at least once every five days to
change the roll or the manufacturers must be encouraged to supply longer rolls.

An alternative to paper tape in the field is magnetic tape. The logging system
would then be modified, though the resulting installation has the advantage of
requiring much less power. It is possible to use ¼-inch standard magnetic tape
which has a very high data packing capacity. For example, a normal 4-inch
casette provides up to three times as long a recording period (i.e. 15 days) as the
1,000-ft. paper tape roll for the same frequency of scan. Unfortunately,
however, magnetic tape records are not generally suitable for direct input to a
computer, so that it is necessary either to obtain expensive additional equipment
to translate such recorded data into the more-conventional punched tape, or to
use the translation service offered by various companies, before any such
experimental data can be inserted into the computer.

Finally, as well as tape records are those of readings printed on to paper rolls
in straightforward numerical form (e.g. in traffic counters) and of traces or
graphs from pen or other analogue field recorders, though in all such cases the
data must be processed and translated (either manually or automatically) before
insertion as input to a computer.

These notes give an introduction to and a basic outline of the problems of
logging experimental data on site and in the field. It is clear that the engineer
must consider the various alternatives open to him most carefully, especially
regarding accuracy, usefulness and cost, before he can establish which is the best
solution to his own data logging problems in his own particular circumstances.

Surveying

Most civil engineers, whilst generally recognising the importance of surveying as
an essential prologue (and perhaps sequel) to any constructional work or similar
undertaking in the field, tend to regard the various activities of the subject as
necessary chores to be borne stoically and executed with diligence and precision
as part of their lot in life.

The three main topics in surveying encountered most frequently by civil
engineers are field measurements, the setting-out of engineering works, and the
computation and plotting of the results, and in view of the above comment it is
therefore not surprising that over the years many minds have endeavoured to
find ways and means of reducing the mental and physical effort required to
accomplish these activities. It is quite outside the scope of this book to give a
detailed account of the various automatic techniques in surveying; what is
possible, however, is to provide the engineer with descriptive notes and an
appraisal of some of the processes which have been, or are being, evolved in each
of the above topics.

Field measurements

Aerial photography and photogrammetry are possibly the automatic techniques par excellence in field measurements. By using aerial photographs the engineer is relieved completely of all physical effort in taking surface measurements and when these are matched in stereoscopic pairs and viewed through the binocular eyepieces in a wide range of photogrammetric instruments, an accurate three-dimensional survey map is obtained, from which measurements can be taken in all three perpendicular directions. The photographs can be taken even at a considerable altitude from aircraft, helicopters and, nowadays, from space craft, and reveal the most detailed information of the terrain and the features thereon.

Tacheometry represents an earlier attempt at measuring lengths and taking levels by sighting on to a staff or target with a theodolite or other instrument and then observing and recording up to three readings from the one setting. This process obviates the need for any physical measurement, though at the expense of eye strain on the part of the engineer. This was a step towards automation but was by no means a complete answer.

The author envisages that further improvements in tacheometry and conventional surveying techniques could be effected using a normal theodolite with optical scales, in conjunction with a suitably adapted 35-mm single lens reflex camera. The optical scales of the theodolite and the readings on the telescope would be photographed on to a half-frame black and white film, so providing either slides or enlarged prints as a permanent record of the actual field measurements. This might be regarded as a semi-automatic process. The engineer would still be required to set up the theodolite and sight on to the target, though he would be relieved of the need to take and record field readings in the usual way.

From this 'half-way house' there is clearly a challenge for manufacturers to design and market instruments to perform the various functions of the theodolite, level and tacheometer as near automatically as possible, with all readings being recorded on tape. Some progress towards meeting this specification has already been made by companies such as Fennel and Kern with the code theodolite FLT and the code tacheometer respectively. Both instruments log the experimental readings in code on film which has to be developed and then translated into punched paper tape by a converter—the ZUSE Z 84. All this equipment is described in a useful paper by van Gent[7] (as well as in manufacturers catalogues), and is obviously very expensive at this early stage of development.

Finally, there is the problem of measuring lengths of from a few to several thousand metres. The readings may be required either to a high degree of accuracy, as in base line measurements, or much less accurately, as in the case of those dimensions normally taken with chains or linen tapes. Different instruments are suitable for these various categories, based on either electro-magnetic or electro-optical systems. Examples are the tellurometer, the geodimeter and the mekometer, and full details can be obtained from manufacturers' catalogues. When the equipment is in position, lengths are measured merely by taking readings on a vernier dial or a digital screen, though no facilities are generally available at present for punched-tape output. With these instruments the approach to some methods in surveying such as traversing and triangulation can now be modified, since the basic triangle can be specified by the 'three sides case' instead of the more usual 'two angles and a corresponding side case', thus eliminating the need to take and record any angular measurements.

Setting-out

This is possibly the section in which least progress has been made towards automation, though the engineer will appreciate that, once setting-out is complete, the work can be checked by using any of the various techniques described above. This suggests that a process of successive approximation could be evolved (as a step towards automation in setting-out) whereby approximate locations (of pegs, or stations) first established using appropriate, perhaps crude, labour-saving methods, are checked by the above techniques, and the errors obtained by computer calculations as described below, so enabling precise corrections to be made to the pegs, etc., by simple, local, adjustments.

There is clearly a need for developing automatic techniques and equipment to carry out the process of setting-out, and it is to be hoped that progress towards this objective will be made in the next few years.

The computation and plotting of results

Much has already been written on the use of computers and ancillary equipment to evaluate results from numerical data. Programs can be written to carry out the various calculations, so eliminating the sheer drudgery of using seven-figure tables and other equally dreary arithmetical processes which have had to be endured in the past. For example, Elliott's have written programs for calculating earthworks quantities,[8] closed traverse surveys,[9] road design,[10] triangulation surveys,[11] calculation of contained volumes,[12] and perspective drawing.[13] These programs have been written chiefly in the obsolescent 803 Autocode, and the problems associated with such a programming language have already been discussed in Chapter 6. (An Algol version of the closed traverse survey program[9] has been written for the author by J. S. Roper.)

With such programs, the engineer in the field can telephone the numerical data he has just obtained to his computer centre, thereby enabling it to be processed immediately by computer. The results can then be telephoned back to him and by this rapid process he can then decide whether or not his practical surveying has been satisfactory, and can make any necessary amendments before leaving the site.

In the case of photogrammetry and aerial surveying, any subsequent calculations can be effected by using a digimeter (see Chapter 6) to obtain co-ordinates from maps on to punched tape for processing by computer.

Automatic drafting machines and map plotters can be used for the reverse process of plotting surveys and drawing maps from data which has been obtained from a computer in the form of punched tape. These techniques are also described in the paper by van Gent[7] to which reference has already been made. (See the notes in Chapter 6 on graph plotters).

Other field measurements

This section is merely intended to indicate some of the other subject areas in
civil engineering to which similar attention could be paid.

   (i) Structural engineering. Field measurements of loads (live, dynamic,
      wind, crane surge, and impact), strains and deflection.

  (ii) Hydrology. Field measurements from rain gauges, flow meters in rivers
      and tidal movements.

 (iii) Public health engineering. Pollution—field measurements from samples
      of atmosphere, water supply and industrial effluent in rivers.

 (iv) Traffic engineering. Field measurements from traffic counters and speed
      meters.

  (v) Materials. Field measurements of cracking of concrete, flow of viscous
      materials (tars, bitumens and asphalt) and hardness and notch ductility
      of steels.

 (vi) Soil mechanics. Field measurements of soil and pore water pressures,
      and moisture content.

 (vii) Astronomy. Field measurements on spacecraft and man-made satellites.

This list is by no means complete, though it extends the range of activities which
can now be tackled by the data-logging techniques described earlier for what
were apparently more restricted applications. Further information can be
obtained from the Bibliography, and some practical illustrations are given in
Chapter 9.

The Processing of Experimental Data

An outline has now been given of how numerical data in the laboratory and on
site can be recorded automatically on to punched tape as a series of data 'words'.
Each 'word' contains a channel identification number followed by the
experimental reading and, if these two components are separated by a comma
and successive 'words' by CRLF, it should be possible to use any tape of data
'words' as direct input to a computer together with an appropriate program.

    The purpose of a useful initial program in all cases would be to check the
validity of the recorded data as far as possible, to ensure, for example, that the
number of readings from, and the identification of, each channel are correct.
The output from the computer indicates whether or not there are any errors or
inconsistencies in the basic data and, if so, their nature, followed by a corrected
data tape—if this is at all possible. In addition, the opportunity can be taken to
eliminate any 'words' which are not required or which have to be ignored for
some reason or other.

    Succeeding programs can then be written to process the experimental data
according to methods based on any particular theory in civil engineering
practice, and the format of the output can be either tabular or graphical. For
example, suitable programs can be written to enable the use of a digital plotter
to plot automatically the co-ordinates of all points on a graph which have been
obtained from the computer by the substitution of experimental data in relevant
formulae. This process is illustrated diagrammatically in Fig. 8.12. The total

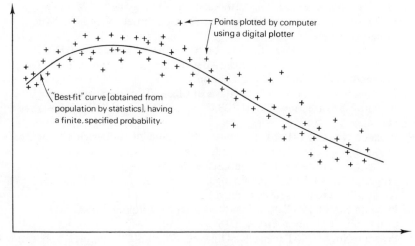

FIG. 8.12

number of points can be quite considerable, thereby producing a significant population for any statistical analysis which may follow. This may take the form of a search for the best-fit curve associated with a specified finite probability (derived from an appropriate correlation function). The best-fit curve can be plotted automatically, with load factors and factors of safety—according to the specified level of probability—being obtained as important results. It may be possible to program this entire process, though the engineer will now require expert knowledge of mathematical theory on statistics, probability and numerical analysis. Such work—restricted as it is to problems of this type—cannot possibly be given the extensive coverage it requires in this book, and reference should be made to the Bibliography.

References

1. Litton, E. and Cole, J. P. Automatic Data Processing Techniques in Civil Engineering Laboratories. Civil Engineering and Public Works Review. Vol. 61, No. 724, November 1966
2. Cole, J. P. The Design and Construction of a 100 way Automatic Scanner. Civil Engineering and Public Works Review. Vol. 62, July 1967, pp. 769-772
3. Cole, J. P. Electronics and the Civil Engineer. Electronic and Radio Technician. Vol. 3, No. 1, March 1969
4. Cole, J. P. Data Logging: making an ally of the 'difficult' art. The Engineer. Vol. 228, No. 5910, 1st May 1969
5. Cole, J. P. A User's Manual for the 2000 way, built-in logging system in the Department of Civil Engineering, Dundee College of Technology. Department of Civil Engineering, Dundee College of Technology
6. Civil Aircraft Accident: Report of the Court of Inquiry into the Accident to Comet G − ALYP on 10th January 1954 and Comet G − ALYY on 8th April 1954. Her Majesty's Stationery Office, London, 1955

7. van Gent. Some remarks on Automization in Surveying and Mapping. Fédération Internationale des Géomètres, Committees V and VI. XIth Congress—Rome 1965

8. Program Nos. LC1 & 1A. Calculation of Earthworks Quantities. Elliott 803 Computer Application Programs. (All enquiries should now be made to International Computers Ltd., London.)

9. Program Nos. LC11 & 11A. Analysis of Closed Traverse Surveys using Bowditch's Correction. Elliott 803 Computer Application Programs. (See 8 above)

10. Program No. LC13—Card—Computer Augmented Road Design. (A package of Algol programs to aid the complete design of roads.) Elliott 803 Computer Application Program. (See 8 above)

11. Program Nos. LC14 & 14A—Triangulation Survey. Elliott 803 Computer Application Programs. (See 8 above)

12. Program No. LC15—Calculation of Contained Volumes. Elliott 803 Computer Application Program. (See 8 above)

13. Program No. LC24—Perspective Drawing. Elliott 803 Computer Application Program. (See 8 above)

Bibliography

Further information can be obtained from the following textbooks, publications and manufacturers' catalogues.

Textbooks

Potma, T. Strain Gauges, Theory and Application. Phillips. 1967
Neubert, H. K. P. Strain Gauges, Kinds and Uses. Macmillan. 1968
Hammond, R. Engineering Structural Failures. Odhams. 1956 (Gives further details of Comet Tank Test)
Rahman, N. A. Exercises in Probability and Statistics. Griffin. 1967
Guest, P. G. Numerical Methods of Curve Fittings. Cambridge U.P. 1961
Fisher, R. A. Statistical Methods for Research Workers. Oliver and Boyd. 1958
Snedecor, G. W. Statistical Methods. 5th ed. Iowa State College Press. 1956

Technical Papers

Royal Aircraft Establishment, Farnborough, England. Structural Test Reports.
Proceedings of a Symposium on Wind Effects on Buildings and Structures. Vols. 1 & 2. April 2-4, 1968. Loughborough University of Technology.
Cole, J. P. Accurate Data from Weaving Traffic. Traffic Engineering and Control. Vol. 8. 1966
Codd, H. A. Instrumentation in Traffic Control. Instrument Review, April 1964

Manufacturers' Catalogues

(a) Electronic Data-Logging Equipment and Measuring Devices.

Advance Electronics Ltd., Bishop's Stortford, England. Chart Recorders and Power Supplies.

Bradley Electronics Ltd., London, England. Digital Voltmeters and Cathode Ray Oscilloscopes.

British Aircraft Corporation, Electronic Systems Group, Bristol, England. Data Logging Systems.

Bryans Ltd., Mitcham, England. X, Y and Chart Recorders.

Cosmos Instruments, Letchworth, England. Multi Channel Chart Recording Systems, Transducers, Power Supplies.

Ether, Ltd., Stevenage, England. Digital Recorders, Chart Recorders and Transducers.

Farnell Instruments Ltd., Wetherby, England. Digital Voltmeters, Power Supplies.

Foxboro-Yoxall, Redhill, England. Chart Recorders and Controllers.

Gage Technique Ltd., Weybridge, England. Structural Instrumentation Engineers.

G.E.C. Elliott Process Instruments, Ltd., London, England. Chart Recorders.

I.D.M. Electronic Ltd., Reading, England. Data-Logging Systems.

Penny & Giles, Blackwood, Mon., Wales. Transducers.

Savage & Parsons Ltd., Watford, England. Strain Recording Equipment.

S.E. Laboratories (Engineering) Ltd., Feltham, England. Digital Voltmeters, Ultra-Violet Recorders, Cathode-Ray Oscilloscopes.

Solartron Electronic Group, Farnborough, England. Digital Voltmeters, Data-Logging Systems, and Cathode-Ray Oscilloscopes.

Southern Instruments, Camberley, England. Ultra-Violet Recorders.

Telequipment, London, England. Cathode-Ray Oscilloscopes.

T.E.M. Sales Ltd., Crawley, England. Chart Recorders.

Tinsley & Co. Ltd., London, England. Strain Gauges and Recorders.

Transducers (CEL) Ltd., Reading, England. Transducers, Power Supplies.

(b) Testing Machines and Systems.

Alfred J. Amsler & Co., Schaffhouse, Switzerland. Testing Machines.

W. & T. Avery, Birmingham, England. Testing Machines.

Denison & Son, Ltd., Leeds, England. Testing Machines.

Enerpac Hydraulic Equipment, Newhaven, Sussex, England. Jacks and Pumps.

Losenhausen (Great Britain) Ltd., Redditch, England. Testing Machines and Jacking Systems.

M.A.N. Augsburg & Augsburg & Nurnberg, Germany. Testing Machines and Jacking Systems.

Schenk, G. m. b. H., A., Schwaebisch Gmuend, Germany. Testing Machines and Jacking Systems.

(c) Automatic Surveying Equipment.

Tellurometer (U.K.) Ltd., Chessington, England. Tellurometers.

Fennel, Kassel, Germany. Code Theodolite FLT.

Kern, Aarau, Germany. Code Tacheometer.

9
Miscellaneous Practical Problems and Their Solution

The engineer should now have rather more than a basic grasp of the subject—in fact he should be capable of undertaking a wide variety of problems using automatic computational techniques. However, there are still many pitfalls awaiting him and, whilst it is possibly useful experience and part of the learning process to allow him to make his mistakes and thereby be a wiser person, it might be more useful for him to first study the following reports on various activities in this field. These are mainly concerned with problems in structural engineering with which the author has been personally involved, though a few have also been selected from other branches of civil engineering (and which have been solved by other workers).

By reading these outlines, the engineer may derive benefit in the following ways.

He may have direct professional interest in the actual problem.
He may appreciate more clearly the type of problems which can be solved by these methods.
He may gain a keener understanding of the processes used, which he can then expand, develop and apply to solve his own problems.
He may devise a completely different—and much better—way of solving any problem.

In most cases the background and circumstances governing the various problems are given, as these inevitably greatly dictate the development of the solution devised.

The Structural Analysis of Large Multistorey Frames Using a Small Store Computer

The author became involved with a variety of problems in analysing large steel-framed multistorey buildings by computer when he was retained as a consultant by Redpath Dorman Long Ltd., when that Company was engaged in designing and constructing a range of such buildings in Hong Kong. Most of the techniques evolved have already been published elsewhere (see below and References), though the reader may find it convenient and useful to peruse the outline of these problems and their solution as now given.

Background to the analyses

This work arose in the early 1960s when the author was a member of the staff of what is now Sunderland Polytechnic, approximately 30 miles from Middlesbrough where the design offices of Redpath Dorman Long Ltd., were located, and the polytechnic had an Elliott 803 machine with an 8,000 word store.

In those days—as previously explained—computer manufacturers tended to provide a supply of library programs free with their computers as a sales promotion technique. Among these in this case was Program LC7[1], which deals with the elastic analysis of plane frames based on the Stiffness Method and was a forerunner to programs such as LC29, A27 and EAPF described previously. The polytechnic computer was the nearest suitable computer to the company's offices and an appreciable amount of machine time was available thereon for industrial applications—particularly so at that time because of its recent installation and the wish on the part of the polytechnic authorities to attract interest from other organisations in the area.

The senior lecturer in computing in the Polytechnic was J. S. Roper—a mathematician—and he collaborated with the author to provide an advisory service to Redpath Dorman Long Ltd., for the structural analysis of various frames devised by the company's design team led by T. V. Thompson (Chief Design Engineer), and K. G. Heward (Chief Designer).

Later, a further Elliott 803 Computer with an 8,000 word store and an additional magnetic tape backing store was installed at the University of Durham, 13 miles from Sunderland and 23 miles from Middlesbrough. J. S. Roper transferred to Durham at the same time and consequently both computer centres became available for all subsequent work on these frames. The company's design engineers were allowed unrestricted access to both computer centres and it was feasible for all personnel concerned to circulate freely between the three locations whenever necessary.

With the availability of these facilities within the same small geographical area as its design offices, it became a fairly attractive proposition for the company to proceed with research and development activities for these computers, followed by the actual production runs. Finally, in common with most engineering undertakings, a strict time limit applied to the entire procedure, so it was impossible to proceed from first principles, devising special methods and then writing lengthy programs for the analysis. Full use had to be made of the existing available software and it was with this background that the following processes were devised.

Details of the frames

These have been fully described in a paper by Thompson.[2] Several such buildings have now been erected in Hong Kong, and it is therefore understandable that many common constructional techniques, involving similar methods of shop fabrication and site erection, etc., were devised and used as far as possible for the supporting structures in each contract, while at the same time meeting the varied aesthetic and functional requirements for each individual building.

The underlying problem throughout was to design and construct supporting structures for high buildings on difficult ground near the sea front subjected to typhoon conditions of wind loading.[3] In addition, an air-conditioning duct system between floors and ceilings and small storey heights had to be provided. Moreover, because of these and other difficulties, neither shear walls nor conventional bracing could be utilised. The solution adopted by the company was to use mild steel, rigid-jointed, welded frames having haunched beams and aluminium curtain walling. The frames were then analysed by computer using elastic theory, and it was possible to erect the resulting structures very rapidly on the restricted sites.

Thus each structure consists essentially of an unbraced steel frame having rigid beam-to-column welded connections on both axes. The main beams are generally in three pieces (Fig. 9.1)–haunched at each end horizontally and also vertically where ducts are required to pass under. All the structures were for office blocks–chiefly for banking organisations–and each had special features or local site problems which presented their own additional difficulties. Examples of these are now given with reference to a particular building.

DETAILS OF HAUNCHED BEAM

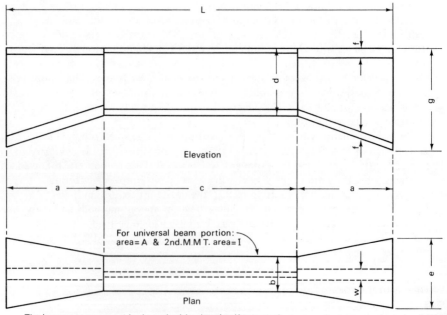

For universal beam portion:
area= A & 2nd.M.M T. area= I

Elevation

Plan

The beam may, or may not, be haunched in elevation. If not g = d.

FIG. 9.1

The Bank of Canton

This building is described by Litton, Roper and Thompson,[4] so that only brief details are given here. The structure is sited on the central district of the City of Victoria and is illustrated in Fig. 9.2. There are 21 main storeys, including the mezzanine, above ground level and a two-storey basement below. The layout of the floors is given in Fig. 9.3 and sectional elevations of two frames in Fig. 9.4, in which the main dimensions are shown.

The frames are essentially in accordance with the general specification given above; frames 3 and 4 each contain 202 members and 125 joints, while frames C and D each have 283 members and 167 joints (Fig. 9.4). The area outlined by the broken line in Fig. 9.3a represents the extent of the Banking Hall at ground level. Fig. 9.4 gives an outline of the frames at Lines 3 and C, from which it is seen that plate girders, one storey deep, span 47 ft. 9 in. over the hall and carry

BANK OF CANTON BUILDING, HONG KONG

FIG. 9.2A

BANK OF CANTON BUILDING, HONG KONG

FIG. 9.2B

the stanchions above, thus providing an uninterrupted floor space of approximately 48 ft. by 60 ft. These details required that special methods had to be devised to solve the various analytical problems presented, apart from and in addition to the other more-general problems which also have to be solved for any frame of essentially this type of construction.

General problems

Elastic analysis of large frames
The first problem was how to analyse a large regular rectangular multi-storey plane frame (Fig. 9.5) having members of uniform section using a small-store computer. (The problem of the haunched beams will be considered later.) As explained earlier, time was critical, and it was desirable to use the existing library Program LC7 if possible. However, this program could only deal with frames having up to approximately 80 joints, and, as each frame here could have up to 300 joints, it was necessary to devise methods of overcoming this difficulty.

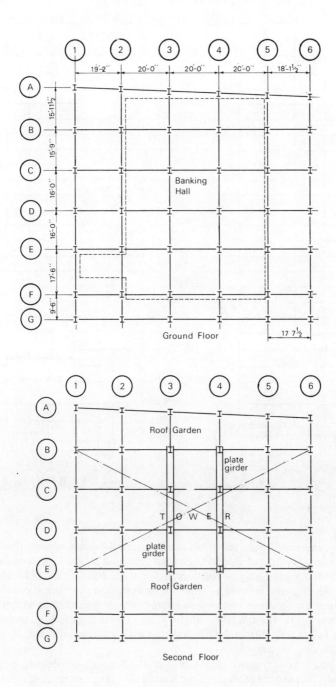

FIG. 9.3

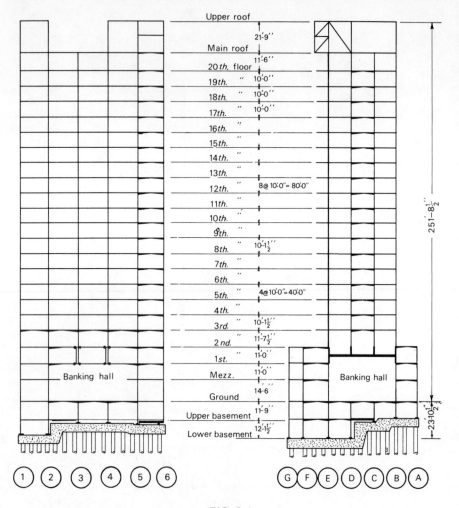

FIG. 9.4

For symmetrical frames subjected to symmetrical or skew symmetrical (e.g. wind) loading, Litton and Roper[5] evolved a method in which one symmetrical half of the frame need only be considered, and if the half-frame was within the limits of the program, it could be analysed by computer. The frame could have an odd or an even number of bays and appropriate supports had to be imposed on members cut by the axis of symmetry. This process is outlined diagrammatically in Fig. 9.6; further details, together with a numerical example, are given in the original paper.[5]

With larger, or unsymmetrical, frames, methods were again devised by Litton and Roper[6] in which each large frame was divided into a series of smaller ones, each subframe so formed being solved separately, with account being taken of compatibility in deflections, forces and moments at joints between frames. An example is illustrated in Fig. 9.7 where the main frame has been divided into

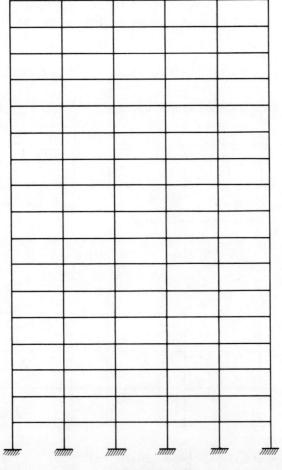

FIG. 9.5

three subframes. It can also be seen that each subframe has been made to overlap those on each side by at least two storeys, so providing an opportunity of obtaining two values for all forces and moments on members in the overlapping zone, which can be used as a check on any calculation.

Each subframe chosen was of such a size that it could be analysed in one piece by LC7, and suitable static supports were then provided to each in turn, except for the bottom subframe already containing the main frame supports. The static supports, or zero deformations, imposed on any subframe, were generally a pin and a roller, providing three reactions altogether which were applied at the bottom joints on the outer columns. This assumed that these two joints were at the same level, which was not necessarily so, but this assumption in no way affected the forces and moments in the members, although it obviously altered the deformations and rotations of all joints in the subframe. (The original paper shows how a Mohr rotational correction can be applied later to eliminate this effect.) The forces and moments acting at all joints on the section cut by the

FIG. 9.6

ELASTIC ANALYSIS OF SYMMETRICAL FRAMES

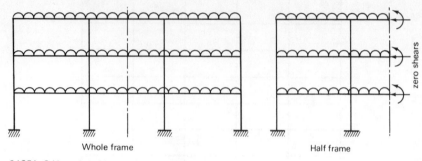

CASE 1. Odd number of bays and symmetrical loading

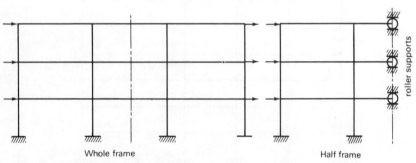

CASE 2. Odd number of bays and skew symmetrical loading

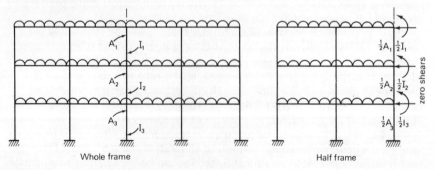

CASE 3. Even number of bays and symmetrical loading

FIG. 9.6 *(Contd.)*

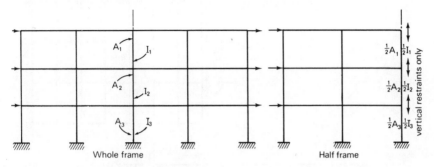

CASE 4. Even number of bays and skew symmetrical loading

subframing process then had to be applied to each subframe to complete the external loading, and the original paper shows that, when these forces and moments are known, the resulting analysis by computer is very accurate indeed.

However, herein lies the crux of the matter. Such forces and moments were not generally known and consequently it was necessary to devise a process of successive approximation which could be continued until the double set of results for common members in the overlapping zone of both subframes were within an acceptable measure of agreement.

Much development work and machine time was expended on this process until a converging series of analyses was devised. The author discovered that the error in the assumed value for any unknown vertical force greatly magnified the error in the resulting vertical deflection output for the same joint, causing the results from successive analyses to diverge or oscillate, rather than to converge to an acceptable measurement of agreement. He therefore deduced that, if vertical deformations could instead be specified as computer input, with vertical forces now being obtained as output, the error in the assumed value for an unknown vertical deformation would produce a greatly reduced error in the resulting vertical force, so causing the results from successive analyses to converge satisfactorily. In view of this, J. S. Roper modified **Program** LC7 (written in 803 Autocode) in a reasonably short period of time, so that deformations could be applied as input, and forces and moments obtained as output, for any joint, so producing **Program DC5/AU/JSR.**[7] This process is described fully in the original publication,[6] and only certain implications arising from its use are discussed here.

When production work started on the actual job, it was soon found that the analysis of one or more subframes had generally to be repeated to achieve satisfactory results. This increased the amount of computer time required and subjected all staff to considerable strain due to the need for complete accuracy and intense concentration at all stages, the dreary painstaking nature of the process, the lengthy format of tabular input and output, and complications and irritations arising from corrections and modifications. Clearly this process could not be used extensively as a typical day-to-day technique, but should be reserved for analysing special urgent problems. It thus became apparent that design staff

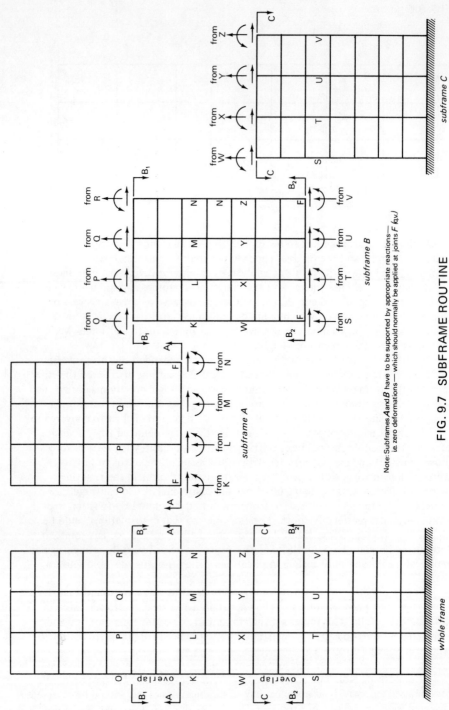

Note: Subframes A and B have to be supported by appropriate reactions—
i.e. zero deformations— which should normally be applied at joints F (q_u)

FIG. 9.7 SUBFRAME ROUTINE

would normally require and expect much more assistance from the computer than given by this technique. This is discussed more fully in Chapter 11, though the author and his associates learned a very valuable lesson from this project. (Later—in the case of the Bank of Canton building—it became essential that each complete frame should be analysed as a whole; this is discussed below.)

Treatment of haunched beams

The detail of a typical haunched beam is shown in Fig. 9.1. This consists of a central portion cut from a universal beam welded to haunches which can be tapered in both plan and elevation. The haunches are symmetrical about the midspan of the beam to reduce the number of parameters, thus simplifying the theoretical analysis, tape punching and programming difficulties.

One method of analysis would have been to subdivide each beam into a connected series of elements, each considered as having constant sectional properties, and then to use Program LC7 in the usual way. However, this would not have been satisfactory, since the number of joints and members, and the maximum difference between member end numbers (the problem with LC7 is similar to that of A27 described in Chapter 4), would then have been substantially increased, thus defeating all the advantages previously gained.

The actual method adopted is described in the paper by Litton, Roper and Thompson,[4] and only a brief outline is given here. Consider the symmetrically haunched beam AB (Fig. 9.8) fixed at B and subjected to forces P_{XA} and P_{YA}

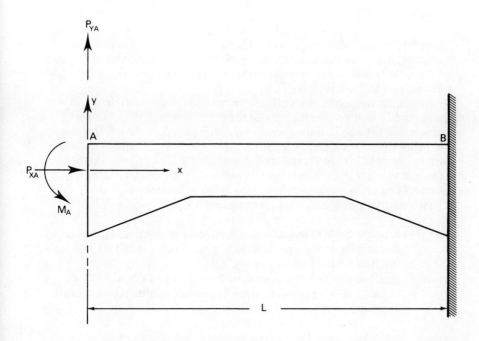

FIG. 9.8

and moment M_A and producing deflections δx_A and δy_A, and rotation θ_A at A. Then it can be shown that

$$P_{XA} = \frac{A_E E \delta x_A}{L}$$

$$P_{YA} = \frac{12EI_D \delta y_A}{L^3} + \frac{6EI_F}{L^2} \theta_A$$

$$M_A = \frac{6EI_D \delta y_A}{L^2} + \frac{4EI_E}{L} \theta_A \,,$$

or in matrix form

$$
\begin{bmatrix} P_{XA} \\ P_{YA} \\ M_A \end{bmatrix} =
\begin{bmatrix} \dfrac{A_E E}{L} & 0 & 0 \\ 0 & \dfrac{12EI_D}{L^3} & \dfrac{6EI_F}{L^2} \\ 0 & \dfrac{6EI_D}{L^2} & \dfrac{4EI_E}{L} \end{bmatrix}
\begin{bmatrix} \delta x_A \\ \delta y_A \\ \theta_A \end{bmatrix}
$$

where E is Young's Modulus, and A_E, I_D, I_E and I_F are constants for a given haunched beam. For a uniform beam A_E is the cross-sectional area of member AB, and I_D, I_E and I_F are all equal to the second moment of area of the section (see Stiffness Method, Chapter 7).

Thus in the haunched beam, A_E can be regarded as the equivalent area and I_D, I_E, and I_F as the equivalent second moments of area of the section of member AB. Because of symmetry, $I_D = I_F$, although I_E is not equal to these, and hence this stiffness matrix for AB is conveniently symmetrical (as for a uniform member). It can also be seen that it is not possible to replace the haunched member by an equivalent uniform one, since two equivalent second moments of area, I_D, and I_E, are required in the simulation.

Two computer programs were then required.

(a)   A program[8],[9] to evaluate A_E, I_D, I_E and fixed end moments and shearing forces for a given haunched beam subjected to the various specified types of loading in Fig. 9.9.

(b)   Modified versions of programs such as LC7 and DC5/AU/JSR to include the two equivalent second moments of area as input for all members.

This was the correct procedure but, because of the shortage of time, an approximation was made in case (b). Only haunched beams having values of I_D

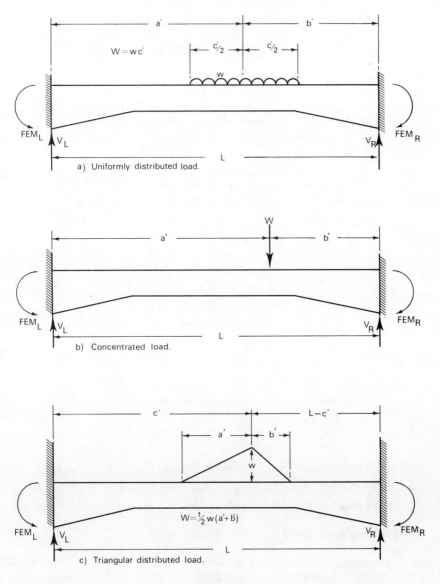

FIG. 9.9

and I_E within 5% of one another were used, and in such circumstances it was possible to dispense with the modification described.

In case (a), Program DC6/AU/JSR [8] was written by J. S. Roper from formulae devised by the author from a method based on column analogy. This program was written for the Elliott 803 computer and is no longer readily available in its original form. However, it has now been re-written by Roper in Algol for an Elliott 4100 computer and a permanent record is now available as Program DCTCES 7.[9]

ANALYSIS OF HAUNCHED BEAMS
PROGRAM NO. DCTCES 7
TYPES OF LOADING

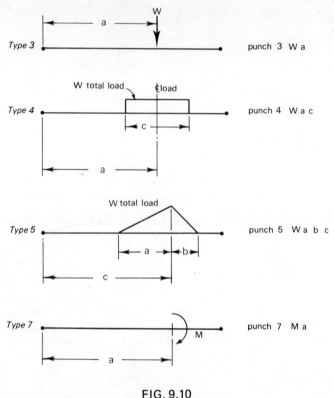

FIG. 9.10

To prepare the input data for Program DCTCES 7

The following items should be drawn up in this order.

1. Descriptive label to be enclosed between / and \.
2. Number of beams (N) to be processed.
3. The values in sequence of e, b, a, d, w, f, c, g, A and I for each beam (i.e. N lines thus) (A and I are the area and inertia of the central beam section.)
4. The number of load cases for each beam, i.e. N numbers.
5. For each load case specified, the details of the loading distribution from the types shown in Fig. 9.10.

For type 3:	punch	3	W	a	
For type 4:	punch	4	W	a	c
For type 5:	punch	5	W	a	b c
For type 7:	punch	7	M	a	

Considering a typical example (Fig. 9.11), the input to the computer is

```
/  TRIAL RUN – UNIVERSAL BEAM 16 x 6   \
   1
  10      6        60      16      0.4375    0.75      240
  35    14.70            647.2
   4
   3     10              100
   4      8              130       96
   5      6               48       24       280
   7    100               40               (CRLF)
```

and the output from the computer is

```
   TRIAL RUN – UNIVERSAL BEAM 16 x 6

         AE     IE     ID     IF    FEM1   FEM2   SH1    SH2
  1     16.48  1148   1257   1257   1173   778.4  4.486  9.514
```

HAUNCHED BEAM – NUMERICAL EXAMPLE

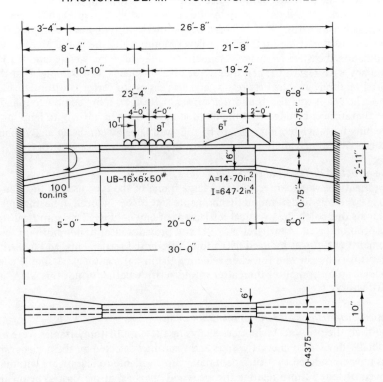

FIG. 9.11

Full details can be obtained from the relevant brochure.[9] As both I_D and I_F are given in the output, a built-in check on the accuracy of the process is provided. These values are derived independently and, of course, should be equal.

The company was then able to draw up a list or table of 'standard' haunched beams for various spans which could be used as required. Once all this data had been obtained, it was possible to analyse the frames and subframes, as described above and below.

An appraisal of some approximate methods of determining bending moments and shearing forces due to wind forces on multi-storey buildings

This investigation was initiated to obtain a good first approximation for the forces and moments acting at all joints on the section cut by the subframing technique. The author reasoned that, if accurate values for such forces and moments could first be obtained, the ensuing analysis of the various subframes by LC7 might well be acceptable at the first attempt. However, as reported above, this did not always turn out to be the case, and eventually another approach was adopted which could be used for both vertical (i.e. dead and imposed) and wind loading.

Another reason for the investigation was curiosity—just how good were the well-known and time-honoured approximate methods? The availability of Program LC7 enabled comparative data to be obtained for various 'test' frames using both the Stiffness Method (accurate!) and various approximate methods.

The work, carried out jointly by the author and W. S. Hedley, has been described elsewhere.[10],[11] Three well-known approximate techniques, the Continuous Portal Method, the Cantilever Method, and the Portal Method, as outlined in the Steel Designers' Manual[12] and elsewhere, were examined for accuracy with regard to their use by practical engineers. It was found that the Portal Method gave the best results and was easiest to perform arithmetically. Nevertheless, it was also found that errors of 100% or more could occur and it was therefore concluded that the method might be used initially to help in selecting tentative section sizes, to be followed later by computer analyses using a successive-approximation process until the sections were satisfactory.

The effect of differential settlement

Investigations were carried out on a large frame of the type used in the Hong Kong contracts to determine the moments and forces induced in beams and columns due solely to a vertical settlement of any column. This empirical study has been reported elsewhere[13], [14] and provides information both on the moments and forces induced by various differential settlements, and how these effects may be allowed for when selecting the initial scantlings. A simple check calculation by computer thereafter would verify whether or not this was satisfactory.

Design proportions and initial selection of scantlings

Whilst all the previous techniques analysing these multistorey frames were generally devised to check the sizes and scantlings selected by the designer, it was now considered that sufficient data were available to investigate whether any consistent pattern relating the areas and 'inertias' of the various beam and column members to one another was emerging. If so, it was hoped it would be possible to lay down some guide lines for design proportions and the initial selection of scantlings, so striking at the heart of the design problem and perhaps leading ultimately to an automatic procedure by computer. The company readily made suitable data available, and, supervised by the author, one of their own student engineers carried out such an investigation at Sunderland Polytechnic. No clear-cut recommendation emerged, though a study of this investigation— filed as a Project Report[15] in the polytechnic—should nevertheless indicate the kind of development work which can be undertaken in this direction.

Special effects

The following problems arose as a result of special difficulties encountered in the case of the Bank of Canton building and full details are given in References 2, 4, 6 and 17.

Elastic Analysis of large frames in one piece
Although a subframe analysis on a small computer had been used successfully for frames of the size used in this building, the presence of the 48 ft girders made a complete analysis of each frame in one operation essential.

The method of analysis in Program LC7 is described in Chapter 7 and consists of setting up and solving a matrix of equations for the frame, there being three equations for each joint, though, in the case of large frames, not all the equations can be stored in the computer at the same time. The maximum number of equations stored at any time depends on u and s, where u is the number of loading cases and s is the maximum value of difference between joint numbers at the ends of any member. Program LC7 was therefore substantially modified to form Program DC1/AU/JSR,[16] intended for use on an Elliott 803 Computer with an 8,000 word store. This proved a major undertaking and took a considerable time. There appeared to be no viable practical alternative, and it could perhaps be argued that LC7 and its (slightly) modified version[7] had already done more than could have been expected in coping thus far.

In the new program, the main frame was automatically divided into a number of subframes, each occupying all the available memory at a given time and so producing an exact analysis, unlike the approximate subframe technique described above. If no backing store such as magnetic tape was available as each subframe was partially analysed, it was overwritten by the next and had therefore to be regenerated later on any subsequent reference to it. The effective number of subframes set up is $\frac{1}{2}N(N+1)$, where N is the number of subframes required for the complete frame. As u and s increase, the number of joints in each subframe decreases, thus increasing N.

In this way it was possible to analyse very large frames, and time, rather than the size of frame, became the limiting factor. The guide to the variation in Elliott 803 computer time required, given in Fig. 9.12, shows the time required for various sizes of frames where s is 5 or 10 and u is 1 or 3. It can be seen that the time required is infinite for a frame of 250 joints when s is 10 and u is 3, thus revealing the theoretical limit in this case. However, if one load case at a time is considered (i.e. if u is 1) the time for each load case is approximately 34.6 hours, and a solution is now theoretically possible, though it is

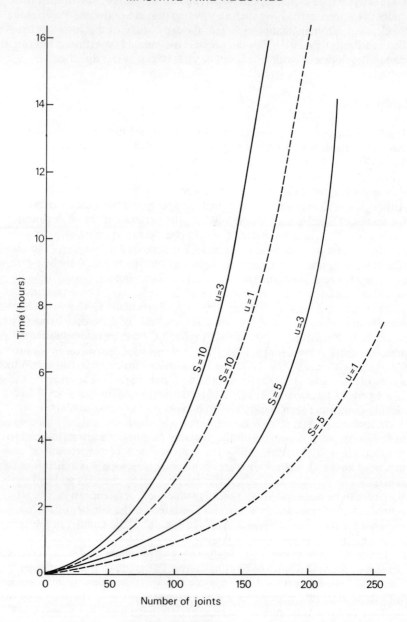

**ELLIOTT 803 COMPUTER
MACHINE TIME REQUIRED**

FIG. 9.12

recommended that not more than 20 hours machine time be taken as the practical upper limit.

The approximate subframe technique described above could now be used very effectively with this new program on very large frames to reduce the computer time required. For example, a frame of 200 joints with $s = 10$ and $u = 1$ required approximately 16 hours' machine time, but when the same frame was considered as two overlapping frames, each of 120 joints, the corresponding time for each subframe was 4 hours (see Fig. 9.12).

This process therefore saves both time and money since charges for computer time are usually quoted as a rate per hour. A similar saving was made with frames having the shape of an inverted T by a two-fold reduction in computer time, firstly by reducing the number of joints in each analysis, and secondly by reducing the value of s in the upper subframe (it being assumed that there were two subframes).

In due course, when the University of Durham computer with its additional storage capacity came into use, Program DC1/AU/JSR was modified to make use of this facility. It then became possible to store the partial analysis of each subframe in the magnetic tape backing store, thereby avoiding the need to overwrite it and regenerate it later. Thus a substantial saving in machine time was achieved. The actual modification was effected by Roper very rapidly, and other refinements and labour-saving techniques were implemented as the complete storage capacity was not now fully utilised. The resulting **Program DC3/AU/JSR**[18] then became virtually the complete answer for analysing all the frames in these contracts—not only for the Bank of Canton—but also for the others. It was also possible to use this new program as an additional check for the subframe technique given above.

Finally it should be noted that Program LC7 and its derivatives[7], [16], [18] have been written in 803 Autocode, and hence if these had to be translated into modern high-level languages such as Algol or Fortran the problem would be quite sizeable. A better approach would be to start with a modern program in **Algol** such as A27 and adapt it in a similar manner to those described above.

Simulation of deep girders

The plate girders 13 ft. deep in the building (Figs. 9.3 and 9.4) necessitated adopting a simulation process to effect a conventional frame analysis by computer. These girders differ from all the other members, which can be satisfactorily portrayed by one-dimensional linear elements, and the problem strictly became one to be solved by the Finite Element Method, in which the girders could be divided into a series of two-dimensional elements, with all other members being treated more simply as linear elements. However, this approach would have made it impossible to use Program LC7 and its derivatives, and writing a completely new program to embrace finite element theory was certainly not considered to be a practical solution.

A more realistic method was to try and effect a simulation which could be analysed by LC7. Accordingly, attempts were first made to substitute either braced or Vierendeel girders of similar inertia as the equivalent structure for the purpose of analysis—but this idea had to be rejected, because the large joint-number differences s became prohibitive for frames of this size on a small-store computer.

The device finally adopted was to represent the girder by a line at mid-storey height between the first and second floors, using the area and 'inertia' values of the girder and assigning greatly enhanced inertia values (i.e. the actual inertia x 100) to the length of the supporting and supported stanchions embraced by the girders. By this means it was considered that reasonable account was taken of the inability of the respective stanchions to deform freely within the 13 ft depth of the girders and also of the tendency of the girders to force the stanchions to follow their own curvature under load. It then became possible to analyse the frames using LC7 and its derivatives.

Allowance for the three-dimensional effects

Sectional diagrammatic plans and elevations of the building in the region of the plate girders are shown in Fig. 9.13. The frames were analysed by a series of two-dimensional analyses of all frames parallel to frame 3, and then followed by a further series of two-dimensional analyses of all frames parallel to frame C (i.e. perpendicular to line 3). However the vertical deflections at joints K, L, M and N in Fig. 9.13 warranted that special treatment should be accorded to the four frames on lines C, D, 3, and 4. It is clear that all members and joints on the vertical columns C3, C4, D3 and D4 are common to the plane frames of lines 3 and C, 4 and C, 3 and D, and 4 and D, respectively, and hence compatibility of vertical deflections and axial column loads should exist for all joints and members in the pairs listed. (All other compatibilities were ignored from a consideration of the practical details of the structure.)

A method of obtaining a close approximation to this state of affairs has been fully described elsewhere,[4] and involves a succession of plane frame analyses of the frames on lines C, D, 3 and 4, using LC7 and its derivatives. This is essentially a unit load method and was actually devised with respect to the space frame and loading given in Fig. 9.14. When successfully developed for that small test frame, it was applied to the Bank of Canton building.

The computer analysis and results for the frame and loading in Fig. 9.14 are given in Fig. 9.15. It is possible to use LC7 for all the analyses of a frame of that size consisting of members having uniform sections. It can be seen from the final results that the axial loads in the members and the vertical deflections of the joints on all the columns common to two frames agree completely with the double set of results given for each of these quantities.

These results do no more than comply with the compatibilities specified above (for vertical deflections and axial column loads on common members), and should not therefore be regarded or interpreted as a complete, three-dimensional elastic analysis (which would involve torsion of the members and further minor compatibilities, etc.). Nevertheless, the method does effectively give results within a designer's authoritative and knowledgeable concept of the behaviour of the structure, and as such, is clearly satisfactory.

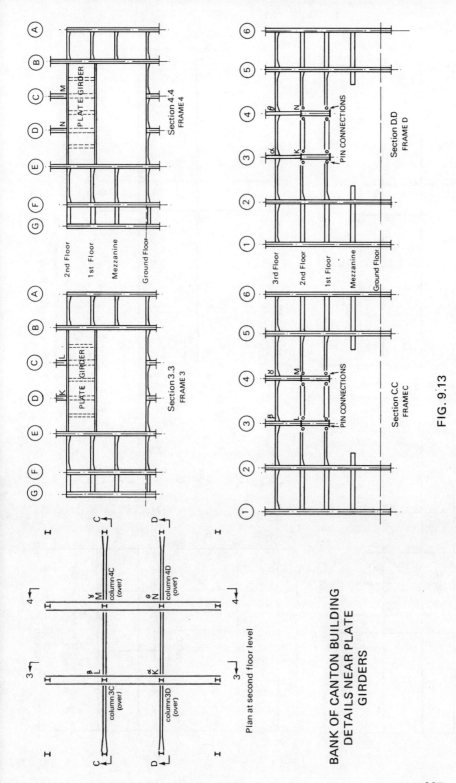

2nd Floor
1st Floor
Mezzanine
Ground Floor

Section 4.4
FRAME 4

PLATE GIRDER

Section 3.3
FRAME 3

PLATE GIRDER

3rd Floor
2nd Floor
1st Floor
Mezzanine
Ground Floor

PIN CONNECTIONS

Section D.D
FRAME D

PIN CONNECTIONS

Section C.C
FRAME C

FIG. 9.13

Plan at second floor level

column 4C (over)
column 4D (over)

column 3C (over)
column 3D (over)

BANK OF CANTON BUILDING
DETAILS NEAR PLATE
GIRDERS

307

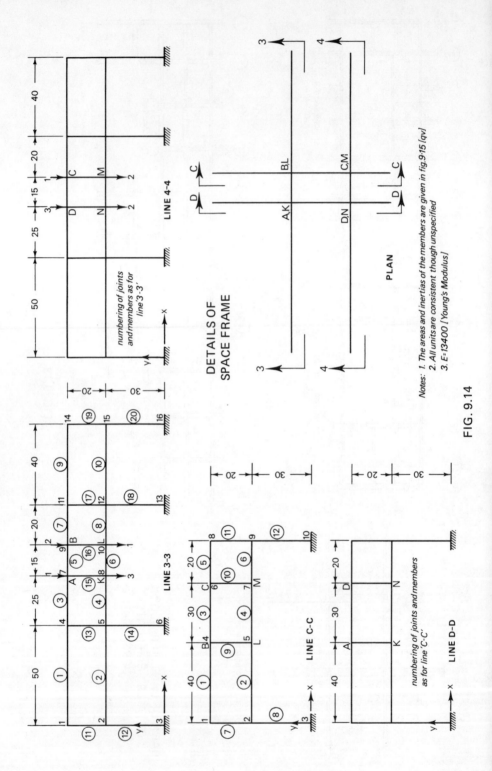

DETAILS OF
SPACE FRAME

Notes: 1. The areas and inertias of the members are given in fig.9.15 [qv]
2. All units are consistent though unspecified
3. E=13400 [Young's Modulus]

FIG. 9.14

LINE 4-4

LINE 3-3

LINE C-C

LINE D-D

PLAN

numbering of joints and members as for line '3-3'

numbering of joints and members as for line 'C-C'

308

FIG. 9.15.
Analysis of Space Frame — Numerical Example

It is required to effect the elastic analysis of the frame shown in FIG. 9.14 in such a way as to take account of the compatibilities between the vertical deflections of the joints K, L, M and N in respect of the plane frames C—C, D—D, 3—3 and 4—4 under the given applied loading.

Sectional Properties of the Members

Line 3-3.   Line 4-4.

Member No.	A	I	Member No.	A	I
1-10 inclusive	10	50	1-10 inclusive	15	60
11,13,15,16,17 & 19	15	60	11,13,15,16,17 & 19	20	70
12,14,18 & 20	20	100	12,14,18 & 20	30	110

Line C-C.   Line D-D

Member No.	A	I	Member No.	A	I
1-6 inclusive	10	50	1-6 inclusive	15	60
7,9 & 12	15	60	7 & 10	20	70
8	20	100	8	30	110
10	20	70	9 & 11	15	60
11	10	50	12	20	100

Analysis of Line 3-3.

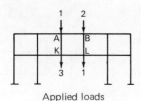
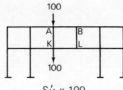
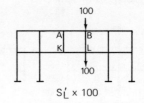

Applied loads $S'_K \times 100$ $S'_L \times 100$

Total load at K = $3 + S'_K$;	Total load at B = $2 + S'_L$,
Total load at A = $1 + S'_K$;	Total load at L = $1 + S'_L$,

where S'_K and S'_L are unknown shears. (See method given in text.) The above three loading cases can be analyzed by computer, using unit values (100) in the 2nd and 3rd load cases.

FIG. 9.15 (*Contd.*)

Analysis of Line 4-4.

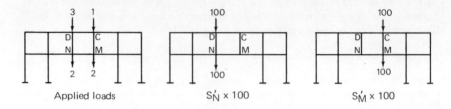

Applied loads $S_N'\times 100$ $S_M'\times 100$

Total load at N = $2 + S_N'$;   Total load at M = $2 + S_M'$,
Total load at D = $3 + S_N'$;   Total load at C = $1 + S_M'$,

where S_N' and S_M' are unknown shears.

Once more these three loading cases can be analysed by computer.

From the results obtained from lines 3-3 and 4-4, the loading to be applied to lines C-C and D-D can be determined using the method and the following analyses effected by computer:—

Analysis of Line C-C.

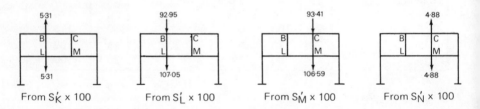

From $S_K'\times 100$ From $S_L'\times 100$ From $S_M'\times 100$ From $S_N'\times 100$

Analysis of Line D-D.

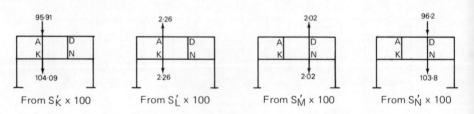

From $S_K'\times 100$ From $S_L'\times 100$ From $S_M'\times 100$ From $S_N'\times 100$

From the results for these cases obtained by computer, account can now be taken of compatibilities between vertical deflections at:

Joint K in lines 3-3 and D-D,
joint L in lines 3-3 and C-C,
joint M in lines 4-4 and C-C, and
joint N in lines 4-4 and D-D.

In this way the following simultaneous equations in S_K', S_L', S_M' and S_N' can be obtained from the computer output:

FIG. 9.15 *(Contd.)*

Vertical deflection at K $= -0.0048 + 0.00153675S_K' + 0.00112485S_L'$ [3-3]

$\qquad\qquad = -0.0000391 - 0.00402195S_K' - 0.0000011S_L'$ [D-D]
$\qquad\qquad\quad + 0.0000001S_M' - 0.0016354S_N'$

Vertical deflection at L $= -0.00415 + 0.0011233S_K' + 0.0012813S_L'$ [3-3]

$\qquad\qquad = \;\; 0.0000293 - 0.0000025S_K' - 0.0048839S_L'$
$\qquad\qquad\quad -0.00201375S_M' - 0.0000002S_N'$ [C-C]

Vertical deflection at M $= -0.00386 + 0.0009137S_N' + 0.0010415S_M'$ [4-4]

$\qquad\qquad = -0.0000068 + 0.S_K' - 0.0020101S_L'$
$\qquad\qquad\quad -0.0021744S_M' - 0.0000016S_N'$ [C-C]

Vertical deflection at N $= -0.0045 + 0.0012559S_N' + 0.0009147S_M'$ [4-4]

$\qquad\qquad = 0.0000178 - 0.0016316S_K' + 0.S_L'$
$\qquad\qquad\quad - 0.0000006S_M' - 0.0017696S_N'$ [D-D]

Hence:

K. $555.86S_K' + 112.59S_L' - \;\; 0.01S_M' + 163.54S_N' = 476.09$

L. $112.58S_K' + 616.52S_L' + 201.37S_M' + \;\; 0.02S_N' = 417.93$

M. $0.S_K' + 201.01S_L' + 321.59S_M' + \;\; 91.53S_N' = 385.32$

N. $163.16S_K' + \;\; 0.S_L' \;\; + \;\; 91.53S_M' + 302.55S_N' = 451.78$

Using a library program to solve these four equations:—

$\qquad S_K' = 0.4777; \qquad S_L' = 0.3710; \qquad S_M' = 0.6725; \qquad S_N' = 1.0322.$

The output deflections, forces and moments for all joints and members can now be obtained, either from the existing results by further manual calculations, or by a completely fresh analysis by computer which would now include the numerical values of S_K', S_L', S_M' and S_N' in the input (so providing the required results, completed by a single load case per line).

FIG. 9.15 (*Contd.*)

The final results for each line by the latter method are:

LINE 3-3:   FINAL RESULTS

DEFORMATIONS

JOINT	X-DIRECTION	Y-DIRECTION	ROTATION
1	-0.00013	0.00002	0.00000
2	-0.00015	0.00001	0.00000
3	0.00000	0.00000	0.00000
4	-0.00010	-0.00040	-0.00006
5	-0.00021	-0.00028	-0.00003
6	0.00000	0.00000	0.00000
7(A)	-0.00026	-0.00357	-0.00003
8(K)	-0.00013	-0.00365	-0.00003
9(B)	-0.00045	-0.00319	0.00006
10(L)	0.00001	-0.00314	0.00008
11	-0.00053	-0.00050	0.00007
12	0.00002	-0.00035	0.00004
13	0.00000	0.00000	0.00000
14	-0.00047	0.00004	0.00001
15	-0.00006	0.00003	0.00001
16	0.00000	0.00000	0.00000

| | MEMBER | | FORCES | | MOMENTS | |
N (	BETWEEN	JOINTS	AXIAL	SHEAR	END 1	END 2
1	1	4	-0.09	-0.06	-0.78	-2.37
2	2	5	0.15	-0.03	-0.17	-1.09
3	4	7	0.85	1.09	12.81	14.42
4	5	8	-0.39	1.33	16.70	16.64
5	7	9	1.72	-0.24	-5.92	2.30
6	8	10	-1.26	-0.38	-7.80	2.10
7	9	11	0.54	-1.32	-13.49	-12.85
8	10	12	-0.08	-1.56	-14.46	-16.80
9	11	14	-0.22	0.15	4.06	2.03
10	12	15	0.27	0.09	2.33	1.17
11	1	2	-0.06	0.09	0.78	0.92
12	2	3	-0.09	-0.06	-0.75	-1.05
13	4	5	1.15	-0.94	-10.44	-8.29
14	5	6	2.51	-0.40	-7.33	-4.58
15	7	8(AK)	-0.81	-0.87	-8.50	-8.84
16	9	10(BL)	0.55	1.18	11.19	12.36
17	11	12	1.47	0.76	8.79	6.38
18	12	13	3.12	0.41	8.08	4.14
19	14	15	-0.15	-0.22	-2.03	-2.34
20	15	16	-0.24	0.05	1.17	0.32

FIG. 9.15 (*Contd.*)

LINE 4-4: FINAL RESULTS

DEFORMATIONS

JOINT	X-DIRECTION	Y-DIRECTION	ROTATION
1	-0.00014	0.00001	0.00000
2	-0.00012	0.00001	0.00000
3	0.00000	0.00000	0.00000
4	-0.00012	-0.00025	-0.00004
5	-0.00015	-0.00017	-0.00002
6	0.00000	0.00000	0.00000
7 (D)	-0.00022	-0.00262	-0.00002
8(N)	-0.00010	-0.00258	-0.00002
9(C)	-0.00034	-0.00220	0.00005
10(M)	-0.00001	-0.00222	0.00006
11	-0.00038	-0.00029	0.00005
12	-0.00000	-0.00020	0.00003
13	0.00000	0.00000	0.00000
14	-0.00034	0.00003	0.00001
15	-0.00005	0.00002	0.00001
16	0.00000	0.00000	0.00000

	MEMBER		FORCES		MOMENTS	
NO	BETWEEN	JOINTS	AXIAL	SHEAR	END 1	END 2
1	1	4	-0.07	-0.06	-0.68	-2.16
2	2	5	0.12	-0.02	-0.14	-0.95
3	4	7	0.80	1.02	11.84	13.63
4	5	8	-0.44	1.15	14.38	14.28
5	7	9	1.51	-0.44	-6.86	0.28
6	8	10	-1.15	-0.33	-6.76	1.77
7	9	11	0.45	-1.07	-10.67	-10.75
8	10	12	-0.09	-1.35	-12.62	-14.48
9	11	14	-0.19	0.13	3.48	1.82
10	12	15	0.23	0.09	2.30	1.22
11	1	2	-0.06	0.07	0.68	0.69
12	2	3	-0.08	-0.05	-0.55	-0.88
13	4	5	1.08	-0.87	-9.68	-7.68
14	5	6	2.24	-0.31	-5.75	-3.62
15	7	8(DN)	0.51	-0.71	-6.77	-7.52
16	9	10(CM)	-0.30	1.06	10.39	10.85
17	11	12	1.20	0.65	7.27	5.65
18	12	13	2.65	0.33	6.53	3.25
19	14	15	-0.13	-0.19	-1.82	-2.06
20	15	16	-0.22	0.03	0.83	0.18

FIG. 9.15 *(Contd.)*

LINE C-C: FINAL RESULTS

DEFORMATIONS

JOINT	X-DIRECTION	Y-DIRECTION	ROTATION
1	-0.00058	-0.00011	-0.00003
2	-0.00042	-0.00008	0.00000
3	0.00000	0.00000	0.00000
4(B)	-0.00073	-0.00320	0.00000
5(L)	-0.00034	-0.00315	-0.00001
6(C)	-0.00096	-0.00221	0.00006
7(M)	-0.00016	-0.00223	0.00007
8	-0.00102	-0.00029	0.00008
9	-0.00013	-0.00021	0.00005
10	0.00000	0.00000	0.00000

	MEMBER		FORCES		MOMENTS	
N C	BETWEEN JOINTS		AXIAL	SHEAR	END 1	END 2
1	1	4	0.50	0.32	6.06	6.93
2	2	5	-0.26	0.37	7.55	7.33
3	4	6	1.04	-0.04	-1.82	0.73
4	5	7	-0.79	-0.01	-1.72	1.46
5	6	8	0.43	-0.57	-6.41	-5.07
6	7	9	-0.19	-0.83	-7.89	-8.63
7	1	2	0.32	-0.50	-6.06	-3.94
8	2	3	0.70	-0.24	-3.62	-3.67
9	4	5(BL)	0.56	-0.54	-5.11	-5.61
10	6	7(CM)	-0.17	0.61	5.68	6.43
11	8	9	0.57	0.43	5.07	3.53
12	9	10	1.40	0.24	5.10	2.19

LINE D-D: FINAL RESULTS

DEFORMATIONS

JOINT	X-DIRECTION	Y-DIRECTION	ROTATION
1	-0.00085	-0.00010	-0.00002
2	-0.00057	-0.00007	0.00000
3	0.00000	0.00000	0.00000
4(A)	-0.00099	-0.00357	-0.00000
5(K)	-0.00050	-0.00365	-0.00001
6(D)	-0.00121	-0.00264	0.00007
7(N)	-0.00033	-0.00259	0.00008
8	-0.00128	-0.00032	0.00009
9	-0.00030	-0.00023	0.00006
10	0.00000	0.00000	0.00000

FIG. 9.15 (*Contd.*)

N O	MEMBER BETWEEN JOINTS		FORCES AXIAL	SHEAR	MOMENTS END 1	END 2
1	1	4	0.70	0.44	8.33	9.16
2	2	5	-0.36	0.52	10.68	10.19
3	4	6	1.46	0.01	-1.80	2.03
4	5	7	-1.12	-0.00	-2.42	2.30
5	6	8	0.64	-0.87	-9.71	-7.74
6	7	9	-0.30	-1.19	-11.05	-12.72
7	1	2	0.44	-0.70	-8.33	-5.72
8	2	3	0.96	-0.34	-4.96	-5.27
9	4	5(AK)	-0.76	-0.76	-7.35	-7.77
10	6	7(DN)	0.66	0.82	7.68	8.75
11	8	9	0.87	0.64	7.74	5.01
12	9	10	2.06	0.34	7.72	2.52

From these results it can be seen that the pairs of values for the vertical deflections at K, L, M and N, which are taken from the results for two appropriate intersecting frames, are in close agreement, while those for the axial forces in members AK, BL, DN and CM are in fair agreement.

Conclusions

The design staff normally require and expect much more assistance from the computer than is given by any of these techniques. The staff were subjected to excessive routine work in which care and concentration were of paramount importance, and it is clearly desirable that these duties too should be programmed, the trend being towards fully automatic design. Towards this goal it would have been helpful to have included more checks on the input data, an extension of the calculations to include stresses as output and indicating whether any section is satisfactory or not, a pictorial representation of input and output enabling the designer to see at a glance errors or whether sections are over- or under-stressed (possibly by writing suitable programs for the digital plotter), and more assistance with initial design steps involving automatic trial-and-error techniques.

Finally it must be admitted that there are other (and perhaps better) ways of undertaking such work, possibly by using bigger and more expensive computers, if these are readily accessible to the engineer. However, invariably the approach to be adopted by a particular engineer will be influenced by his answer to the question, 'With due regard to my own special circumstances, what is the best approach to my own computational problems?'

Welded Plate Girders

Details of a welded plate girder are shown in Fig. 9.16. A common design problem is, given the span and loading, to find suitable values for all dimensions in the cross section. The girder may, or may not, have lateral restraints and be of uniform section. This is a real design problem this time (as compared with the techniques for check calculations and analysis given above for multistorey

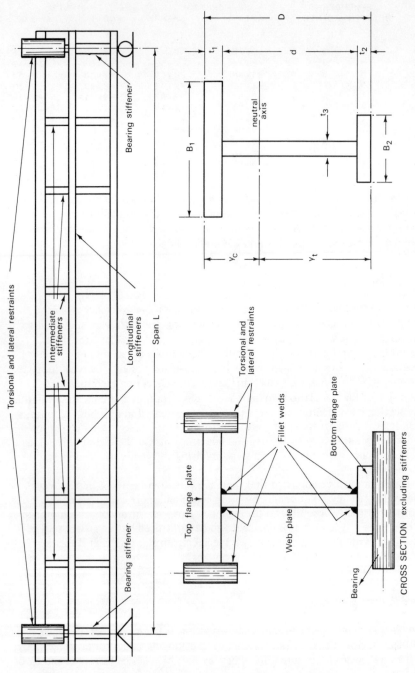

FIG. 9.16

316

frames) and the author, in common with other civil engineers, has battled with it many times using a trial and error approach until arriving at a satisfactory solution. It therefore seemed desirable to try to establish a general method for expediting the design calculations for obtaining the optimum cross sectional dimensions of such a girder in accordance with the requirements of various design codes and standards, and accordingly the author carried out a series of design investigations in which all numerical calculations were accomplished by computer. The work is described in detail elsewhere,[19],[20] and only a brief outline—with notes on more recent developments—is given here. The main investigation was of welded mild steel plate girders with a uniform I-shaped section throughout, simply supported and with effective lateral and torsional restraints only at the ends of the span. The calculations were carried out in imperial units in accordance with BS 449:1959[21] and minimum weight was taken as the criterion of efficiency. If

M_R = maximum moment of resistance of section.

f_{bc}, f_{bt} = maximum compressive stress in compression flange and maximum tensile stress in tension flange under applied loading and self-weight of girder, respectively

p_{bc}, p_{bt} = maximum permissible compressive and tensile stresses, respectively, as given in BS 449.

y_c, y_t = distance from X–X axis to extreme fibres of compression and tension flanges, respectively.

BM_{max} = maximum bending moment on span due to applied loading and self-weight of girder.

K_1 = function given in Table 5 of BS 449.

λ = quantity depending on nature of support of compression flange given in BS 449, Clause 26.

I_x = second moment of area of section about X–X axis.

S_{max} = maximum shearing force that can be applied to any given section.

A' = cross-sectional area.

Then in the limit when $BM_{max} = M_R$

$$M_R = \frac{f_{bc}I_x}{y_c} \quad \text{and} \quad M_R = \frac{f_{bt}I_x}{y_t}$$

where $f_{bc} \leqslant p_{bc}$ and $f_{bt} \leqslant p_{bt}$.

In order to identify and consider all possible stress distributions on a given section, the stress ratio R is introduced, where

$$R = \frac{p_{bc}}{p_{bt}} \Big/ \frac{f_{bc}}{f_{bt}} = \frac{f_{bt}}{p_{bt}} \Big/ \frac{f_{bc}}{p_{bc}}$$

The following conditions can then be deduced.

If $R < 1$, the tension flange is understressed and the compression flange is fully stressed; therefore $f_{bc} = p_{bc}, f_{bt} = (y_t/y_c)p_{bc} < p_{bt}$, and $M_R = (p_{bc}I_x)/y_c$.

If $R = 1$, both the tension and compression flanges are fully stressed simultaneously; therefore $f_{bt} = p_{bt}, f_{bc} = p_{bc}$, and $M_R = (p_{bc}I_x)/y_c$ or $M_R = (p_{bt}I_x)/y_t$.

If $R > 1$, the tension flange is fully stressed and the compression flange is understressed; therefore $f_{bt} = p_{bt}, f_{bc} = (y_c/y_t)p_{bt} < p_{bc}$ and $M_R = (p_{bt}I_x)/y_t$.

Thus the moment of resistance M_R, and hence the maximum bending moment BM_{max} which may be applied, can be obtained for a given span and a given section. For any particular section, p_{bt} and p_{bc} may be obtained according to the regulations in BS 449.[21]

It is also possible to specify four criteria which must be satisfied so that a specified shearing force might be applied, from which S_{max} can be determined. All the calculations were carried out by computer using the specially written programs now described.

Details of the computer programs

Program ABEL 1[22]

This basic program provides, inter alia, M_R for a given section and was written in Algol for an Elliott 803 Computer by A. Bergson from information obtained or devised by the author. The computer input consists of span L, sectional dimensions B_1, t_1, B_2, t_2, D and t_3 (Fig. 9.16), together with constants λ and K_1. The output includes $M_R, A, p_{bc}, f_{bc}, f_{bt}$, slenderness ratio $1/r_y$, and stress ratio R, together with section identification. The program follows the design procedure given in BS 449[21] and has the following features. The three cases of section (1, 2, and 3 on pages 29 to 31 of BS 449) are automatically allowed for and the output indicates which case has been used. All sections failing to comply with the restrictions in BS 449 are automatically rejected, the reason being given in the output. The output indicates which of four categories of web stiffening is required. and is given tabularly. This is essentially a check calculation for any proposed section of girder. (Numerical examples and further details are given in references 19 and 20.)

Program ABEL 1A[23]

This is a modified version of the previous program giving greater accuracy in R as required in some of the design investigations.

Program ABEL 2[24]

When B_2 is varied from its minimum to maximum values while all other sectional dimensions are kept constant, the graph of M_R against A' consists of a line defined as a contour. There are certain significant points on a contour and this program obtains the various values of B_2 for such points, so enabling Programs ABEL 1 or 1A to be used precisely for these values.

Program ABEL 3[25]

A significant point on any contour occurs where R equals 1. This is the fully stressed section and features prominently in all design optimisation studies. This program, therefore, based on ABEL 1A, locates the section on a given contour at which R equals 1.

Program ABEL 4[26]

This program provides S_{max} from a consideration of four different criteria based on average shear, local intensity of shear, and welding of the compression and tension flanges to the web.

Use of the programs

These programs have all been used extensively with a range of numerical data and a general method was obtained of finding the sectional dimensions of the girder having the minimum weight necessary to resist a given moment on any specified span. This method can be used to prepare large-scale charts and tables for use in a design office, and as the basis for automatic design at minimum weight. (In the latter case a special, all-embracing computer program might be written.) These two practical applications have yet to be implemented—the delay being due to the introduction of SI units, with the need for new or modified British Standards and plate sizes, and to the departure of the author from the centre used to obtain these calculations. These difficulties are now considered.

Recent developments and practical difficulties

In 1968 the author transferred from Sunderland Polytechnic to Dundee College of Technology and in the process experienced many of the difficulties outlined in Chapter 6. Any engineer might be subjected to such problems on changing from one computer installation to another, either in the same or a different establishment, and he has then to decide what to do about the software and development work he has been evolving—possibly at great expense and with considerable effort over many years—for a particular machine or system. The author considers that insufficient attention has been paid to such problems, and is convinced that benefit will be derived from studying the method adopted in this particular case.

The ABEL series of programs, written in Algol for an Elliott 803 machine, was available on five-channel punched tape in 803 teleprinter code, whereas the installation at Dundee College of Technology consisted of an Elliott 4100 machine with facilities for eight-channel punched tape in 4100 teleprinter code. The use of library translation programs was considered and rejected (see Chapter 6), and since the flow diagrams and print-outs of the existing programs were not readily available (this can frequently be a problem), these were written afresh by A. Q. Agnew[27] and A. R. McNicol[28] under the supervision of the author. The opportunity was taken to rectify any errors and inaccuracies, to alter some of the subroutine techniques used, and to modify the layout of input and output data to use the line-printer now available. Thus the WPG series of programs (in Algol) was compiled.

Program WPG 1[29]

This basic program is essentially a translation of ABEL 1.[22] However, the opportunity was taken to improve the subroutine obtaining p_{bc} from the critical stress C_s by interpolation, to arrange to print the output data on a line printer, and to re-arrange the format of both input and output data.

Program WPG 2 and 2A[30]

These programs perform the same task as ABEL 3[25] but include several refinements. Program WPG 2A uses a slightly different technique which generally reduces the machine time required.

Program WPG 4[31]

All the versions of this program have been devised to study the effect of lateral buckling from different types of lateral restraint and of a change of section. This is achieved by varying λ and N. The output can be either graphical (using a digital plotter), or tabular, according to which of the five versions is used.

Program WPG 5[32]

Again this is essentially a translation, though this time of ABEL 4,[26] and provides the critical shearing force which may be applied to any section of girder.

Use of the programs

These programs have been used to carry out further investigations and full details are recorded elsewhere. Agnew[27] attempted to make some progress towards automatic design, while McNicol[28] investigated the use of lateral restraints and non-uniform sections, and studied critical shearing forces in detail.

Introduction of SI units

Whilst all previous work (including both the ABEL and WPG programs) had been carried out in imperial units, the author, in spite of having just established the WPG series of programs at his new computer centre, nevertheless found he could deal reasonably favourably with the introduction of SI units to the design of these members.

There are four distinct time periods for which provision must be made during the transition from imperial to SI units.

1. Imperial units in operation. BS 449:1959 (imperial units) in use.
2. Optional units in operation, i.e. either imperial or SI units may be used, though only one version of BS 449:1959 (imperial units) is available.
3. SI units in operation. An interim issue of BS 449:1959 Part 2[38] is now available (being essentially a straightforward conversion of the existing Standard with some rounding-off, but not constituting a major revision).
4. SI units in operation. A major revision of BS 449 in use.

At the time of writing the third period has just been reached, and the first period has already been fully catered for.

In the second period, it was first necessary to adapt the WPG series of programs to SI units in accordance with the current state of affairs. Only minor modifications were necessary, involving the inclusion of conversion factors at the beginning and end of the existing programs, together with a slight re-arrangement of the print statement. The basic procedure was as follows.

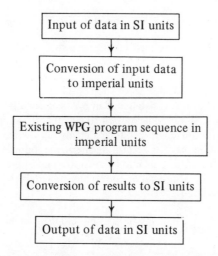

This process can be readily applied to the entire WPG series, and has been used by D. A. H. Smith[33] to modify Programs WPG1 and 2A, resulting programs being WPG1M[34] and WPG2AM[35] respectively. In the third period a complete revision of the basic programs was necessary, and has been done by Smith in respect of Programs WPG1 and 2A, the resulting programs being WPG1MM[36] and WPG2AMM.[37] respectively. (These latter programs are in accordance with BS 449: 1969 Part 2.[38] This does not constitute a major technical revision, and either version of the Standard may be used provided that one set of values is used consistently in any particular design.)

At present, plate sizes in SI units are not readily available and hence it was not considered desirable to prepare design charts or an automatic-design program at this stage, it possibly being wiser in any case to await the introduction of the major revision of BS 449 before proceeding with such major undertakings.

Design studies in SI units

This work by Smith has been reported elsewhere,[33] and includes an introduction to the design procedures adopted by several other countries, which are compared with the UK procedure. The grades of steel available in each country were then investigated and a comparative study made of minimum-weight girders for short, medium, and long spans, with particular reference to the USA and UK codes. (Two programs were written for designing girders to the USA Code,[39] using essentially the techniques of the previous work.[19], [20])

The ultimate goal

The target at which the author has been aiming is a system in which the automatic design program will be linked to an assembly jig, plate-rolling mills and furnace, within a single complex to form a fully integrated process, so that a request for a girder to a specified span and load can be rapidly and economically fulfilled, thereby extending the range already covered by universal beam and autofab sections and encouraging a greater everyday use of larger spans. This work is being continued.

Analysis of Readings From Electrical Resistance Strain-Gauges

Among the first problems to be tackled in processing experimental data recorded by an automatic logging system (such as described in Chapter 8) are possibly those analysing the results from electrical resistance strain-gauges (see Chapter 8, and references 1 and 2, for details of such strain-gauges).

Consider a strain-gauge of resistance R in a Wheatstone Bridge circuit of the type shown in Fig. 8.2. If a load is applied to a gauge, the resulting increase ΔR of the resistance of the gauge will produce a change in the reading of the digital volt meter, which can be used to determine the strain induced in the specimen. Ideally the initial DVM reading should be zero, though in practice it is more realistic to allow for a non-zero value due to any unavoidable original lack of balance in the bridge. Then

$$\text{strain} = \frac{\text{change in reading on DVM screen}}{S \times (\text{GF})} \hspace{2em} (1)$$

where S is the scaling factor and GF the gauge factor.

$$(\text{Note:} \quad \text{strain} = \frac{\Delta R}{R} \div \text{GF})$$

If the DVM forms part of a data logging system and the readings are recorded on punched tape, a simple program based on equation (1) will provide the values of strain directly from the data tape, together with the results of any subsequent calculation. The format of such programs depends greatly on the logging system employed and the results required. The following example illustrates the procedure.

Strain-gauge rosettes

In a two-dimensional stress system the conditions of stress and strain at any point can be completely defined if the values of the strains in three different directions at the point are specified, together with Young's Modulus and Poisson's Ratio. This can be achieved experimentally by positioning three strain-gauges as in Fig. 9.17 to deal with the condition at point X, such an arrangement being known as a strain-gauge rosette. The three linear gauges are inclined at angles α and β to one another, typical values being 60° and 120°, or 45° and 90°, though other values can also be used.

For any given loading condition the strains at X in the three directions can be obtained from the gauges, and from this information the two principal strains and the two principal stresses at X, and the inclinations of these to a reference direction (generally taken through the axis of strain-gauge no. 1) can then be determined.

Program DCTCEL 1: Strain-gauge rosette analysis
For the logging system in the Department of Civil Engineering, Dundee College of Technology, Program DCTCEL 1[40] written in Algol by J. R. Thorpe for an Elliott 4100 computer provides these results directly from the data tape for

STRAIN GAUGE ROSETTE

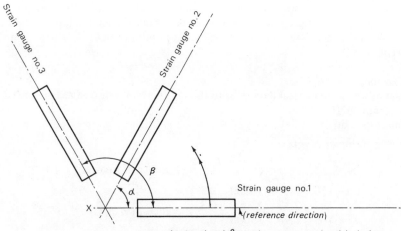

note: Angles α and β are always measured anticlockwise from the reference direction

FIG. 9.17

several rosettes in a single operation. The program accepts as input data 'words' on punched tape directly from the logging system. The computer input consists of:

  (i) A descriptive title between / and \,
 (ii) The number of rosettes,
(iii) For each rosette in turn, the values of α, β, gauge factor, Young's Modulus and Poisson's Ratio, in that order; and
 (iv) The three data 'words' for each rosette—these being given in a specified sequence (the engineer will doubtlessly recall that each data 'word' consists of the channel identity number, followed by the DVM reading and separated by a comma). (See Chapter 8 for full details.)

The output from the computer includes:

  (i) The program number and title;
 (ii) The descriptive title;
(iii) A print out of input data;
 (iv) The following results in tabular form.

Rosette number	Reference gauge channel identity number	Angle (deg.)	Principal strain	Principal stress	Angle (deg.)	Principal strain	Principal stress

Notes

In this particular system facilities exist for adjusting the initial readings of the DVM to zero, so that only one set of three data 'words' is required from each rosette. The scaling factor S is constant for the system, and has therefore been incorporated in the program so that it can be excluded from the input data. Full details of the formulae used and the program print out can be obtained from the brochure.[40]

Example

The following numerical data refer to the results from four rosettes processed by Program DCTCEL 1.

Input to computer

/ TEST CASE \

4   Number of rosettes

60	120	2.1	13000	0.3	
45	135	2.17	13000	0.3	Values of α, β, GF, E, σ
45	105	1.98	13000	0.3	
45	60	2	13000	0.3	

01,+0605   data "words"
02,−0063   for
03,+0425   rosette 1

04,+0228   data "words"
05,+0000   for
06,+0134   rosette 2

07,+0031   data "words"
08,−0214   for
09,−0302   rosette 3

10,−0077   data "words"
11,−0185   for
12,−0122   rosette 4

Output from computer

PROGRAM NO. DCTCEL 1: STRAIN GAUGE ROSETTE ANALYSIS
J. THORPE 1970

TEST CASE

ROSETTE DATA

ROSETTE NO.	ALPHA	BETA	GAUGE FACTOR	YOUNG'S MODULUS	POISSON'S RATIO
1	60.0	120.0	2.1	13000	0.30
2	45.0	135.0	2.2	13000	0.30
3	45.0	105.0	2.0	13000	0.30
4	45.0	60.0	2.0	13000	0.30

ROSETTE NUMBER	REFERENCE GAUGE CHANNEL IDENTITY NUMBER	ANGLE (DEG.)	PRINCIPAL STRAIN	PRINCIPAL STRESS	ANGLE (DEG.)	PRINCIPAL STRAIN	PRINCIPAL STRESS
1	1	−22.453317	.00068708	9.5020977	67.546683	−.00007311	1.9001699
2	4	−11.297296	.00022247	2.7540387	78.702704	−.00009897	− .46042452
3	7	− 7.6245709	.00003841	−1.0143178	82.375429	−.00036471	−5.0454813
4	10	33.735842	−.00020025	−2.0063213	−56.264158	.00019935	1.9896064

The reference gauge, from which angles α and β and those to the direction of the principal strains and stresses are all measured anti-clockwise, is always the first to which any reference is made in the set of three for any rosette.

Conclusion

Two different kinds of programs can be identified and used in such work:
a special program only applicable to a given installation using the experimental results directly in their punched tape form, (though it is perhaps desirable to use another special program first of all to check the 'crude' data for errors or inconsistencies) (Program Number DCTCEL 1 is such an example in this category); or
a more general library program apply to experimental results from any installation, (though normally much preparatory work would first be necessary here to obtain the input data in the required form).

This is only a beginning. It should now be possible to refine and extend DCTCEL 1 to obtain, for example, the maximum shear stress and its direction, strain energies, etc., if required. As experimental laboratory techniques are developed and improved, so will it become possible to modify this program further.

Automatic Recording of Experimental Measurements

Further examples of the automatic or semi-automatic recording of experimental or operational data from other electrical measuring devices such as transducers, are now given.

Collapse tests on model portal frames and beams

The author has described elsewhere[41],[42] the series of experiments which he has carried out on apparatus of his own design for testing to collapse various model steel portal frames and beams. The equipment used was simple, with loads applied and experimental data read manually. This restricted the range of experiments to fairly fundamental issues, though with data logging equipment and displacement transducers (both linear and rotational), and the development of an automatic loading device (consisting essentially of a suspended tank filled with water from a source controlled automatically in various specified ways by a valve), it is possible to study the effects of the rate of loading, stepped loading, and continuous loading, etc. Moreover, the joint rotations and linear displacements of significant points on the specimens can be measured by transducers and the results recorded, either as a continuous trace using a pen or ultra-violet recorder, or on punched tape at finite time intervals. These experiments will be reported in due course though it is now possible to illustrate a load test on a model portal frame with a fixed base (Fig. 9.18). The loading is vertical (being applied manually at midspan by steel weights), the vertical deflection at the point of loading is measured by a deflection transducer, and the results therefrom (and from strain-gauges) are recorded either digitally (on the DVM or on paper tape) or as a continuous trace (on an ultra-violet or pen recorder). These results can then be processed further using special programs for the computer and the graph plotter.

LOAD TEST ON A FIXED BASED, MODEL STEEL PORTAL FRAME

FIG. 9.18A

LOAD TEST ON A FIXED BASED, MODEL STEEL PORTAL FRAME
(Close-up View of Specimen)

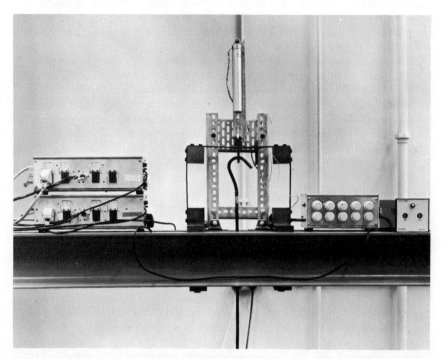

FIG. 9.18B

Investigations into the design and construction of welded crane jibs

The author became involved with this project when in charge of research in the
Department of Civil Engineering at Sunderland Polytechnic. The work,
sponsored by the British Crane and Excavator Corporation Ltd.,* was carried
out in the laboratories of the Polytechnic. Full details will be given elsewhere[43]
in due course.

Typical welded jib specimens were fabricated by the company in lattice
construction using rectangular hollow tubular steel sections; these were
approximately 15 ft long and had four main booms at 2-ft centres tied together
by square, diagonal and skew bracing to form a jib of square section. Each
specimen was placed horizontally in the loading frame and subjected to a stepped,
compressive loading up to a maximum of 100 tons by a jacking system, the
loading being applied either at, or eccentric to, the centroid of the jib section and
acting along, or parallel to, the longitudinal axis of the specimen.

The test rig was equipped with electrical measuring and logging devices, and
the results recorded on five-channel punched tape. Among the physical

* Now Coles Cranes Ltd.

quantities measured were the oil pressure in the jacks (using a pressure transducer), the applied load (using a load cell), the linear deflections in three perpendicular directions (using displacement transducers), and the strains in various directions (using electrical resistance foil strain-gauges). Up to 100 channels were available at a single time for these purposes in the logging system designed and constructed by J. P. Cole.

The loading was applied manually up to each threshold, where it was held constant until a complete set of measurements were recorded. A special computer program is to be written to process these results and finally, a theoretical space frame analysis will be effected using a suitable library program (e.g. one from the STRESS System—Reference 3, Chapter 6) to compare the theoretical and experimental results. (The author was obliged to withdraw from this project before its completion due to his departure from Sunderland.)

The Clyde tidal model

This was a scale representation of the river and estuary from the tidal limit several miles above Glasgow to a line between Innellan and Wemyss Bay (see Fig. 9.19) and was constructed by George Wimpey & Co. Ltd. for the Clyde Port Authority. Much of the design of the model and the collection of data necessary for its construction was carried out by the Civil Engineering Department of the University of Strathclyde, under the direction of the late Professor William Frazer. All variations in the bed of the river and firth were carefully reproduced. The inflowing water at the seaward end was a brine solution of the correct density, this flow being continually pumped at a rate in excess of the maximum flood tide demand, as is the usual practice in the controlled weir method of tidal generation, while 13 fresh water sources were accurately metered to represent the major tributaries true to scale. Tides were introduced according to any pre-determined programme using fully automatic equipment which actuated and controlled the rise and fall of the tide control gate (Fig. 9.20) in response to error signals from comparison of measured and programmed water level. An investigation of the river itself to collect data necessary for the design and operation of the model was carried out continuously over a period of three years. The Clyde Port Authority's Hydrographer, using the survey launch 'Crannog', studied depths in the river and firth and the levels of the sand and mud flats below Bowling. The University of Strathclyde, with the survey vessel 'Strathclyde', measured currents, temperature, and salinity variations, and the suspended silt content throughout the area. A mass of information on tides was also available, since automatic tide gauges have been operating at various points from Gourock to the Broomielaw for more than 70 years. Electronically operated and automatic recording tide gauges were provided on the model at locations corresponding to those of the automatic gauges on the river. The information from the model tide gauges, and also details of model velocities and salinity were then recorded for comparison purposes.

It has been established that the model reproduced the conditions known to exist on the river and estuary, and investigations were carried out on several projects which could be vital to the development of the Clyde, including the long projected full reclamation of the tidal flats. The results from some of these

CLYDE TIDAL MODEL

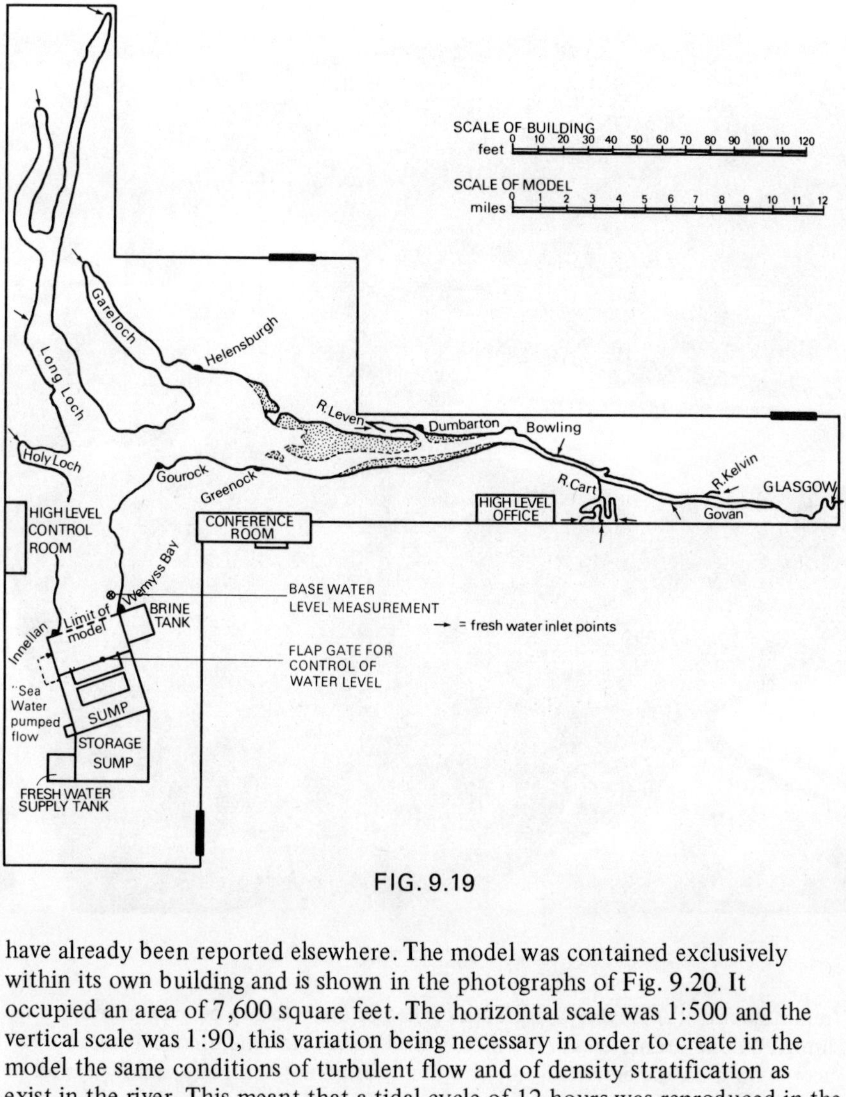

FIG. 9.19

have already been reported elsewhere. The model was contained exclusively within its own building and is shown in the photographs of Fig. 9.20. It occupied an area of 7,600 square feet. The horizontal scale was 1:500 and the vertical scale was 1:90, this variation being necessary in order to create in the model the same conditions of turbulent flow and of density stratification as exist in the river. This meant that a tidal cycle of 12 hours was reproduced in the model in approximately 15 minutes and one year's tides were simulated in a week's continuous operation.

Engineering installations

A brief mention must be made of the use of telemetry in operating, controlling, or monitoring some engineering installations such as hydro-electric power stations, reservoirs, water supply systems and sets of traffic signals. (Telemetry includes both telemetering and telecontrol, meaning 'remote metering' and

CLYDE TIDAL MODEL

FIG. 9.20A

'remote control'.) Most hydro-electric schemes are possibly best operated as an integrated, co-ordinated complex of several power stations, since advantages can accrue both on the electrical side (by the interlinking of electrical power) and on the hydraulic side (by having a single policy for water storage and flow, especially for all power stations served by the same river).

An example in this category is the North of Scotland Hydro-Electric Board 'Strathfarrar and Kilmorack' project. This 102 MW network, which includes four main power stations, two compensation stations and an area switching centre, is completely monitored and controlled from the master station at Fasnakyle.

The display of system parameters at Fasnakyle is extremely detailed. Two large mimic type panels cover electrical and hydraulic values respectively, while the four sub-panels on the master control console show the main generator readings for each of the four remotely-controlled main power stations.

CLYDE TIDAL MODEL

FIG. 9.20B

Telemetering presentation on the hydraulic side includes water levels at each of the dams serving the four unattended main generating stations, together with the flow rates at appropriate points along the river system (which has a catchment area of 350 square miles). The corresponding level and flow rate values for Fasnakyle itself are also shown on the hydraulic diagram. Relevant alarm/annunciator indication is given on each of the three main display sections, the hydraulic panel being arranged to cover excessive 'screen loss'. (The method of detecting this condition depends on the accurate measurement of the levels on the upstream and downstream sides of the screens interposed in the loch dams, these measurements of level being obtained from vibrating cylinder pressure transducers placed on opposite sides of the screen.) Further details of this interesting application on the use of telemetry are available.[44], [45]

In addition, the Sunderland District Water Board uses some telemetering and

CLYDE TIDAL MODEL
AUTOMATIC RECORDING EQUIPMENT

FIG. 9.20C

telecontrol for operating and monitoring its domestic and industrial water supply
system. The Dee and Clwyd River Authority, Chester, has also developed a fairly
advanced system of river gauging by telemetry, and the Wye River Authority,
Hereford, use telemetry to provide flood warnings. Sets of traffic signals in the
city of Glasgow are controlled by telemetry.[46]

Conclusions

By their very nature many of the applications of automatic techniques in civil
and structural engineering mentioned have had to be described in highly
subjective terms. The author is certainly not trying to dictate to the reader the

correct way, but rather has tried to illustrate typical problems and how these might be tackled in various circumstances. Of course, there are other, perhaps better, solutions and it is hoped that the reader will now be better equipped to cope with whatever assignment confronts him in this field.

References

1. Program No. LC7. Structural Analysis of a Plane Framework. Elliott 803 Applications Group Program. Sept. 1962
2. Thompson, T. V. 'Building high in a typhoon zone—Hang Seng Bank, Hong Kong'. The Structural Engineer, Vol. 44, No. 5, May 1966, pp 171-181
3. Hong Kong Ordinance No. 68 1955 (revised 1960). 'Buildings'
4. Litton, E., Roper, J. S. and Thompson, T. V. Computer Calculations for High Buildings in Hong Kong. Int. Symp. on the Use of Electronic Digital Computers in Structural Engineering, University of Newcastle-upon-Tyne, July 1966, Working Session No. 5, Paper No. 5
5. Litton, E. and Roper, J. S. Symmetrical Frame Solution by Computer. Engineering, Vol. 197, p 82, 1964
6. Litton, E. and Roper, J. S., The Structural Analysis of Large Multi-Storey Frames Using a Small-Store Computer. Proceedings of the Institution of Civil Engineers, Vol. 34, pp 201-214, Paper No. 6945, June 1966
7. Roper, J. S. Program No. DC5/AU/JSR. University of Durham 1965
8. Litton, E. and Roper, J. S. Program No. DC6/AU/JSR. University of Durham 1965
9. Litton, E. and Roper, J. S. Program No. DCTCES 7. Department of Civil Engineering, Dundee College of Technology, 1968
10. Litton, E. and Hedley, W. S. An Appraisal of Some Approximate Methods of Determining Bending Moments and Shearing Forces due to Wind Forces on Multi-Storey Buildings. Proc. of the Symp. on Wind Effects on Buildings and Structures, Loughborough University of Technology, Vol. 2, Paper No. 37, April 1968
11. Hedley, W. S. Wind Force Analysis of Multi-Storey Frames. B.Sc. Honours Project Report, Department of Civil Engineering, Sunderland Polytechnic, 1965. (Supervised by E. Litton)
12. The Steel Designers' Manual. 4th ed. Crosby Lockwood, 1972
13. Litton, E., and Buston, J. M. The Effect of Differential Settlement on Steel Framed Multi-Storey Buildings. The Structural Engineer, Vol. 46, No. 11, Nov. 1968
14. Buston, J. M. Structural Analysis due to the Effects of Settlement on Multi-Storey Buildings. B.Sc. Honours Project Report, Department of Civil Engineering, Sunderland Polytechnic, 1966. (Supervised by E. Litton)
15. Hutchinson, K. T. An Investigation into the Design Proportions of Multi-Storey Buildings. B.Sc. Honours Project Report, Department of Civil Engineering, Sunderland Polytechnic 1967. (Supervised by E. Litton)
16. Roper, J. S. Program No. DC1/AU/JSR. University of Durham, 1965
17. Murray, N. M.—Litton, E. Published Correspondence on Paper No. 6945 (see 6 above). Proceedings of the Institution of Civil Engineers, Vol. 35, p. N9. November 1966

18. Roper, J. S. Program No. DC3/AU/JSR. University of Durham, 1965
19. Litton, E. Design Investigations on Some Welded Mild Steel Plate Girders. M.Sc. Thesis, University of Durham, 1967
20. Litton, E. Design Investigations on Some Welded Mild Steel Plate Girders. Proceedings of the Institution of Civil Engineers, Vol. 39, Page 583, Paper No. 7086 S, April 1968
21. British Standard 449:1959. The Use of Structural Steel in Building. British Standards Institution, 1959
22. Litton, E. and Bergson, A. Program No. ABEL 1. Sunderland Polytechnic and the University of Durham, 1965
23. Litton, E. and Bergson, A. Program No. ABEL 1A. Sunderland Polytechnic and the University of Durham, 1965
24. Litton, E. and Bergson, A. Program No. ABEL 2. Sunderland Polytechnic and the University of Durham, 1965
25. Litton, E. and Bergson, A. Program No. ABEL 3. Sunderland Polytechnic and the University of Durham, 1966
26. Litton, E. and Bergson, A. Program No. ABEL 4. Sunderland Polytechnic and the University of Durham, 1966
27. Agnew, A. Q. Initial Techniques for an Automatic Design Process for Welded Plate Girders. B.Sc. Honours Project Report, Department of Civil Engineering, Dundee College of Technology, 1969. (Supervised by E. Litton)
28. McNicol, A. R. Miscellaneous Investigations on Some Welded Mild Steel Plate Girders. B.Sc. Honours Project Report, Department of Civil Engineering, Dundee College of Technology, 1969. (Supervised by E. Litton)
29. Agnew, A. Q. and McNicol, A. R. Program No. WPG1, Department of Civil Engineering, Dundee College of Technology, 1969
30. Agnew, A. Q. Program Nos. WPG 2 and WPG 2A. Department of Civil Engineering, Dundee College of Technology, 1969
31. McNicol, A. R. Program No. WPG 4. Department of Civil Engineering, Dundee College of Technology, 1969
32. McNicol. A. R. Program No. WPG 5. Department of Civil Engineering, Dundee College of Technology, 1969
33. Smith, D. A. H. Comparative Design Studies on Plate Girders from Various Countries. B.Sc. Honours Project Report, Department of Civil Engineering, Dundee College of Technology, 1970. (Supervised by E. Litton)
34. Smith, D. A. H. Program No. WPG 1M. Department of Civil Engineering, Dundee College of Technology, 1970
35. Smith, D. A. H. Program No. WPG 2AM. Department of Civil Engineering, Dundee College of Technology, 1970
36. Smith, D. A. H. Program No. WPG 1MM. Department of Civil Engineering, Dundee College of Technology, 1970
37. Smith, D. A. H. Program No. WPG 2AMM. Department of Civil Engineering, Dundee College of Technology, 1970
38. British Standard 449: Part 2: 1969. Metric Units. The Use of Structural Steel in Building. British Standards Institution, 1969
39. Specification for the Design, Fabrication and Erection of Structural Steel for Buildings. American Institute of Steel Construction, 1963

40. Thorpe, J. R. Program No. DCTCEL 1. Department of Civil Engineering, Dundee College of Technology, 1971
41. Litton, E. Laboratory Experiments on Plastic Theory. The Engineer, Vol. 219, page 754, 1965
42. Litton, E. Collapse Tests on Model Steel Portal Frames. The Engineer, Vol. 221, page 801, 1966.
43. Buston, J. M. Investigations into Crane Jib Construction and Design. M.Sc. Thesis. University of Durham. (To be published)
44. Young, R. E. Telemetry Engineering. Iliffe Books, 1950
45. 'Strathfarrar Hydro-Electric Scheme', Elec. Rev., 172, No. 21, Iliffe ; Electrical Publications, Ltd., London, (May 24th 1963)
46. Holroyd, J. and Hillier, J. A. Area Traffic Control in Glasgow—a summary of results from four control schemes. Traffic Engineering and Control, Vol. 11, pp 220-223. September 1969

10
The Approach to New Problems

Chapter 9 reports some actual engineering problems and their solution in various circumstances using automatic computational techniques. These include examples in which an organisation was not as adequately equipped as it would have wished to be to deal with the 'enquiry' suddenly thrust upon it, and while the solution then devised might reveal considerable ingenuity, it must inevitably be dominated by what is expedient and possible, rather than by what is ideally desirable (in the widest possible sense).

It is part of an engineer's job to cope with such situations, and the challenge can be both exciting and stimulating. Nevertheless, clearly a happier state of affairs is to be prepared for such eventualities if possible, and thus there is a need for much more preliminary research and development in all branches of civil engineering affected by automation. In fact, some allowance to prepare for introducing new techniques in advance of undertaking actual 'production' work is essential.

How would a particular project be tackled if the engineer has a free choice of time and resources? Can any lesson be learned from the previous examples? It is from this background that the following comments to assist the beginner in his approach to new problems in selected fields of activity are made.

General Notes

An attempt should first be made to specify what is involved in or required of each topic in a particular context, and a method then devised to make the optimum use of the basic resources to achieve that end. The 'basic resources' consist chiefly of:

(a) staff—deployment and attitudes being two important aspects;
(b) equipment and facilities—with particular reference to availability and location;
(c) time—how much or how little?; and
(d) methods used at each stage of the process.

Major changes 'in the order of things' involving a considerable financial outlay may have to be effected to implement the new techniques to any great extent. However, there is a wide choice, from the renting or hiring to the outright purchase, of appropriate software and hardware, and it is most

336

important for the engineer to decide in good time which approach he should recommend his own organisation to adopt.

A possible ideal is ready access to a computer or computers, for which programs are available to deal with the size and type of undertaking generally designed, constructed, or investigated by the organisation concerned. The input and output format should be readily able to be prepared, checked, modified, scrutinised, or used by the staff in a convenient, pleasing form, possibly involving graphical displays with coloured traces. The programs should ensure that as much as possible of the arithmetical manipulation is performed by the computer, thus avoiding a burden of routine, manual work which would, in fact, nullify all the advantages gained and would introduce errors, cause delays, create staffing problems, and so forth.

Since items (a), (b) and (d) have already been discussed in previous chapters, it is now only necessary to highlight some problems which the engineer might encounter when choosing which way to achieve the specification, when setting up the resulting system or process, or when operating the system on actual 'production' work.

Cost analysis of the various alternatives can be investigated with other considerations such as time, convenience, reliability and accessibility. Some of the possibilities, with their advantages and disadvantages, are given in the list below.

Process	*Advantages and disadvantages*
1. Setting up own exclusive system (equipment, staff, and methods).	*Advantages* Complete control and accessibility. *Disadvantages* High costs; capital outlay; overheads; problem of under-utilisation; maintenance; training of staff.
2. Buying machine time and using own staff, (e.g. computer 'workshop').	*Advantages* Much less expensive; virtually no capital outlay; no responsibility for installation or its operation. *Disadvantages* Uncertain service which can be withdrawn or changed at any time. No guarantee of machine time, especially when urgently required. Problem of geographical location. (If lucky, this can be an advantage.) Great dependence on many services, equipment, and persons, over which one has no control.

Process	Advantages and disadvantages
3. Asking 'experts' to under-take all duties.	*Advantages* No capital outlay; no 'tie-up' of own resources; no problems on systems and methods as these are left to experts. Quick results can be expected (time can be specified.) *Disadvantages* High dependence on experts, their methods and equipment. (One might be virtually totally ignorant of all aspects of their work, seeing the experts as 'black boxes' giving the necessary results!) No 'feel' or commitment in methods used (one has to accept experts' results at their face value—their credibility possibly being enhanced by persuasive 'sales talk'!) Lack of permanency—problems exist if service is withdrawn or changed, leaving one heavily committed to techniques about which, as explained, one may have little knowledge.
4. Setting up team involving part-time consultants.	*Advantages* Inexpensive; capital outlay can be held down; expert advice readily available; reasonable amount of control over all personnel; training on methods used can be provided for own staff. *Disadvantages* Possible doubts on reliability and lack of permanence. (Precautions must be taken on this score!)
5. Writing own programs for own or other computers and loggers.	*Advantages* Complete control over methods; relatively small capital outlay. *Disadvantages* Heavy staff commitment; problem of training staff; time delays; uncertainties regarding problems and development work which might arise; difficulties of access to other computers (in respect of their availability and geographical locations).

Process	Advantages and disadvantages
6. Having programs written by experts for own or other computers.	*Advantages* No capital outlay and low commitment from own staff (in this respect); prompt results can be expected (can be specified, though realistically, not always achieved). Own methods can be programmed. *Disadvantages* Great dependence on many services, persons, and equipment over which one has no real control. Cost of service can escalate, e.g. if experts and their computers are fairly inaccessible.
7. Drawing from, and contributing to, resources on a national basis, e.g. engineering institutions, Governmental research laboratories, national computer centres, or university computer centres.	*Advantages* Concentration of skills, methods, equipment and expert staff and services in single system to provide wide range of hardware and software for one's use. No capital outlay; low commitment from own staff. Standardisation of computer data—this is most useful. *Disadvantages* No control over system; great dependence on many services, persons and equipment over which one has no control. Lack of permanency on some equipment and methods (e.g. others can decide to effect changes directly against own interests.) Time delays; lack of priority for own project. General geographical and physical inaccessibility (though one can be lucky in this respect.)
8. Forming a local group to set up joint system shared by several users in same area.	*Advantages* Some control possible over operating system and providing resources; good accessibility; capital outlay shared by several organisations; pooling of software between participants. *Disadvantages* By joining such a group, one's business competitors might obtain information which can be used against one's own interests; possible strife between rival factions for machine time, subsequent developments or policy.

Process	Advantages and disadvantages
9. Use of own console in time-sharing computer complex.	*Advantages* Exclusive control of, and accessibility to, own input and output terminal. Capital outlay shared by several organisations; availability of much software written by experts; opportunity to use wide range of hardware. *Disadvantages* Lack of permanence (changes might be decided by others); problem of priorities; dependence on others.

This list can readily be extended—perhaps even by the reader, now—though its objective—to help the engineer make up his mind which approach to adopt in his own particular circumstances—has possibly been achieved. The author sees all the above possibilities (and others) as eminently satisfactory methods of obtaining a solution in certain instances, and does not intend malicious criticism of any process by listing the disadvantages. If the table enables the reader immediately to appreciate and overcome any difficulties in his own context, then that more than justifies the making of the point.

The day may not be too far distant when a systems analyst will devise a program to appraise all the possible alternatives open to a particular organisation, and by taking account of various relative values awarded for different considerations (by the organisation concerned according to its priorities), will indicate which process is the best 'buy'.

The above notes can apply to virtually any subject area, and attention will now be paid to special points in selected fields of interest.

Structural Analysis

This topic is dealt with extensively throughout the book and is therefore only mentioned briefly here. Most calculations are to check or prove a structure for strength, stability, deformation or some other limit state. The ideal arrangement here is possibly to have ready access to a computer for which a range of programs is available for the size and type of structures to be designed, constructed or analysed. The engineer who finds such needs met on his own doorstep is very fortunate, and it is more usual for him to have to set up a system either from scratch or from existing facilities which have been in no way created specially for his present needs.

A greater use of experimental work and load tests in structural analysis seems likely in the coming years, and this possibility has been mentioned briefly in Chapter 7, where it was suggested that the co-ordination of theory and practice to 'prove' a structure might become a standard modern procedure. Thus the engineer may wish to provide for the introduction of logging devices and for the instrumentation of structures when planning his approach to automatic computational techniques in respect of structural analysis. Such a decision means that analytical methods must be devised to use these facilities in this way, either by the engineer's own staff or by others. Thus a considerable amount of forward planning, research and development and staff training, is desirable and perhaps essential before the new system is used on actual production work.

Feasibility Studies

The introduction of automatic computation provides ideal opportunities to investigate many more possible solutions to problems than could have been contemplated before. If an appropriate program is first prepared, it is relatively simple to try out a wide range of values—some perhaps being quite unconventional—to see whether a more economical solution can be obtained.

For example, in the design of welded plate girders it might be feasible to use such girders to carry greater loads over longer spans, exceeding all present ideas of maximum dimensions. The methods already developed[1] are particularly suitable for large deep girders and efficient design dimensions could be obtained from the existing range of special programs, and the resulting practical engineering problems analysed. The steel manufacturers and welding technologists could then be presented with specific problems and a clear incentive to obtain a solution. This, then, is the type of approach which could be applied to a variety of problems in different subject areas.

Design

The specification here, for example, might be the automatic design of a complete scheme or of some basic elements, or the provision of some aids to design. Both specifications are now considered in respect of the design of structures, though again the widest possible interpretation and application of these notes is intended.

(A) Automatic design

To illustrate a possible approach to automatic design the multistorey frame example from Chapter 9 is first used.

Multistorey steel framed buildings
By studying the various problems relating to the design of the multistorey frames as outlined, the engineer will deduce that it may be feasible to write a program to provide the scantlings of all sections in a frame fabricated according to this type of construction. This would then virtually be the automatic design of the structural steelwork for such a building.

This program has not been written, nor will it be here. However, some ideas on the synthesis of a method which might be adopted are provided. The essential constituents would include:

(i) Basic data on loading and layout dimensions, with criteria for efficient or acceptable design;

   (ii) A table giving all practical beam and column sections which might be used;

  (iii) An initial choice of scantlings for all sections;

  (iv) Programs (now relegated to subroutines) to analyse frames of such construction;

  (v) Programs (or subroutines) to obtain stresses or stress ratios (of the form $f_{bc}/p_{bc} + f_c/p_c$) from the structural analysis;

  (vi) Decision taking (e.g. Are the sections satisfactory or not? If not, are they over- or under-stressed in any individual case?);

  (vii) Subsequent choices for all sections resulting from decision taking;

 (viii) Repetition of (vi) and (vii) until all sections are satisfactory;

  (ix) Print-out of recommended sizes for all sections;

   (x) Structural analysis of final design for all loadings; and

  (xi) Stresses and stress ratios for the final design and all loadings,

---

together with a series of systems programs to carry out these steps.

Such an ambitious undertaking could clearly only be contemplated for a computer with a large backing store.

Compiling such a system would be a mammoth task for a team of systems analysts. Some such projects having wider more-general applications have already been tackled in the USA (e.g. see the notes on STRESS, FRAN and ICES STRUDL 1 in Chapter 6, and references 3 to 5 in that chapter. Another system—GENESYS—is now being developed in Britain by R. J. Allwood at the University of Technology, Loughborough).

Welded plate girders

Reference is made in Chapter 9 to design investigations[1] by the author on welded plate girders, which resulted in a method for determining the dimensions of the girder of minimum weight required to support a given load on a given span. Although, for the reasons explained in Chapter 9, such a program has not yet been written, the writing of a program for automatic design becomes a much simpler task in this case, and should be capable of use on a typical small-store computer. This approach to automatic design can be within the grasp of the average engineer.

(B) Aids to design

Item (A) above dealt essentially with the 'true' structural design problem, part of which consists of the fairly creative task of determining member sizes from a loose overall specification. Complementing this task are the more mundane duties such as check calculations, design of joints, taking-off quantities, load tests and testing of materials, and it is clear that programs could readily be devised to solve, or assist in solving, these problems. Still further aids to design can be provided by charts, tables and standard details prepared by computer to suit the needs of any particular organisation.

Moreover, if the designer, a most important person, can be relieved of dull routine duties and assisted in his more-creative work, he will be more efficient, his morale will be raised, and he will have the time and opportunity either to consider many more alternative solutions or to undertake additional projects.

Experimental Work

Having identified and specified the various duties, their nature (temporary or permanent), the equipment required, and whether this is to be bought, borrowed or hired, should be established. The engineer may well initiate these advances in his own organisation and have to plead the case for introducing the new methods (in experimental work) and to justify the financial outlay by showing the savings and benefits which would accrue by their adoption.

Obtaining approval for introducing automatic techniques and devices for experimental work might be difficult in some organisations, especially in those that have not used such methods before. This can be quite an innovation, perhaps more so on the site and in the laboratory than in the design office. The development of more such techniques as a standard procedure will strike at the 'grass roots' of many problems in civil and structural engineering, and the author is convinced that such an approach is worth serious consideration.

Costing and Management Problems in Construction

The emphasis of this book is more towards 'scientific' problems arising in civil and structural engineering—chiefly during design—than to those encountered in construction. This is intentional, not because of the insignificance of or lack of challenge from constructional problems but because of his own restricted experience on such matters recently when substantial strides have been made in introducing these techniques to all aspects of construction. Nevertheless, the book would be incomplete without some reference to this field. These introductory remarks are supplemented by a bibliography enabling a more detailed study of this important subject area to be made if required.

Construction is the process of deciding how to effect and then actually effecting civil engineering works in an efficient and economical way, taking acccount of special requirements in respect of time and other restrictions. Among the resources available are 'time', 'money', 'manpower', 'plant', 'materials', 'instruments', and 'methods', and the considerations include 'quality', 'cost', 'aesthetics', 'function', 'progress', and 'completion date', together with the 'welfare' of all personnel.

Each of these items is a complex issue with many subdivisions and ramifications, and their permutation, combination, control and use to complete the works, is the art of construction. Automatic computational techniques such as, for example, preparing and analysing data by computer, can aid this process. Many of the library programs—especially those on financial matters—are written in COBOL (see Chapter 6) and input and output data are frequently on punched cards, enabling the data to be readily altered as required by substituting one card for another. In this way it is possible to prepare and analyse Gantt and bar charts,[2] cost-time curves and planning networks, employing Critical Path Methods[3] (CPM) and the Program Evaluation and Review Technique[3] (PERT) with these latter diagrams.

CPM consists essentially of planning the execution of a project by drawing up a network showing the sequence and interrelation of all the activities to be undertaken (a typical simple network being shown in Fig. 10.1), and

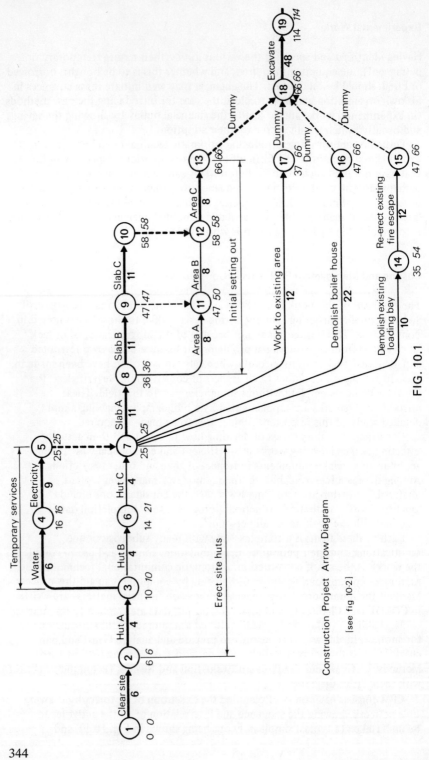

FIG. 10.1

Construction Project Arrow Diagram

[see fig 10.2]

analysing and manipulating these to determine the best overall method of completing the work. It is thus possible to compute the earliest and latest starting and finishing times for each activity, so determining the key operations which control the quickest efficient execution of the works. The trace on the network from the start to the completion of the project over these key operations in sequence, is then known as the critical path.

Various library programs are available to process numerical data using either CPM or PERT, among these being L01[4] and L06[5] for the Elliott 803 or 503 computers, and PERT 1[6] for the Elliott 4100 system. The function of program L01—project planning of control—is to analyse the interrelationships and duration of the activities making up a project, and to compute the earliest and latest starting and finishing times for these activities. This indicates the critical path which is shown by zero float (Fig. 10.2).

The data to be input consists of: the first event in the project; the starting time; the total time allowed; a list of the activities, each defined by the preceding event, the succeeding event and its duration. The program then outputs: the earliest and latest time for each event; the critical path; the earliest and latest times for starting and finishing each activity; and the total float time for each activity.

The maximum size of the project is 700 activities. All data must be in integer form, each number being greater than 0 and less than 8191. The project length must not exceed 8191 time units. For the largest possible project, the machine time taken is ten minutes with an Elliott 803 computer. A numerical example using Program L01 is shown in Fig. 10.2.

This type of CPA (Critical Path Analysis) by computer can be repeated several times on the same project as work progresses, thereby updating the results and catering for any unavoidable changes in the work schedule.

Such techniques can also possibly be used to assist: work and method study; financial, operational and administrative control; stock and materials control; and to measure work completed. The reader must however appreciate that these techniques are mere tools to assist the engineer with his managerial duties, and can in no way control, direct, or make decisions on, the operation.

Operating the System

When giving advice there is always a danger of being misunderstood or misinterpreted, with the recipients being perhaps not very pleased with the information offered to them. This, of course, can apply to engineers (of which the author is one!) all of whom pride themselves on their ingenuity in effecting solutions or devising and operating some system once they have studied the basic principles involved. This explains why the subject matter in so many textbooks is almost exclusively devoted to a description of the basic method or issue and nothing more, leaving the reader to use it as he will. For example, many books on structural analysis give only the most fleeting reference as to how the methods might be applied to computers or practical applications. On the other hand, it is certainly not helpful to provide the average engineer with vague and incomplete numerical data, evaluated by an incomprehensible program on a rare, inaccessible and expensive computer, as the sole introductory illustration to the

TYPICAL EXAMPLE *(USING PROGRAM LOI).*

CONSTRUCTION PROJECT   MAY 1964. *(SEE FIG. 10.1.)*

RESULTS

C R I T I C A L   P A T H   A N A L Y S I S

E.T.   EARLIEST TIME
L.T.   LATEST TIME
P.E.   PRECEDING EVENT
S.E.   SUCCEEDING EVENT
*   CRITICAL PATH

EVENT	E.T.	L.T.		P.E.	S.E.	DURATION	START E.T.	START L.T.	FINISH E.T.	FINISH L.T.	TOTAL FLOAT
1	0	0	*	1	2	6	0	0	6	6	0
2	6	6	*	2	3	4	6	6	10	10	0
3	10	10		3	6	4	10	17	14	21	7
			*	3	4	6	10	10	16	16	0
6	14	21		6	7	4	14	21	18	25	7
4	16	16	*	4	5	9	16	16	25	25	0
5	25	25	*	5	7	0	25	25	25	25	0
7	25	25		7	14	10	25	44	35	54	19
				7	16	22	25	44	47	66	19
				7	17	12	25	54	37	66	29
			*	7	8	11	25	25	36	36	0
14	35	54		14	15	12	35	54	47	66	19
15	47	66		15	18	0	47	66	47	66	19
16	47	66		16	18	0	47	66	47	66	19
17	37	66		17	18	0	37	66	37	66	29
8	36	36		8	11	8	36	42	44	50	6
			*	8	9	11	36	36	47	47	0
9	47	47		9	11	0	47	50	47	50	3
			*	9	10	11	47	47	58	58	0
11	47	50		11	12	8	47	50	55	58	3
10	58	58	*	10	12	0	58	58	58	58	0
12	58	58	*	12	13	8	58	58	66	66	0
13	66	66	*	13	18	0	66	66	66	66	·0
18	66	66	*	18	19	48	66	66	114	114	0
19	114	114	*								

FIG. 10.2

usefulness of a particular method. The author trusts that he himself will not be accused of having committed the same offence! Nevertheless, he is convinced that a few notes on operating a system (consisting of the modern techniques and devices) on actual production work might be useful.

At this stage, staff deployment and attitudes are most important. The former problem is discussed in Chapter 3, and attitudes can only be won over by the system being manifestly seen to be an efficient, labour-saving way of accomplishing the actual production work of the particular organisation. The prior training of staff (see Chapter 11) in respect of technical 'know-how' and attitudes should reap its reward on production work, and provision should be made for this. Due attention should always be paid to the arrangements made for processing data by computer and it is recommended that:

1.   A register for each engineering project should be kept by the engineer, giving every tape, card and paper a coded serial number for identification. For each number the register entry should include a descriptive title, the date and time, a mark or modification number, the machine time taken to provide the results, and notes indicating whether the results obtained are satisfactory, with a reference being given if they are not, as to the next appropriate action.

2.   All program and data tapes should have a length of blank tape at the outer end, on which the program or serial number, the mark or modification number, the date and a brief heading are written. These tapes should be stored in rolls, preferably in an appropriate filing cabinet having small shallow drawers.

3.   In addition, duplicate or master copies of each program tape should be kept separately to enable replacement tapes to be made in the event of accidental damage to the original tape.

4.   The data preparation sheet and diagrams, the print-out of the actual input, and the output from the computer for each run should all be appropriately labelled, folded to a standard size, clipped together and filed.

5.   Copies of the program print-outs, program brochures, and blank data preparation sheets should always be available.

6.   The entire operation should be systematic and the register kept up to date. The cumulative machine time, hence cost of machine time to date, could also be recorded in the register.

Similar arrangements should be made with card input and output of data. During production work on the computer the engineer should be prepared for any of the following events.

Errors occurring, either in the data, the program, the compiler or the systems programs (the latter two regrettably being possibilities!) and the need for their subsequent rectification.

The need for subsequent or additional runs due to modifications or errors, all of which must generally be paid for.

The breakdown of the computer.

The non-availability of the computer for a time due to maintenance or servicing.

The need to check the data at all stages.

Readiness and ability to make a decision—perhaps involving financial outlay— as a result of a turn of events. (Valuable time on production work may be lost if the engineer does not have this authority.)

The many miscellaneous administrative and routine duties to be discharged include:

the packing of papers and tapes to catch the post;

the printing-out of additional copies of results;

attendance at awkward hours; and

transmitting results or instructions by telephone, and so on.

The engineer should always be prepared for various phenomena occurring when processing numerical data by logger or computer, ill-conditioning possibly being the most likely cause of trouble in this respect (see Chapter 5). (There are various ways of overcoming ill-conditioning, the most drastic perhaps being to re-write a subroutine.)

Other such difficulties during production runs might include: scaling problems (for example, where the denominator in a complex expression can unexpectedly have a value approaching zero, so producing very large numerical quantities); the computer stopping unexpectedly during a run (provision for the printing out of error indicators on such occasions being most useful in helping to locate the trouble); and the manifestation of 'errors' in a library program (perhaps because the engineer is expecting too much of a particular program or applying it to a use not envisaged by its originator).

Concluding Remarks

The suggestions and ideas given in this chapter are intended as a memorandum to the engineer, indicating how the development and exploitation of the methods described might be initiated and accomplished, and are supplemented by a list of practical difficulties which could well arise in the process, and of which account should be taken. Some of the points made, and others, are considered further in Chapter 11.

References

1. Litton, E. Design Investigations on Some Welded Mild Steel Plate Girders. Proceedings of the Institution of Civil Engineers, Vol. 39, Page 583, Paper No. 7086S, April 1968
2. Clark, W. The Gantt Chart. Pitman
3. Antill, J. M. and Woodhead, R. W. Critical Path Methods in Construction Practice. Wiley
4. Program Number L01. Project planning of control mk I. Elliott 803 and 503 Applications Group Program for Civil Engineering. (All enquiries should now be made to International Computers, Ltd., London)
5. Program Number L06. Project planning of control mk III. Elliott 803 and 503 Applications Group Program for Civil Engineering. (see 4 above)
6. Program Number PERT 1. Elliott 4100 Application Program. (see 4 above)

Bibliography

This is a list of textbooks on management techniques in the construction of civil engineering works using the modern developments.

Antill, J. M. and Woodhead, R. W. Critical Path Methods in Construction Practice. Wiley. 1965
Lockyer, K. G. An Introduction to Critical Path Analysis. 3rd ed. Pitman. 1969

Lockyer, K. G. Critical Path Analysis. Problems and Solutions. Pitman. 1966

Brennan, J. Applications of Critical Path Techniques. E.U.P. 1968

Modder, J. J. and Phillips, C. R. Project Management with CPM and PERT. Reinhold. 1964

Barnetson, P. Critical Path Planning. Newnes. 1968

O'Brien, J. J. CPM in Construction Management. McGraw Hill. 1965

Kelley, J. E. and Walker, M. R. Critical Path Planning and Scheduling. (Proceedings of the Eastern Joint Computer Conference 1959)

Pilcher, R. Principles of Construction Management. McGraw Hill. 1967

Battersby, A. Network Analysis for Planning and Scheduling. Macmillan. 1967

Woodgate, H. S. Planning by Network. Business Publications Ltd. 1967

Buffa, E. S. Production-Inventory Systems: planning and control. Homewood. 1968

Anthony, R. N. Planning and Control Systems. A Framework for Analysis. Harvard Business School. 1965

Clark, W. The Gantt Chart. Pitman. 1952

11
Implications Arising from the Development and Use of Automatic Computational Techniques

Perhaps it is now profitable to consider what might happen to all sectors of the civil and structural engineering industry due to the introduction of modern methods. Already, in the words of the song writer, 'fings ain't what they used to be!' This outcome is the logical extension of the process described in Chapter 1 as the Age of Automation, whereby the pick and shovel have been replaced by the bulldozer and scraper, the slide rule by the computer, and now, in addition, manually operated laboratory and field instruments by automatic devices.

It is first important to realise that a new philosophy has been created and all engineering problems—both old and new—can now be appraised from an entirely different viewpoint. Many old prejudices or difficulties can now be cast aside, e.g. a reluctance to embark on a certain undertaking because of the sheer size of the job, or a disinclination to try other feasible possibilities due to a shortage of time, and so on. Engineering projects can also now be regarded more as one continuous integrated process with much closer collaboration between office, laboratory and site.

From these concepts, a completely new technological world can emerge where the total benefits can greatly exceed those obtained by summating the individual ones from each separate subject. The author believes it would be wrong when thinking along these lines to adopt the attitude fostered by George Orwell in his novel '1984'. Instead, a more positive altruistic outlook should be fostered, though of course it is foolish to regard all the warnings of any Jeremiah as insignificant. Indeed, much 'newspeak' has encroached already into the everyday parlance of engineers, though in itself this is not perhaps a bad thing. ('Doubletalk' and 'doublethink', of course, are!) The engineer must always try—especially in this Age of Automation—to think beyond his machines, equipment and techniques, and consider the effect of the new philosophy on mankind, asking himself the question 'Are these changes good or bad for civilization?'

Returning now to the day-to-day world of the engineer, the effects of introducing modern methods on well-established, conventional engineering practice are now discussed.

350

Responsibility

Possibly the most important implication of all is the question of responsibility—who is responsible for the validity of the results obtained by the new techniques? This matter is discussed in Chapter 4, and it is only necessary to reiterate briefly here that the author believes this responsibility rests fairly and squarely with the engineer. It is therefore vital that he should be completely satisfied with whatever program, method or process is used, and should always carry out his own checks and trial runs before ever attempting to make serious use of any program or technique on actual 'production' work.

The Drawing Office

Drawings will always be needed as this is how an engineer can best depict what he wishes to have done. Therefore there should be no need for Luddite attitudes to change in the drawing office. Nevertheless, major changes will doubtless take place due to the introduction of automatic drafting machines and map and graph plotters, leading to the redeployment of staff to other duties. Some staff will clearly need to be retrained in terms of technical 'know-how' and, equally important, in attitudes of mind. This problem is considered further in a wider and more-general context later.

A range of automatic drafting machines has been described by van Gent,[2] from the E–51 machine (of Concord Control Inc.) which can draw lines and print names and numbers photographically, to a smaller machine such as the DIP–360–365 (of California Computer Products Inc.) which is only suitable for drawing profiles and the like, and finally to machines developed by Dennert and Pape, Zuse, Haag Streit and Coradi, of which the Z64 model of Zuse is a fine example.

While graph plotters are now used fairly widely, automatic drafting machines are not at present, possibly due to the magnitude and nature of the changes which their introduction would produce in the traditional drawing office regarding staffing, equipment and methods. The financial and sociological implications of such a change must be considered very carefully—particularly because of the high cost of the machines and their peripheral equipment at this early stage of development. However, as such instruments become more widely used, the cost of each will doubtlessly be reduced. As these machines are introduced, it should be possible to transfer the drawing-office staff from their previous duties to the challenging new activities involving maintaining, servicing and operating the new equipment (after suitable training). In addition, much software will be required to use the new system most efficiently, and the technocrats who devise the appropriate programs will be leaders in their field.

The automatic measurement of areas on drawings and graphs can also now be accomplished by photo-planimeters and electro-planimeters such as those developed by Becker and Zuse.[1] Thus graphical integration can be effected automatically so that this technique could be utilised as a subroutine in a design program.

Finally, the introduction of automatic drafting machines, planimeters, and analogue-to-digital converters clearly provides ideal opportunities for linking many other diverse engineering activities such as those of the design and commercial offices, the laboratory and the site, so pointing the way to the achievement of a single integrated process.

Design

From the discussion in Chapters 9 and 10, the implications arising from the development and use of the new techniques in design should be apparent. The engineer will be relieved of the need to accomplish many complicated design calculations personally, thus being freed to think more creatively in terms of the overall design scheme, involving aesthetics, function, cost and construction, and examining many more alternatives—some perhaps quite unconventional—than was previously possible.

If the physical and mental well-being of the design office staff is duly considered as the new system is devised and used, so that, for example, results are given in a concise satisfactory form (e.g. pictorially, graphically, or as a small table with critical values clearly marked), from which the essential information can be readily and effortlessly extracted, the morale and efficiency of the staff will be high and their attitude right. Modern methods provide an opportunity to extend the design process to the site where, from the results of loading tests or experimental measurements, it would be possible for extra haunches, reinforcing bars, flange plates, or shear walls to be specified and added after the erection of the structure has commenced. This could become a standard practice and would be relatively simple to effect, especially if due allowance has previously been made for such optional strengthening.

Estimating

It is now particularly opportune for an engineering organisation, if it so wishes, to keep records and statistics of individual costs incurred and difficulties encountered, together with details of the time taken and the processes used in constructing all the works for which it has been responsible. These records are perhaps best kept on punched cards so that they can be grouped and re-grouped in various ways, and then processed by computer to obtain financial and other results. It is thus possible first to compare the actual results with those estimated, noting the differences and seeing whether any lessons may be learned for future occasions. Secondly, a cumulative dossier of information can be drawn up to enable the subsequent estimating of similar work to be better and more realistic. Thirdly, a systematic analysis of successive constructional undertakings may reveal where savings or improvements can be made on subsequent works. Analysing all network diagrams and critical path analyses from start to completion of any project may well provide useful information on the entire construction process, and is worth considering.

To undertake such investigations and analyses, it will be essential for the organisation concerned systematically to obtain and record the necessary information, to devise and write (or to have written) appropriate computer programs, and to possess, or have access to, appropriate staff, equipment and facilities, all of which must be paid for—perhaps out of the savings achieved by the exercise!

On Site

On site, there will be more measuring equipment and supporting staff for 'scientific' work, and time-and-motion observers backed by clerical staff for work on the commercial side. Whilst this may cause some 'interference' and have 'nuisance value', nevertheless it will ensure that much more attention is focussed on the site by the design office, committing it to much more-active 'real' participation in the project than that possible with the 'remote control' approach of head office which has persisted in the past. Indeed, it is likely that more and more contractors—particularly those working in consortia—will establish design offices on large sites, so obviating the gulf which has so long existed between designer and site engineer. Closer communication can thus be achieved between design and site staff, thereby reducing the opportunity for confusion.

Loading Tests

The proving of structures by loading tests in lieu of calculations is already permitted. For example, subclause 9c of B.S. 449: Part 2:1969[2] (dealing with structural steel in building), states:

> '*Experimental basis*: Where, by reason of the unconventional nature of the construction, calculation is not practicable, or where the methods of design given . . . above are inapplicable or inappropriate, loading tests shall be made, . . . to ensure that the construction has:
> 1. adequate strength to sustain a total load equal to twice the sum of the dead load and the specified superimposed load; and
> 2. adequate stiffness to resist, without excessive deflection, a total load equal to the sum of the dead load and 1½ times the specified superimposed load.'

Again, subclause 605 of CP114: Part 2:1969[3] (dealing with reinforced concrete in building) states:

> '*Load testing of structures*: Load tests on a completed structure should be made if required by the specification or if there is reasonable doubt as to the adequacy of the strength of the structure. Such tests need not be made until the expiry of 56 days of effective hardening of the concrete.
> In such tests, the structure should be subjected to a superimposed load equal to one and a quarter times the specified superimposed load used for design, and this load should be maintained for a period of 24 hours before removal. During the tests, struts strong enough to take the whole load should be placed in position leaving a gap under the members.
> If within 24 hours of the removal of the load, the structure does not show a recovery of at least 75 per cent of the maximum deflection shown during

the 24 hours under load, the test loading should be repeated. The structure should be considered to have failed to pass the test if the recovery after the second test is not at least 75 per cent of the maximum deflection shown during the second test.

If during the test, or upon removal of the load, the structure shows signs of weakness, undue deflection or faulty construction, it should be reconstructed or strengthened as necessary.'

Finally in the draft Unified Code of Practice[4] for concrete in which the concept of limit states of collapse, excessive deflection, excessive local damage, excessive vibration, fatigue, durability, and fire resistance, is introduced, load tests are once more envisaged, together with the interesting possibilities of model testing or development testing of prototypes. The text states:

'Designs based on model testing or development testing of prototypes are acceptable, subject to the engineers responsible being suitably experienced.'

The implication of such clauses could be profound. The introduction of electrical measuring devices and loggers brings the opportunity of making loading tests a standard procedure in the design or acceptance of a structure. Such tests could be carried out on a model or a prototype, or on the actual structure itself. One possible approach embracing these principles would be first to undertake rough design calculations, making provision in detailing for optional site strengthening, and then to carry out loading tests to decide whether the structure is satisfactory or must be strengthened. Transducers, strain gauges and other devices with appropriate wiring and pick-up points could be built into the structure and sited beneath foundations, behind dams and retaining walls, etc., to provide records during the life of the structure. Telemetering (see Chapter 9) could also report (factored!) critical values of strain, deflection and soil pressure, etc., sustained at key locations in a structure, and so forth.

Finally, low-level loading tests could assist in analysing a structure, being used to obtain flexibility influence coefficients, the elements of a stiffness matrix and stress concentrations, in difficult unconventional designs.

Organising the Work

Planning a system

The engineer may well find himself the motivating and initiating force trying to introduce modern methods into his own organisation. Not only may he have to plan the system, or at least its basic concept, he may also have to 'sell' the idea to those who make policy decisions and hold the purse strings. If he envisages the extension in due course of automatic computational techniques 'right across the board' from the drawing, design and commercial offices to the laboratory and site, and the monitoring of completed works, it is essential that the entire undertaking should first be thought out as a whole, even although the full proposals are not all implemented at once. Planning the system thus ensures the effective co-ordination and integration of the various services, equipment and staff involved, rather than producing a piecemeal approach in isolated sectors which later become a liability when attempts are made to match together what are then incompatible units.

Its installation

In parallel with the work of setting-up and installing the system, should be the preparation of software and the training of staff (see below). Also, if production by traditional methods is to continue during this period, a phased operation should be planned, culminating in a complete transfer to the modern methods in due course. Some rather indirect financial implications are associated with installing the system, perhaps involving the provision of additional buildings, teleprinter lines and other communicating links, amenities for night time operations, and so forth, all of which must be taken into account when planning, costing, and constructing the scheme.

Its use

Time and money must be allowed for the commissioning of the system. This involves a time allowance for staff to become accustomed to using the new techniques, and provision for 'de-bugging' the equipment and eliminating snags which may only be revealed as the system comes into operation.

An operating plan should be devised to include a means of allocating priorities, facilities for checking the data provided, and an emergency procedure to be adopted in the event of any breakdown in the system. Every effort should then be made to operate the system efficiently by ensuring if at all possible that all expensive equipment is fully utilised on production work during a significant part of each day. This depends both on the modus operandi and on the deployment and training of the staff.

Research and Development

When any organisation first decides to partly or wholly use automatic computa-tional techniques, it also virtually commits itself to a programme of research and development (R & D) in that particular field of automation. It is important that top management appreciates and accepts the need for R & D and should regard it not as a liability or overhead but as a useful essential part of the production work, so that an allowance for these activities is always included as normal practice in all quotations, etc. Properly controlled and directed R & D certainly makes an organisation more efficient and more competitive, giving far reaching benefits all round. However, top management may well not view R & D at all in this way (for example, if they are accountants or businessmen), and it is then the engineer's 'duty' to act as advocate, counsellor and diplomat so as to establish the correct attitudes in that sector of influence and power. For example, it is quite unrealistic to imagine that existing software, written by others, will meet all the requirements of any organisation without at least some adaptation or modification. What is more likely is the need to prepare new programs and techniques for the specialist activities undertaken, which must therefore be allowed for when estimating.

It should be possible to undertake R & D at slack times when staff and equipment are not fully utilised, though obviously such activity must always be carried out in advance of any application to production work. R & D provides an ideal opportunity for maintaining staff interest, and also may be a useful means of retaining their services! Care should be taken to ensure that all such work has as wide an application as possible, so as to deal with many types of problems at the one time. Attempts should also be made for such work to have some lasting usefulness, contributing to the general or specialist pool of knowledge within the organisation. Lastly, it is essential that all the results and techniques derived are carefully and permanently recorded in a central source, that news of all the developments is systematically communicated to all staff and management, and that full details for their practical use are readily available in a properly edited and written form.

Training of Staff

Proper facilities for the training of staff must exist. Whilst many colleges, manufacturers and research stations provide useful services in this respect, such instruction cannot possibly be tailor-made for any particular organisation with its own special problems unless, of course, it has been designed for that very purpose. A very large organisation could, however, set up its own training school, and even offer these facilities to others for an appropriate fee! Nevertheless it is much more feasible for the organisation—however small—to engage an expert for a short period to advise on how to make provision for the training of staff to deal with its own particular problems.

Nowadays most universities and colleges give instruction in the use of automatic computational techniques in their degree courses, so that the next generation of engineers will generally have a better knowledge of the modern methods from the outset. This more fundamental approach can also be supplemented by short specialist and post-graduate courses on various aspects of the subject to provide the concentration of expert knowledge required in any particular field.

For those organisations whose staff have little or no experience in modern methods, a training programme could be evolved to include: attending a computer programming course; demonstrations of equipment and techniques by experts; attending carefully selected short courses; the use of textbooks and technical papers; visits to research establishments; attendance at conferences and symposia; and developing correct staff attitudes to automation.

Attending a computer programming course
Perhaps the best possible introduction to the modern methods for all staff is provided by their attending a fairly general course on basic computer programming and appreciation. It may well be that those attending will never have occasion, in their daily duties, to make direct use of this knowledge by having to write programs themselves. Nevertheless such instruction is most useful, as it embodies the very essence of the new system and should provide a valuable grounding in the basic concepts, especially if practical program writing, tape punching, computer operation and so forth are included.

A high-level programming language such as ALGOL, FORTRAN or COBOL should be used and, just as with a foreign language, competence and skill in its application can only be attained and maintained by daily study and use (which, of course, may not always be possible!)

Demonstrations of equipment and techniques by experts
The staff should first of all become familiar with the new equipment and techniques as soon as possible by means of demonstrations and question-and-answer sessions, so that when they subsequently attend any course of lectures they have a satisfactory prior knowledge of the nature, appearance and function of the hardware and software to which reference will inevitably be made.

Attending carefully selected short courses
It is very difficult to study a subject exclusively from a textbook and it may therefore be helpful to attend a suitable short course at a university, college, or research centre. These generally consist of either three to five days' study at the one time—perhaps on a residential basis—or of a series of weekly meetings each of about two hours' duration. The former enables participants to devote their entire effort to the course for a few days, to meet and converse with others (thereby exchanging notes and ideas), and really to develop an interest in the subject, while the latter is a means of acquiring knowledge steadily over a longer period. Moreover, if these latter classes are held outside normal working hours, the participants can continue to discharge their normal professional duties as usual.

Typical short courses include:

Computer programming (ALGOL, FORTRAN or COBOL)
Computer appreciation.
Mathematics for computers (e.g. numerical analysis, matrix algebra, Boolean algebra, logic, statistics, and probability).
The use of data-logging devices and systems in civil engineering laboratories and sites.
Design, servicing and maintenance of electrical equipment, instrumentation and installations for civil engineering laboratories and sites.
The design and layout of systems in civil engineering embodying automatic computational techniques.
Computers for structural engineers.
Structural analysis by computer (e.g. matrix methods, finite element methods for elastic, plastic or dynamic analyses, etc.)
Design studies by computer.
Modern method of construction management (PERT, CPA, networks, etc.)
Structural testing and loading tests using electrical logging devices (together with the analysis and processing of the results).
Traffic analysis and surveys.
Modern techniques in automatic surveying.
Water engineering problems.
Automatic recording of field measurements (e.g. pH values, wind velocity, river pollution and flow, structural quantities, etc).
Supervised practical projects in civil engineering using automatic computational techniques.

It is worth noting that universities and colleges may be prepared to organise and operate any short course on request, provided they are guaranteed sufficient support for the venture.

Use of textbooks, etc.
Textbooks such as this and those listed in the bibliographies can be used, not only to obtain a suitable background and basic knowledge of the various subjects, but also as reference manuals. As the reader's personal knowledge increases, he may find visits to research establishments, conferences and symposia useful. (If he does this at too early a stage, he may well be discouraged!)

Developing correct staff attitudes to automation
Almost as important as the technical knowledge of modern methods is the attitude of the staff concerned towards the introduction of the new techniques. Only when the staff are convinced that 'new means better'—and is manifestly seen to be so—will the entire new technology have any chance of functioning as efficiently as it should. This matter should therefore be given considerable attention and a public relations campaign promoted throughout the organisation concerned to 'sell' the new ideas.

Standardisation and Permanent Records

Finally, there is an obligation on all users to standardise their data and maintain permanent records of their work for posterity. Just as Whitworth standardised the screw thread throughout the engineering world in his day, so should computer data and program formats likewise be unified. It is obviously most desirable to be able to use a library program, specially written for one computer, on a different one, if the latter is sufficiently large, and details are given in Chapter 6 (q.v.) of how such problems have been greatly simplified by the introduction of high-level languages such as Algol and Fortran. Nevertheless, many programs still cannot be used on other computers even when this appears to be straightforward, perhaps due to the existence of machine-code blocks, to the need for some obscure subroutine which is not clearly specified or available, or simply because the program originators did not foresee the possibility of this extended use. Also with data there would be much less opportunity for confusion or errors if all engineers were to adopt a standardised procedure. An Institution of Structural Engineers' committee for the 'Standardisation of input information for computer programmes' has existed at least since 1966/67, and has published two technical reports.[5], [6] In addition a joint committee on the 'Use and development of computers in the civil engineering industry' was set up in May 1969 by the Institution of Civil Engineers, the Association of Consulting Engineers and the Federation of Civil Engineering Contractors. It is conceivable that a British Standard might conceivably be prepared in due course and clearly the matter is being taken seriously by these various bodies.

Finally the author believes it is desirable for a national library of computer programs to be established by an appropriate organisation to record for all time details of the special programs that have been written. Each organisation could be invited to send full particulars of its programs (brochures, print-outs, tapes or cards) for registration and deposition, and arrangements could be made for copies to be provided as required. (The author is currently establishing a small 'software bureau' in the Department of Civil Engineering at Dundee College of Technology as part of his contribution towards such a venture and has included in this book, as far as possible, full details of the programs for all to use as required. He hopes that others will be encouraged to do the same!)

References

1. van Gent. Some Remarks on Automatization in Surveying and Mapping. Federational Internationale des Geometres, Committees V and VI. XIth Congress—Rome 1965.
2. British Standard 449: Part 2: 1969. The Use of Structural Steel in Buildings. British Standards Institution.
3. British Standard Code of Practice. CP 114: Part 2: 1969. The Structural Use of Reinforced Concrete in Buildings. British Standards Institution.
4. "Proposed New Unified Code of Practice" Reference: BLCP/7, BLCP/8, BLCP/35 30th September 1969. Draft British Standard Code of Practice for the Structural Use of Concrete. (to supersede CPs 114, 115 & 116).
5 Technical Report: 1967. The use of Digital Computers in Structural Engineering (2nd Edition). The Institution of Structural Engineers.
6. Technical Report: 1967. Standardization of Input Information for Computer Programs in Structural Engineering. The Institution of Structural Engineers.

12
Conclusion

This is the final chapter, and, before making the concluding remarks, it might be useful to first of all summarise in general terms what has been attempted in these pages.

Summary

An attempt has been made to provide a 'bird's eye view' of the problems of automation in the various sectors of civil and structural engineering, particularly those arising from the introduction and use of automatic computational techniques within these fields of interest. A basic grounding on the important aspects was first given to ensure that the reader could become immediately involved, and appreciate the reason for and nature of each topic quickly and clearly. This work was then consolidated and co-ordinated to form a fully-integrated treatment of the subject, and several examples were given of practical applications to illustrate these techniques, particular attention being paid to the background circumstances and to how the various difficulties were overcome. Finally some suggestions were made as to how an appropriate system might be set up to utilise the automatic computational techniques in office, laboratory and site, with due consideration being shown to a wide range of implications arising therefrom.

Concluding Remarks

The reader will appreciate that, whilst particular computers, loggers, programming languages, etc., have of necessity had to be used to devise the various techniques, these techniques are not in general restricted to such software and hardware, but have much wider applications. Again, whilst most attention has been paid to problems in structural engineering, the techniques might conceivably be once more extended into much wider applications in other branches of engineering. This is the 'approach', and the reader should now be equipped to make a more detailed and knowledgeable study of whichever subject or item he chooses from within this large complex of technology. He should no longer feel like an 'enfant terrible', but should have acquired the confidence of his own convictions of how to set about tackling any problem in this field. The book might be regarded as a bridge between the practising engineers and the computer 'experts', and the author trusts that he has made this connecting link both wide enough to allow free passage and of sufficiently robust material for it to endure. This, then is the 'approach'—perhaps it might be possible in due course to report on 'modern advances'!

360

Appendix 1: Program DCTCES 3: Elastic Analysis of Various Single-Span Beams with either a Concentrated or Uniformly Distributed Load

Details of Formulae Used (see Chapter 3)

For all cases:

Reaction at end 1	= F [1]
Reaction at end 2	= F [2]
F.E.M. at end 1	= M [1]
F.E.M. at end 2	= M [2]
B.M. at load point	= M [3]
Max. positive B.M.	= M [4]
Max. negative B.M.	= M [5]
Distance of max. positive B.M. from end 1	= X [1]
Distance of max. negative B.M. from end 1	= X [2]
Distance of max. deflection from end 1	= X [3]
Deflection of end 1	= D [1]
Deflection of end 2	= D [2]
Deflection of load point (if applicable)	= D [3]
Max. deflection on span	= D [4]

$Z(1), Z(2)$, and $Z(4)$ are all unstable structures.

$Z(3)$:— Cantilever (fixed at end 2) $Q = 1, R = 3$.

For $P = 1$: $F[1] = 0, F[2] = W, M[1] = 0, M[2] = WB, M[3] = 0, M[4] = WB,$

$M[5] = \text{N.A.}, X[1] = L, X[2] = \text{N.A.}, X[3] = 0, D[1] = \dfrac{WB^2}{8EI}\left(L + \dfrac{A}{2}\right), D[2] = 0,$

$D[3] = \dfrac{WB^3}{3EI}$, and $D[4] = \dfrac{WB^2}{3EI}\left(L + \dfrac{A}{2}\right)$.

For $P = 2$: $F[1] = 0, F[2] = W, M[1] = 0, M[2] = \frac{1}{2}WL, M[3] = \text{N.A.}$

$M[4] = \frac{1}{2}WL, M[5] = \text{N.A.}, X[1] = L, X[2] = \text{N.A.}, X[3] = 0, D[1] = \dfrac{WL^3}{8EI},$

$D[2] = 0, D[3] = \text{N.A.}$, and $D[4] = \dfrac{WL^3}{8EI}$.

Z(5): Simply-supported beam, $Q = 2$, $R = 2$.

For $P = 1$: $F[1] = \dfrac{WB}{L}$, $F[2] = \dfrac{WA}{L}$, $M[1] = 0$, $M[2] = 0$, $M[3] = \dfrac{-WAB}{L}$,

$M[4] = $ N.A., $M[5] = \dfrac{-WAB}{L}$, $X[1] = $ N.A., $X[2] = A$, $X[3] = \sqrt{\dfrac{A(L+B)}{3}}$

when $A \geqslant \frac{1}{2}L$, and $X[3] = \sqrt{\dfrac{B(L+A)}{3}}$ when $A < \frac{1}{2}L$, $D[1] = 0$, $D[2] = 0$,

$D[3] = \dfrac{WA^2 B^2}{3EIL}$, $D[4] = \dfrac{WAB(L+B)}{27EIL} \sqrt{3A(L+B)}$ when $A \geqslant \frac{1}{2}L$, and

$D[4] = \dfrac{WAB(L+A)}{27EIL} \sqrt{3B(L+A)}$ when $A < \frac{1}{2}L$,

For $P = 2$: $F[1] = \frac{1}{2}W$, $F[2] = \frac{1}{2}W$, $M[1] = 0$, $M[2] = 0$, $M[3] = $ N.A.,

$M[4] = $ N.A., $M[5] = \dfrac{-WL}{8}$, $X[1] = $ N.A., $X[2] = \frac{1}{2}L$, $X[3] = \frac{1}{2}L$, $D[1] = 0$,

$D[2] = 0$, $D[3] = $ N.A., and $D[4] = \dfrac{5WL^3}{384EI}$.

Z(6): Propped cantilever (fixed at end 2, prop at end 1) $Q = 2$, $R = 3$.

For $P = 1$: $F[1] = \dfrac{WB^2}{2L^3}(2L+A)$, $F[2] = \dfrac{WA(3L^2 - A^2)}{2L^3}$, $M[1] = 0$,

$M[2] = \dfrac{WAB}{2L^2}(L+A)$, $M[3] = \dfrac{-WAB^2}{2L^3}(2L+A)$, $M[4] = M[2]$, $M[5] = M[3]$,

$X[1] = L$, $X[2] = A$, $X[3] = \sqrt{\dfrac{L(WAB - L \cdot M[2])}{WB - M[2]}}$ when $B < A\sqrt{2}$, $X[3] = A$

when $B = A\sqrt{2}$, and $X[3] = L - \dfrac{2L \cdot M[2]}{WA + M[2]}$ when $B < A\sqrt{2}$, $D[1] = 0$,

$D[2] = 0$, $D[3] = \dfrac{WA^2 B^3}{12EIL^3}(3L+A)$, $D[4] = \dfrac{WAB^3}{3EI} \cdot \dfrac{(L+A)^3}{(3L^2 - A^2)^2}$ when $B > A\sqrt{2}$,

$D[4] = \dfrac{WL^3}{102EI}$ when $B = A\sqrt{2}$, and $D[4] = \dfrac{WAB^2}{6EI} \sqrt{\dfrac{A}{(2L+A)}}$ when $B < A\sqrt{2}$.

For $P = 2$: $F[1] = \dfrac{3W}{8}$, $F[2] = \dfrac{5W}{8}$, $M[1] = 0$, $M[2] = \dfrac{WL}{8}$, $M[3] = $ N.A.

$M[4] = M[2]$, $M[5] = \dfrac{-9WL}{128}$, $X[1] = L$, $X[2] = \frac{3}{8}L$, $X[3] = 0.4215L$,

$D[1] = 0$, $D[2] = 0$, $D[3] = $ N.A., and $D[4] = \dfrac{WL^3}{185EI}$.

Z(7) Cantilever (fixed at end 1) $Q = 3$, $R = 1$.

For $P = 1$: $F[1] = W$, $F[2] = 0$, $M[1] = WA$, $M[2] = 0$, $M[3] = 0$, $M[4] = WA$,

$M[5] = $ N.A., $X[1] = 0$, $X[2] = $ N.A., $X[3] = L$, $D[1] = 0$, $D[2] = \dfrac{WA^2}{3EI}\left(L + \dfrac{B}{2}\right)$,

$D[3] = \dfrac{WA^3}{3EI}$, and $D[4] = \dfrac{WA^2}{3EI}\left(L + \dfrac{B}{2}\right)$.

For $P = 2$: $F[1] = W$, $F[2] = 0$, $M[1] = \dfrac{WL}{2}$, $M[2] = 0$, $M[3] = $ N.A.,

$M[4] = \dfrac{WL}{2}$, $M[5] = $ N.A., $X[1] = 0$, $X[2] = $ N.A., $X[3] = L$, $D[1] = 0$,

$D[2] = \dfrac{WL^3}{8EI}$, $D[3] = $ N.A., and $D[4] = \dfrac{WL^3}{8EI}$.

Z(8): Propped cantilever (fixed at end 1, prop at end 2) $Q = 3$, $R = 2$.

For $P = 1$: $F[1] = \dfrac{WB(3L^2 - B^2)}{2L^3}$, $F[2] = \dfrac{WA^2(2L + B)}{2L^3}$, $M[1] = \dfrac{WAB(L + B)}{2L^2}$,

$M[2] = 0$, $M[3] = \dfrac{-WA^2B(2L + B)}{2L^3}$, $M[4] = \dfrac{WAB}{2L^2}(L + B)$,

$M[5] = \dfrac{-WA^2B}{2L^3}(2L + B)$, $X[1] = 0$, $X[2] = A$, $X[3] = \dfrac{2L \cdot M[1]}{WB + M[1]}$ when

$A > B\sqrt{2}$, $X[3] = A$ when $A = B\sqrt{2}$ and $X[3] = L - \sqrt{\dfrac{L(WAB - L \cdot M[1])}{WA - M[1]}}$

when $A < B\sqrt{2}$, $D[1] = 0$, $D[2] = 0$, $D[3] = \dfrac{WA^3B^2}{12EIL^3}(4L - A)$,

$D[4] = \dfrac{WA^3B}{3EI}\dfrac{(L + B)^3}{(3L^2 - B^2)^2}$ when $A > B\sqrt{2}$, $D[4] = \dfrac{WL^3}{102EI}$ when $A = B\sqrt{2}$,

and $D[4] = \dfrac{WA^2B}{6EI}\sqrt{\dfrac{B}{(2L + B)}}$ when $A < B\sqrt{2}$,

For $P = 2$: $F[1] = \tfrac{5}{8}W$, $F[2] = \tfrac{3}{8}W$, $M[1] = \dfrac{WL}{8}$, $M[2] = 0$, $M[3] = $ N.A.,

$M[4] = \dfrac{WL}{8}$, $M[5] = \dfrac{-9WL}{128}$, $X[1] = 0$, $X[2] = \tfrac{5}{8}L$, $X[3] = 0.5785L$, $D[1] = 0$,

$D[2] = 0$, $D[3] = $ N.A., $D[4] = \dfrac{WL^3}{185EI}$.

Z(9): Built-in beam $Q = 3$, $R = 3$

For $P = 1$: $F[1] = \dfrac{WB^2}{L^3}(L + 2A)$, $F[2] = \dfrac{WA^2}{L^3}(L + 2B)$, $M[1] = \dfrac{WAB^2}{L^2}$,

$M[2] = \dfrac{WA^2B}{L^2}$, $M[3] = \dfrac{-2WA^2B^2}{L^3}$, $M[4] = \dfrac{WA^2B}{L^2}$ when $A \geqslant B$, $M[4] = \dfrac{WAB^2}{L^2}$

when $A < B$, $M[5] = \dfrac{-2WA^2B^2}{L^3}$, $X[1] = 0$ when $A > B$, $X[1] = L$ when $A < B$,

$X[1] = 0$ and L when $A = B$, $X[2] = A$, $X[3] = \dfrac{2LA}{L + 2A}$ when $A > B$, $X[3] = \frac{1}{2}L$

when $A = B$, $X[3] = \dfrac{L^2}{L + 2B}$ when $A < B$, $D[1] = 0$, $D[2] = 0$, $D[3] = \dfrac{WA^3B^3}{3EIL^3}$,

$D[4] = \dfrac{2WA^3B^2}{3EI(L + 2A)^2}$ when $A > B$, $D[4] = \dfrac{WL^3}{192EI}$ when $A = B$, $D[4] =$

$D[4] = \dfrac{2WA^2B^3}{3EI(L + 2B)^2}$ when $A < B$.

For $P = 2$: $F[1] = \frac{1}{2}W$, $F[2] = \frac{1}{2}W$, $M[1] = \dfrac{WL}{12}$, $M[2] = \dfrac{WL}{12}$, $M[3] = $ N.A.,

$M[4] = \dfrac{WL}{12}$, $M[5] = \dfrac{-WL}{24}$, $X[1] = 0$ and L, $X[2] = \frac{1}{2}L$, $X[3] = \frac{1}{2}L$, $D[1] = 0$,

$D[2] = 0$, $D[3] = $ N.A., $D[4] = \dfrac{WL^3}{384EI}$.

Notes   N.A. denotes 'not applicable'.
         Hog bending moments are positive.

Appendix 2: Program DCTCES 4: Elastic Analysis of a Single Bay, Pitched Roof, Portal Frame with Pinned Bases

Details of Formulae Used (see Chapter 3)

Class of loading $\qquad\qquad H_A$

1. $\dfrac{WL(3+5m)}{16Nh}$

2. $\dfrac{WL(3+5m)}{15Nh}$

3. $\dfrac{WL(3+5m)}{16Nh}$

4. $\dfrac{-Wf(C+m)}{8Nh} - \dfrac{W}{2}$

5. $\dfrac{Wf(C+m)}{8Nh} - \dfrac{W}{2}$

6. $\dfrac{W}{8N}\left[2(B+C)+k\right] - W$

7. $\dfrac{-W}{8N}\left[2(B+C)+k\right]$

8. $\dfrac{Pc}{2Nh}\left[B+C-k\left(\dfrac{3a^2}{h^2}-1\right)\right]$

9. $\dfrac{Pc}{2Nh}\left[B+C-k\left(\dfrac{3a^2}{h^2}-1\right)\right]$

10. $-P$

11. $\dfrac{PLC}{4Nh}$

12. $\dfrac{-P(B+C)}{2N}$

13. $\dfrac{-P(B+C)}{2N}$

14. $\dfrac{-P}{2}$

Notes: (i) Since there is only one redundant reaction (taken as H_A) in each case considered, it is only necessary to list the various values of H_A for each class of loading. All other results can then readily be obtained by simple statics.

(ii) For the meaning of the symbols used, see Fig. 3.6. In addition:

$$k = \frac{I_2 h}{I_1\sqrt{\frac{1}{4}L^2 + f^2}}$$

$$m = 1 + \frac{f}{h}$$

$$B = 2(k + 1) + m$$

$$C = 1 + 2m$$

$$N = B + mC$$

Index